Lecture Notes in Artificial Intelligence   5989

Edited by R. Goebel, J. Siekmann, and W. Wahlster

Subseries of Lecture Notes in Computer Science

Luc De Raedt (Ed.)

Inductive
Logic Programming

19th International Conference, ILP 2009
Leuven, Belgium, July 02-04, 2009
Revised Papers

 Springer

Series Editors

Randy Goebel, University of Alberta, Edmonton, Canada
Jörg Siekmann, University of Saarland, Saarbrücken, Germany
Wolfgang Wahlster, DFKI and University of Saarland, Saarbrücken, Germany

Volume Editor

Luc De Raedt
Katholieke Universiteit Leuven, Department of Computer Science
Celestijnenlaan 200a, 3001, Heverlee, Belgium
E-mail: luc.deraedt@cs.kuleuven.be

Library of Congress Control Number: 2010929279

CR Subject Classification (1998): F.4.1, H.3, H.2.8, F.1, I.2.6, I.2.3

LNCS Sublibrary: SL 7 – Artificial Intelligence

ISSN   0302-9743

ISBN-10   3-642-13839-X Springer Berlin Heidelberg New York
ISBN-13   978-3-642-13839-3 Springer Berlin Heidelberg New York

Typesetting: Camera-ready by author, data conversion by Scientific Publishing Services, Chennai, India
Printed on acid-free paper   06/3180

Preface

The ILP conference series has been the premier forum for work on logic-based approaches to machine learning for almost two decades. The 19th International Conference on Inductive Logic Programming, which was organized in Leuven, July 2-4, 2009, continued this tradition but also reached out to other communities as it was colocated with SRL-2009 – the International Workshop on Statistical Relational Learning, and MLG-2009 – the 7th International Workshop on Mining and Learning with Graphs. While these three series of events each have their own focus, emphasis and tradition, they essentially share the problem that is studied: learning about structured data in the form of graphs, relational descriptions or logic. The colocation of the events was intended to increase the interaction between the three communities.

There was a single program with joint invited and tutorial speakers, a panel, regular talks and poster sessions. The invited speakers and tutorial speakers were James Cussens, Jason Eisner, Jure Leskovec, Raymond Mooney, Scott Sanner, and Philip Yu. The panel featured Karsten Borgwardt, Luc De Raedt, Pedro Domingos, Paolo Frasconi, Thomas Gärtner, Kristian Kersting, Stephen Muggleton, and C. David Page. Video-recordings of these talks can be found at www.videolectures.net. The overall program featured 30 talks presented in two parallel tracks and 53 posters. The talks and posters were selected on the basis of an extended abstract. These abstracts can be found at http://dtai.cs.kuleuven.be/ilp-mlg-srl/. In addition, as in previous years, a selection of the papers of ILP 2009 have been published in a volume in the *Lectures Notes in Artificial Intelligence* series and in a special issue of the *Machine Learning Journal*. From the initial 54 extended abstracts (6 pages in LNCS format) that were submitted to ILP-2009, 5 papers were invited for the special issue, 10 papers were published as a long paper and 14 more as a short paper in the proceedings. These papers were prepared after the conference. Many of the other abstracts were accepted for poster presentation.

This event would not have been possible without the help of a large number of people: I would especially like to thank Bernhard Pfahringer for suggesting this colocation, the invited and tutorial speakers, the panelists, the Chairs of MLG (Hendrik Blockeel, Karsten Borgwardt and Xifeng Yan) and SRL (Pedro Domingos and Kristian Kersting), Filip Železný for handling the Leuven papers, the Program Committee, the additional reviewers, the authors that submitted papers, the participants, the videolectures organization, and the extensive local organizing team (listed herein) as well as the sponsors (the BNVKI – the Benelux Association for Artificial Intelligence, Pascal 2 – the Network of Excellence, FWO – the Research Foundation Flanders) for their financial support. In preparing this volume, the support of Jan Struyf (who managed Easychair) and Laura Antanas (who managed the proceedings) was essential.

February 2010   Luc De Raedt

Organization

ILP-SRL-ML 2009 was organized by the Department of Computer Science, Katholieke Universiteit Leuven, Belgium.

Chairs

MLG Program Chairs   Hendrik Blockeel
     Karsten Borgwardt
     Xifeng Yan

SRL Program Chairs   Pedro Domingos
     Kristian Kersting

ILP Program Chair   Luc De Raedt
General Chair   Luc De Raedt

Program Committee

Annalisa Appice
Kurt Driessens
Alan Fern
Paolo Frasconi
Tamas Horvath
Kristian Kersting
Stefan Kramer
Niels Landwehr
John Lloyd
Stan Matwin
Stephen Muggleton
David Page
Bernhard Pfahringer
Taisuke Sato
Jude Shavlik
Christel Vrain
Akihiro Yamamoto
Filip Zelezny

James Cussens
Sašo Džeroski
Peter Flach
Lise Getoor
Katsumi Inoue
Ros King
Nada Lavrač
Francesca Lisi
Donato Malerba
Raymond Mooney
Ramon Otero
David Poole
Celine Rouveirol
Vitor Costa Santos
Prasad Tadepalli
Stefan Wrobel
Gerson Zaverucha

Additional Reviewers

Laura Antanas
Nicola Di Mauro
Aneta Ivanovska
Ken Kaneiwa
Lily Mihalkova
Barbara Pieters
Lothar Richter
Joerg Wicker

Michelangelo Ceci
Andreas Hapfelmeier
Yoshitaka Kameya
Dragi Kocev
Louis Oliphant
Ingo Thon
Lisa Torrey

Local Organization

Submissions	Jan Struyf
Registration and Hotels	Siegfried Nijssen
	Martijn van Otterlo
	Christophe Costa Florencio
Web	Fabrizio Costa
	Christophe Costa Florencio
Publicity	Kurt Driessens
Proceedings	Kurt Driessens
	Laura Antanas
Logistics	Daan Fierens
Social Program	Celine Vens
Administration	Karin Michiels
Sponsoring	Karsten Borgwardt
	Jan Ramon
Video	Beauregart Piccard
Advice	Maurice Bruynooghe
Volunteers	Laura Antanas
	Eduardo Costa
	Tom Croonenborghs
	Anton Dries
	Robby Goetschalckx
	Fabian Guiza
	Bernd Gutmann
	Angelika Kimmig
	Parisa KordJamshidi
	Theofrastos Mantadelis
	Wannes Meert
	Dhany Saputra
	Leander Schietgat
	Nima Taghipour
	Ingo Thon
	Vinicius Tragante do O

Sponsoring Institutions

BNVKI, the Benelux Association for AI
PASCAL2, the Network of Excellence for Statistical Modelling and Computational Learning
FWO, Research Foundation Flanders, through its Networks "Declarative Methods in Computer Science" and "Machine Learning for Data Mining and its Applications"

Table of Contents

Knowledge-Directed Theory Revision

Kamal Ali, Kevin Leung, Tolga Konik, Dongkyu Choi, and Dan Shapiro

Cognitive Systems Laboratory
Center for the Study of Language and Information
Stanford University, Stanford, CA 94305
kamal3@yahoo.com, kkleung@stanford.edu, konik@stanford.edu,
dongkyuc@stanford.edu, dgs@csli.stanford.edu

Abstract. Using domain knowledge to speed up learning is widely accepted but theory revision of such knowledge continues to use general
syntactic operators. Using such operators for theory revision of teleoreactive logic programs is especially expensive in which proof of a top-level
goal involves playing a game. In such contexts, one should have the option
to complement general theory revision with domain-specific knowledge.
Using American football as an example, we use Icarus' multi-agent teleoreactive logic programming ability to encode a coach agent whose concepts correspond to faults recognized in execution of the play and whose
skills correspond to making repairs in the goals of the player agents.
Our results show effective learning using as few as twenty examples.
We also show that structural changes made by such revision can produce performance gains that cannot be matched by doing only numeric
optimization.

1 Introduction

Teleoreactive logic programs (TLPs) hosted in systems such as Icarus [1] are
programs that are goal-oriented yet able to react when an external factor may
cause already achieved subgoals to become false. A TLP consists of a graph
(hierarchy) of first-order rules and a corresponding graph of skills. Proof of a
top-level goal starts by forward chaining from a set of facts - the *perceptual
buffer* - to compute its transitive closure. Next, proof of the top-level goal is
attempted using backward chaining. When reaching a leaf in the proof tree
which is currently false, the teleoreactive framework will switch to a skill tree,
back-chaining until it reaches a leaf skill which can be executed.

Theory revision of skills is different than revision of concepts in that many
candidate revisions cannot be repeatedly evaluated against a static training set.
As a skill is changed, it forces the environment to react. Evaluation of a candidate revision of a skill requires the game to be played again - several times
for stochastic domains. Thus it becomes imperative to reduce the number of
training examples by orders of magnitude. This leads to the central thesis of
this paper: that a skill revision engine be capable of reading and using domain-
specific revision rules when available, before falling back to general syntactic
revision operators.

L. De Raedt (Ed.): ILP 2009, LNAI 5989, pp. 1–8, 2010.

This paper does not present a sophisticated theory revision algorithm - the whole point is to show that for revision of skills, domain-specific theory revision rules are *necessary* in order to keep down the number of required training examples, and that only a simple revision engine is needed as long it can take advantage of such powerful revision rules. By its sophistication, a skill revision algorithm cannot make up for its lack of domain knowledge if it is to lead to revision in a small number of training examples. The situation is analogous to the transition in AI from general learning algorithms to expert systems, where it became accepted that to speed up learning, the learner should have the capacity to use and learn with domain-specific expert-provided rules before falling back to general learning methods.

For skill revision in American football, for example, it is easy to elicit domain-specific theory revision operators because coaches are well aware of domain-specific ways in which they can perturb their play designs to improve performance. As a very simple example, in a skill where a ball receiver fumbled reception of the ball, the domain-specific revision knowledge base could exploit the knowledge that there are usually multiple receivers to try another receiver. The revision knowledge embodied here is that there is a subset of players that can be thrown to, and that one need not only throw to the player currently specified in the logic program.

To implement domain-specific theory revision, we take advantage of a recent extension to Icarus [1] - multi-agent capability. A natural and elegant way to exploit the multi-agent ability for theory revision is to create a "coach agent" whose concepts correspond to faults observed while the program is in play and whose skills correspond to fixes. As the game proceeds, each agent uses its inferences to decide which skills to execute. At each time tick, the coach agent also makes inferences too - some of them corresponding directly to faults, others used in support of faults. After the game is over, skills of the coach agent are applied - these skills are implemented by rules whose preconditions are arbitrary combinations of faults and whose actions involve modifying programs of the other agents.

Since our thesis is that for skill revision which needs expensive execution of plays, the revision engine should be able to accommodate domain-specific revision rules, we chose American football as a domain because its plays are intricate coordinations chosen before the play by the coach and communicated to all the players. In addition, it has the property that once play begins, the plan may get modified due to efforts of the opposing team, so it also requires the agents to be reactive. These properties are a good fit for the teleoreactive capabilities of Icarus. Our empirical validation in section 5 show that using a proof-of-concept small revision theory using about a dozen faults and ten fixes, domain-specific revisions can produce significant performance improvement using only twenty training examples.

With regard to prior work, these "debugging rules" are similar in sprit to declarative theory refinement operators other ILP systems use, but unlike theory refinement operators that are not particularly designed for temporal events, the debugging rules of our system detect and accumulate problems over to the end of the game. Other authors [2] have something equivalent to debugging knowledge in the form of a *critic agent* but this agent does not rely on execution to find faults - it does its work in the planning phase prior to execution. Gardenfors [3] notes a similarity between belief revision and nonmonotonic logic - this is analogous to our work except at a "meta" level: we are revising theories (not beliefs) so the nonmonotonicity applies between revisions.

The rest of the paper is organized as follows: section 2 presents representation, inference and execution of teleoreactive programs in Icarus. Section 3 explains how we modeled American football in Icarus. Section 4 presents DOM-TR: our theory revision algorithm that inputs domain-specific rules. Section 5 gives results showing how domain-specific theory revision allows effective learning using just a few tens of examples.

2   Teleoreactive Logic in Icarus

Representation -   The ICARUS architecture (details in [1]) represents conceptual and procedural knowledge using first-order logic. Concepts consist of a generalized head, perceptual matching conditions in a :percepts field, tests for constraints in a :tests field, and references to other concepts in a :relations field. For example, the concept *cch-near* shown in table 1 matches against two objects of type *agent*, and binds values to variables (indicated with ?). Skills (right column in table 1) are also represented with generalized heads and bodies - the body can refer directly to executable procedural attachments in an actions field or to other skills in a subgoals field.

Table 1. Some sample concepts and skills in the football domain

```
((cch-near ?a1 ?a2)                           (RWR-play 11-1-013-1)
  :percepts ((agent ?a1 x ?x1 y ?y1)            :percepts ((agent ?N role RWR
             (agent ?a2 x ?x2 y ?y2))                       startx ?X starty ?Y))
  :tests    ((<= (sqrt (+ (expt (- ?x1 ?x2) 2)  :actions  ((*startAt RWR ?X ?Y)
                          (expt (- ?y1 ?y2) 2)))            (*PassRouteCornerLeftAtYard
              *threshold*)))                                 RWR *RWR-ydsb4diagLeft*)
                                                            (*finish ?N)))
((cch-covered-recipient ?a1 ?a2)
  :percepts ((agent ?a1 team offense recipient t) ((QB-play 11-1-013-1)
             (agent ?a2 team defense))             :percepts ((playState 11-1-013-1)
  :relations ((cch-near ?a1 ?a2)))                            (agent ?N role QB
                                                             startx ?X starty ?Y closestRcvr ?R))
((cch-open-recipient ?agent)                       :subgoals ((startAt QB)
  :percepts ((agent ?agent team offense recipient t))         (QBFallback 11-1-013-1)
  :relations ((not                                            (waitForReceivers 11-1-013-1)
              (cch-covered-recipient ?agent ?a2)))) (pass ?R)))
```

Interpretation -   Icarus accepts a user-provided top-level goal per agent, the set of agents being pre-defined by a user. It operates in distinct cycles, and it infers the current state of the world at the beginning of each cycle. Based on the inferred state, the system finds an executable path through its skill hierarchy, which starts with a currently executable skill for the top-level goal and terminates with a primitive skill with procedural attachments. ICARUS repeats the process for all the agents, and executes all implied actions simultaneously in a single cycle.

For example, to achieve a goal, (QB-play 11-1-013-1), the QB agent must execute a corresponding skill with the same head – like the second skill in the right column on table 1 – that entails making true *in sequence* the four sub-goals specified in its subgoals clause. Teleoreactivity in the system mainly results from disjuntive skills it has for a particular goal, which provide alternatives to the system based on the current state of the world.

3   American Football – An Example Domain

For the American football domain, Icarus controls the offensive team. We define one agent per player on the offensive team and an additional agent for the coach. The agents execute their skills against an "environment" that is managed by the RUSH football simulator [4]. RUSH controls the defense, physics of the ball and stochasticity of the environment.

We use the system in [5] to transform video of real games into a sequence of segments per player. Each segment consists of a symbolic label and numeric parameters - for example, slantLeft(253,283,RWR,10) indicates the Right Wide-Receiver should run diagonally for ten yards from time ticks 253 to 283. To translate this into a logic program for Icarus, each sequence is translated into an Icarus skill as in the right column of table 1. Constants in the skills are replaced by variables (parameters) that have those constants as their default values. For example, 10 becomes the variable (parameter) *RWR-yardsb4diagLeft*.

4   Theory Revision

Table 2 gives a few examples of the revision knowledge which was elicited from an expert. Note that it is generalized to first order and can either make changes that are local to an agent or changes that involve multiple agents. By execution, we may find, as shown in rule 1 in Table 2 that X manages to get more open and thus the ball should be thrown to X. The revision operators can change variables into constants and vice versa, replace predicates, introduce new variables and

Table 2. Examples of domain-specific faults and fixes

Fault: cch-farther-open-recipient(X,Y)
Description: X is an open receiver farther downfield than Y
Fix: Change intended receiver to X from Y
Fault: cch-crowded-recipients(X,Y)
Description: Receivers X and Y are too close to each other
Fix: Move the starting locations of each receivers 2 yards apart

introduce negation. Not only is the primary theory first-order, but the revisions themselves are first order. For instance, the skill in the lower right of table 1 was modified by theory revision which introduced a new variable (?R) whose value is picked up from the :percepts clause and used in the :subgoals clause.

Table 3. DOM-TR: Structural Learning Breadth-First Search

```
learn-structure(MaxDepth)
1   Depth = Depth + 1
2   For N (4) times:
3       Simulate the play, record faults into Faults
4   If Faults is empty then exit
5   For fix in fixes(Faults) do
6       Temporarily apply fix
7       Simulate N (4) times, recording rewards
8   AR = reward from best fault/fix pair
9   Improvement = AR - BestReward
10  if Improvement ≤ 0 then exit
11  else commit fix; BestReward = AR
12  If depth ≤ MaxDepth then go to 1
```

Table 3 presents DOM-TR - a breadth-first algorithm that inputs a domain-specific debugging theory to do theory revision. The domain-specificity of the debugging theory greatly reduces the breadth of the search - for our current knowledge base, the breadth factor is usually only one or two. Note that the core step of the algorithm - evaluating a fix - involves playing several ($N = 4$) games[1] because of the stochasticity of this domain.

Table 3 shows that the algorithm begins by playing N games, recording all occuring faults. The other purpose of playing these games is to establish a baseline reward value. After playing the N initial games, it iterates over fixes whose preconditions match the observed faults. For each fix, it applies the fix, plays N games to compute average reward and then undoes the fix. If the best fix so found produces an improvement compared to the baseline, it is permanently applied, otherwise the search terminates. We do not check explicitly for cycles (where a revision undoes a previous revision), opting instead to limit the depth of the revision tree to some small value such as $MaxDepth = 5$.

5 Results

For our experimental methodology, we randomly select an offense from our library of videos and a defense from RUSH's library. In each such trial, DOM-TR is used to produce a learning curve. The process is repeated twenty times. Figure 1 shows the results - application of DOM-TR to our small prototype revision theory improves performance from a baseline figure of 1 yard to 4 yards. Although this gain may appear numerically small, a repeatable three-yard gain using only 20 plays (and given that our revision theory is just a proof-of-concept) is significant in American football.

---

[1] Early experiments showed that values of N higher than 4 did not yield substantially greater benefits.

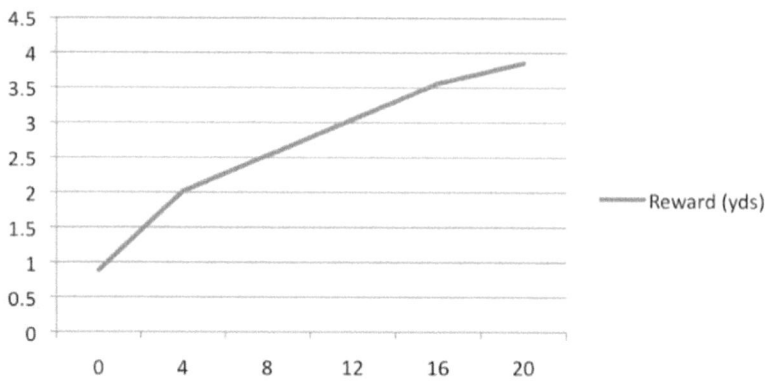

Fig. 1. Learning curve for theory revision

In a second experiment, we wanted to compare structural learning using DOM-TR to an algorithm that only did numeric optimization (by hill-climbing over all parameters mentioned in the theory). In figure 2, the leftmost points of the curve correspond to the end of the theory revision stage ($x = 20$) and show that DOM-TR achieves a performance of six yards by using the domain-specific theory revision rules. The middle curve corresponds to taking the same Icarus program but disallowing structure learning - only using parameter learning. The results show that without the benefit of the powerful domain-specific theory revision rules, to reach the six yard performance mark takes parameter optimization almost an order of magnitude more examples ($x = 160$ versus $x = 20$). Furthermore, even asymptotically, the algorithm that only does parameter optimization

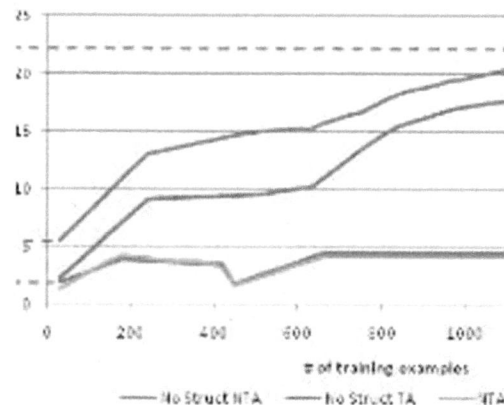

Fig. 2. Uppermost curve: structural and parameter learning; middle curve: parameter learning only; bottom curve: learning for a "knowledge-free" agent

(middle curve) cannot match the reward obtained by the hybrid algorithm (upper curve) that combines structure learning with numeric optimization.

So far, we have compared two learning algorithms, both on a rich domain-specific theory. Now we vary the program itself to answer "what is the impact of this domain-specific theory"? - can we compare it to a "knowledge-free" theory to serve as a control? It is somewhat problematic to define a knowledge-free program that achieves non-zero performance. Should the knowledge-free program be just the null program? Should it correspond to an agent that knows about team games but not American football? Should it correspond to expert-level knowledge from rugby that should serve as a starting point for adaptation to American football?

In order to achieve non-zero performance, we defined the control theory to be one that knows all the rules of American football but none of the strategies that coaches possess in their toolkits. We defined such a "naive" agent to be one that places all players that are eligible receivers on the scrimmage line under the instructions to run as fast as possible towards the touchdown line. The quarterback's behavior is defined to throw immediately (he does not have the notion of a protective fallback) to the receiver that is furthest downfield. Thus this program lacks notions of coordinated and distracting plays, notions of running downfield and then cutting across, and so forth. The Icarus program that corresponds to this definition has a non-zero set of rules so the theory-revision change-knowledge rules we used earlier can also be applied to this program. It also has parameters but they parameterize a space that consists simply of starting x and y coordinates for the players. This contrasts with the informed program whose parameters correspond to semantically rich and relevant constructs such as the length of time the quarterback should wait before throwing the ball, or how far he should fall back from the scrimmage line to protect his throw.

Figure 2 shows that after theory revision ($x = 20$) this program achieves a performance of two yards, in comparison to the six yard gain achieved when theory revision was applied to the informed theory (some theory revision rules such as switching receivers are applicable even to this control theory). Thus, applying domain-specific theory revision operators to the informed theory produce a 200% gain over application of the same revision rules to the uninformed theory.

Figure 2 also shows that the control agent is only able to take minimal advantage of the greedy hill-climbing parameter-learning algorithm - probably because its parameters correspond to low-level x and y coordinates. In American football, it is important for there to be coordination between multiple players which may not be learnable by a greedy method operating over a set of raw, low-level parameters. such as coordinates. By contrast, the greedy hill-climbing algorithm is able to significantly improve the informed theory since its parameters implicitly correspond to degrees of coordination between multiple players.

6   Conclusions

In domains where learning from training examples is orders of magnitude more expensive than usual and where skills need to be revised, general syntactic-only theory revision needs to be augmented by domain-specific revision knowledge. This paper demonstrates theory revision in the context of multi-agent teleo-reactive logic programs by the implementation of a "coach" agent whose concepts correspond to faults and whose skills correspond to fixes that modify those agents' programs. Using American football as an example, we have demonstrated that domain-specific theory revision can produce meaningful gains using as few as twenty examples and that this affords an order of magnitude faster learning that that realized by doing only parameter optimization.

Acknowledgements

This paper reports research sponsored by DARPA under agreement FA8750-05-2-0283. The U. S. Government may reproduce and distribute reprints for Governmental purposes notwithstanding any copyrights. The authors' views and conclusions should not be interpreted as representing official policies or endorsements, expressed or implied, of DARPA or the Government.

References

1. Langley, P., Choi, D.: A unified cognitive architecture for physical agents. In: Proceedings of the Twenty-First National Conference on Artificial Intelligence, Boston. AAAI Press, Menlo Park (2006)
2. Wilkins, D., Myers, K., Lowrance, J., Wesley, L.: A multiagent planning architecture. In: Proceedings of AIPS 1998 (1998)
3. Gardenfors, P.: Belief revision and nonomotonic logic: Two sides of the same coin? In: Proceedings of the ninth European Conference on Artificial Intelligence. Pitman Publishing (1990)
4. Sourceforge: Sourceforge.net - rush 2005 (2005),
   http://sourceforge.net/projects/rush2005/
5. Hess, R., Fern, A.: Discriminatively trained particle filters for complex multi-object tracking. In: IEEE Conference on Computer Vision and Pattern Recognition (2009)

Towards Clausal Discovery for Stream Mining

Anton Dries and Luc De Raedt

Department of Computer Science,
Katholieke Universiteit Leuven
{firstname.lastname}@cs.kuleuven.be

Abstract. With the increasing popularity of data streams it has become time to adapt logical and relational learning techniques for dealing with streams. In this note, we present our preliminary results on upgrading the clausal discovery paradigm towards the mining of streams. In this setting, there is a stream of interpretations and the goal is to learn a clausal theory that is satisfied by these interpretations. Furthermore, in data streams the interpretations can be read (and processed) only once.

1 Introduction

The data base and data mining community has devoted a lot of attention recently to dealing with data streams. A data stream consists of a continuous, often high speed, flow of data points that is so large that each data point can be processed just once. Often these data streams are also susceptible to continuously evolving underlying patterns [1]. Special purpose querying and mining techniques have been developed to cope with such streams in the literature [2]. However, to the best of the authors' knowledge, such techniques have not yet been used in a logical and relational learning setting and there are several interesting questions that arise in this context: What is an appropriate formalization of stream mining in a logical and relational learning setting? and if we have such a setting, can we still develop efficient learning techniques? and for what purposes can this setting be employed?

In this note, we provide an initial answer to these questions. More specifically, we show that the learning from interpretations setting introduced by Valiant [3] in the propositional case and upgraded by De Raedt and Dzeroski [4] for the relational case, constitutes an appropriate setting for read-once stream mining. Furthermore, we also show how the algorithms used in these PAC-learning results can be adapted for use in an incremental read-once setting. The resulting algorithm computes a *jk*-clausal theory that is satisfied by all examples. While the learning from interpretations setting has been quite popular in inductive logic programming, cf. Claudien [5], Tertius [6], Logan-H [7], it is – to the best of the authors' knowledge – the first time that a read-once logical and relational learning algorithm is developed. The framework we present extends and generalizes that of Dries et al. for the propositional case of mining mining k-CNF theories from streams of item-sets [8]. Dries et al. have also shown that the induced propositional theories can be used for a variety of purposes, such as classification, missing value imputation and detection of concept drift [9].

L. De Raedt (Ed.): ILP 2009, LNAI 5989, pp. 9–16, 2010.
© Springer-Verlag Berlin Heidelberg 2010

2   A Logical and Relational Stream Mining Definition

We shall assume that examples are (Herbrand) interpretations, that is, sets of ground atoms. We shall induce clausal theories consisting of a set of (range-restricted) clauses, where each clause is a formula of the form $c : h_1 \vee \ldots \vee h_n \leftarrow b_1 \wedge \ldots \wedge b_m$ where the h_i and b_i are logical atoms. We call the disjunction of h_is the head of the clause and the conjunction of b_is the body of the clause[1]. An example e *satisfies* or is *covered by* a clause $c : H \leftarrow B$, notation: $e \models c$, if and only if $\forall \theta : B\theta \not\subseteq e \vee H\theta \cap e \neq \emptyset$. When there exists a substitution θ such that $B\theta \subseteq e$ and $H\theta \cap e = \emptyset$ we say that the example *violates* the clause and call θ a *violating substitution*.

As a generality relation we shall employ the usual notion of θ-subsumption. Basically, a clause s θ-subsumes a clause g, notation $s \preceq g$, if and only if there exists a substitution θ such that $s\theta \subseteq g$. In the learning from interpretations setting, the direction of generality is reversed as compared to the learning from entailment setting, cf. [10]. Therefore, the θ-subsumption is also monotonic, that is, if a clause s is satisfied by an example e, and $s \preceq g$ then e will also satisfy g. This property will be important for the stream mining setting, because once an example is covered by a clause, it is guaranteed that all refinements under θ-subsumption (applying substitutions or adding literals to the head or the body) will also cover the example, and hence, the example can be forgotten.

Systems such as Claudien [5] and Tertius [6] now start from a given set of examples E, a language of clauses $\mathcal{L}$ and compute the theory

$$\mathcal{T} = \{c \in \mathcal{L} | E \models c \text{ and } \forall s \in \mathcal{L} : s \prec c \rightarrow E \not\models s\}.$$

This theory only computes the maximally specific clauses covering the examples, as the others ones are entailed by $\mathcal{T}$, and, hence, logically redundant. Claudien and Tertius compute – in a way – $\mathcal{T}$ and realize this by treating the examples in batch using top-down search. However, this does not satisfy the read-once requirement of stream mining. We therefore propose to compute $\mathcal{T}$ incrementally and requiring that each example is read just once. This leads to the following problem that has to be tackled once for each example:

Given:

- a language of clauses $\mathcal{L}$ with language bias
- a theory of clauses $\mathcal{T}_{n-1} \subseteq \mathcal{L}$
- an interpretation e

Find $\mathcal{T}_n = \{c \in \mathcal{T}_{n-1} | e \models c\} \cup$
$\{g \in \mathcal{L} | \exists s \in \mathcal{T}_{n-1} : e \not\models s \text{ and } s \prec g \text{ and } e \models g \text{ and}$
$\neg \exists g' \in \mathcal{L} : s \prec g' \prec g : e \models g'\}$

---

[1] A clause is range-restricted if all variables appearing in the head of the clause also appear in its body.

It should be clear that the final theory that is computed in this way coincides with the theory that is computed directly, that is, by processing the examples in batch using a non-heuristic, complete algorithm, and, that the final theory is independent of the example order.

The problem setting can also easily be extended to take into account background knowledge in the usual way when learning from interpretations. It corresponds to employing a definite clause theory and computing for each example the least Herband model of the background theory and the set of facts describing the clauses, cf. [10,5].

This setting is also natural for dealing with streams, because in many applications such as robotics and video analysis every state can be described in a natural way by an interpretation and over time streams of such interpretations would be generated. Although it remains an open question, it seems harder to adapt the usual notion of learning from entailment to mining streams because that setting assumes information about the examples resides in the background theory and has – mainly – been focussing on heuristic classification and not on description.

Finally, as shown for the propositional case by Dries et al. [8], the setting above can be used for classification and for missing information imputation. For example, when confronted with an unlabeled example a score is calculated for each possible class label. This score is based on the number of clauses that are violated by the example and the class label. The optimal class label is then chosen as the one with the best score, i.e., the one that violate the least amount of clauses. Because the learning algorithm does not make a distinction between a class attribute and any other attribute, the same method can be used to predict an arbitrary number of missing values. In the first-order case these missing value problems correspond to finding suitable completions for partial interpretations, or in other word, what information do we need to add to an interpretation to make it satisfy the current theory?

3   Algorithm

The algorithm that we introduce is based on the candidate-elimination algorithm of Mitchell [11], which computes border-sets of hypothesis consistent with the data. The set $\mathcal{T}$ that we compute can be considered the border of maximally specific clauses consistent with the examples. The resulting algorithm is shown below. The initial theory contains the maximally specific clause, that is, the empty one, which is not satisfied by any example.

For each example in the stream of examples, we then repeat the following three steps: find clauses in $\mathcal{T}_n$ that are not satisfied by the current example, refine these clauses, and possibly compress the resulting theory to remove those clauses that are not maximally specific, and, hence, logically redundant. Because the first step and third step are straight-forward, we focus on the second step.

The most interesting step is that of applying the refinement operator $\rho(c, e)$ to generate refinements of the clause c with respect to e, that is, $\rho(c, e) = \{c' | c \prec c'$ and $e \models c'$ and not $\exists d : c \prec d \prec c', e \models d\}$. This corresponds to computing the minimally general generalizations c' of c such that $e \models c'$.

As usual in logical and relational learning, the refinement operator is governed by a language bias that defines which literals can be added to the head and which literals can be added to the body of the clause, and that also specifies mode restrictions on the variables in the new literals. These mode restrictions specify whether a literal can use constants, existing variables, or whether it can introduce new variables. The refinement operator generates a complete set of refinements within the restrictions set out by the bias. This completeness guarantees that all possible (minimal) clauses are generated and that the final theory is independent of the order of the examples in the data stream.

$$\mathcal{T}_0 = \{\Box\}$$
for $e_n \in \mathcal{E}$ **do**
    for $c \in \mathcal{T}_{n-1}$ **do**
        if $e_n \models c$ **then**
            add c to $\mathcal{T}_n$
        else
            add all $c' \in \rho(c, e_n)$ **to** $\mathcal{T}_n$
        end if
    end for
    compress$(\mathcal{T}_n)$
end for

Fig. 1. Incremental clausal discovery algorithm

Because there are typically many possible refinements of a clause, we want to minimize the number of generated refined clauses by eliminating redundant ones already during the refinement phase. Here, the read-once property comes in handy because the refined clause should only add those literals that are necessary for the current example. To this aim, we employ the set of substitutions Θ_F that violate the clause. We can distinguish between two types of refinements. On one hand we have literals that only use variables that are already part of the clause. Finding a minimal set of such literals is equivalent to solving a set covering problem where the chosen literals should form a minimal set that covers all violating substitutions. We do this by first listing the possible literals that can be added to the clause and the violating substitutions they resolve. This gives us a set of the following form $\{l_1 : [\theta_1, \theta_2, \theta_3], l_2 : [\theta_3, \theta_4], l_3 : [\theta_1, \theta_2, \theta_5], l_4 : [\theta_1, \theta_2, \theta_3, \theta_5]\}$, which indicates that, for example, l_1 resolves substitutions θ_1, θ_2, and θ_3. From this we find all minimal sets of literals that cover all substitutions. In this case $\{l_2, l_3\}$ and $\{l_2, l_4\}$ are the only such sets, and $\{l_1, l_2, l_3\}$ is not because l_1 is redundant.

On the other hand we have literals that add new variables to the clause. Avoiding redundancy for these literals is harder, because they cause dependencies between the new literals. For example, it would only be possible to add a literal that uses variable X to the head of a clause if a literal creating variable X is already added to the body of the clause. However, it is still possible to reduce the number of redundant refinements by taking into account the violating substitutions of the clause for the current example.

Finding an exact solution to this problem is hard because there are many cases to consider. Therefore, we use a heuristic approach that uses a scoring function on new literals based on the set Θ_F and the current example. It is important to note that this heuristic only reduces the number of non-minimal refinements, and has no impact on the final theory of the algorithm.

Our approach is based on a simple and effective way of avoiding the generation of permuted versions of the same clause. These permutations can be easily avoided by using a fixed order in which new literals are added. This approach is used in most refinement operators found in ILP. We modify this approach by using a dynamic order on the literals that allows us to greedily expand the best refinements first. This greedy breadth-first approach requires a scoring function that assigns a value to (partial) clauses based on how well they cover the current example. This scoring function is based on the fraction of all possible substitutions (for the current example) that violate the clause. This score can be easily calculated based on the set of violating substitutions Θ_F and the constants present in the example using the formula

$$Score(c, e) = \frac{|\Theta_F(c, e)|}{\prod_{Var \in c} |Values(e, Var)|}$$

In this formula $|Value(e, Var)|$ is the number of distinct values, present in example e, that can be assigned to Var (i.e. that have same type). This function has three interesting properties: (1) it can only decrease (or stay equal) when a literal is added to the clause, (2) it can deal with the introduction of new variables in the clause, and (3) it reaches zero if-and-only-if the clause is satisfied by the example and does not require further refinement. Furthermore, if a new literal does not decrease the value of this function, then it is redundant (unless it introduces a new variable).

4   Extended Example

Consider the following simple language for describing scenes containing objects

$$\mathcal{L} = \{square(id), triangle(id), circle(id), in(id, id)\},$$

and its associated background knowledge

$$polygon(X) \leftarrow triangle(X). \qquad polygon(X) \leftarrow square(X).$$

We use this language to represent the following two scenes.

$$e_1 = \{triangle(a), circle(b), in(a, b)\} \quad \triangle$$

$$e_2 = \{square(x), triangle(y), in(x, y)\} \quad \triangle$$

We set our initial theory $\mathcal{T}_0$ to contain only the empty clause $\square$. After processing example e_1 the system learns 10 different clauses.

$triangle(A) \leftarrow in(A, B)$	(1)		$\leftarrow polygon(A) \wedge circle(A)$	(6)
$circle(B) \leftarrow in(A, B)$	(2)		$\leftarrow polygon(A) \wedge in(B, A)$	(7)
$polygon(A) \leftarrow in(A, B)$	(3)		$\leftarrow square(A)$	(8)
$triangle(A) \leftarrow polygon(A)$	(4)		$\leftarrow circle(A) \wedge in(A, B)$	(9)
$\leftarrow in(A, A)$	(5)		$\leftarrow in(A, B) \wedge in(B, A)$	(10)

These are all the minimal clauses that can be learned from this limited example. There is still some redundancy in this set, for example clauses (1) and (6) combined with background knowledge imply clause (9). However, we cannot remove clause (9) because clauses (1) and (6) might not remain true for future examples. Indeed, when we update our theory with respect to example e_2 we see that clauses (3), (5), (6), (9) and (10) still apply, but the other, including clause (1), no longer do. If we had removed clause (9), it would no longer be part of the theory, and since it is not a refinement of any existing clause, it would not be reconsidered as part of the theory.

We can now repeat the process for example e_2. Clause (7) $\leftarrow polygon(X) \wedge in(Y, X)$, for example, is violated by example e_2 because there is a substitution $\theta_F = \{X/b, Y/a\}$ which violates the clause. There are several refinements that resolve this substitution, for example $triangle(X)$, $square(Y)$, or $\neg triangle(Y)$. In total, theory $\mathcal{T}_2$ contains 45 clauses.

5   Experiments

Implementation. We implemented a preliminary version of our algorithm in Sicstus Prolog. It uses a specialized trie-based data structure to store the theory of clauses, which enables fast discovery of violated clauses and clause subsumption. In its current version this implementation has two limitations: (1) it does not test for background subsumption, i.e., it does not remove clauses that are subsumed by the background knowledge, and (2) it enforces inequality constraints on variables, i.e., it forces different variables to take distinct values. Both these limitations lead to a final theory that contains more clauses than strictly necessary, but which is still minimal under these restrictions.

Poker dataset. For our experiments we primarily used the Poker dataset to analyze the behavior and performance of our algorithm. This dataset was introduced by Blockeel et al. [12] and contains a descriptions of poker hands and their classification. We used a dataset containing 10.000 interpretations of the form,

$$\{card(R_1, S_1), card(R_2, S_2), card(R_3, S_3), card(R_4, S_4), card(R_5, S_5), class(C)\},$$

where R_i and S_i indicate rank (ace through king) and suit (spades,hearts,clubs or diamonds) respectively. This dataset contains seven classes: nothing, pair, double pair, full house, three of a kind, flush, and poker.

Rule learning behavior. First we ran our algorithm directly on this dataset without any additional background knowledge or length restriction on the learned clauses. Processing the entire dataset took approximately 125 seconds and produced 154 rules. Within these clauses several subgroups can be discovered. For example, there a 60 different clauses to express that there can not be 6 cards in a hand, or there are 8 different rules to express the elements of class *pair*. Some of these groups could be simplified by using more flexible variable inequality constraints. Interestingly, only 32 examples contributed to the theory; the other 9.968 examples did not violate any clauses, and consequently did not update the theory.

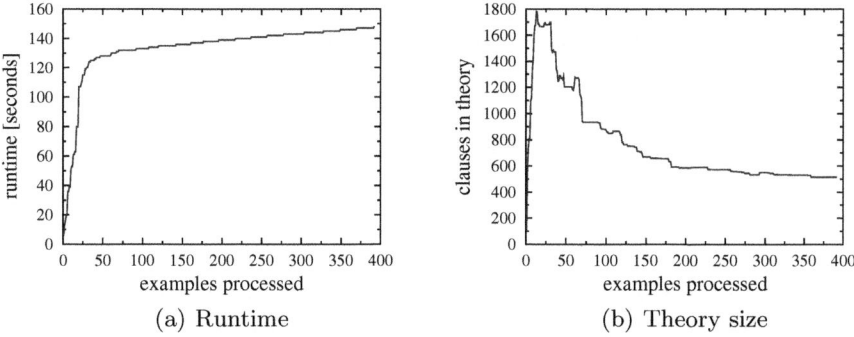

(a) Runtime   (b) Theory size

Fig. 2. Evolution of runtime and theory size for the Bongard dataset

For the second run, we defined two background predicates: *same_rank* and *same_suit* that count the number of cards of the same rank and suit respectively, for example, $same\_rank(2, R) \leftarrow card(R, A) \land card(R, B) \land A \neq B$.

We modified our language bias to use these new predicates instead of the *card* atom. Running the algorithm with this setting took 25 seconds and produced 95 rules. In this case, only 12 examples were responsible for the entire theory. Interestingly, three of the classes (full house, flush and poker) were only represented by a single example in this reduced dataset.

We also repeated these experiments for a dataset containing Bongard problems [13] (of which a simplified example was used in the previous section). We restricted the theory to clauses of length 5. The results of these experiments are summarized in Figure 2. For this dataset the final theory was based on 94 out of 392 examples. Figure 2(b) shows clearly the effect of pruning due to the maximal length constraint.

Classification and completion of partial interpretations. We also performed experiments to see whether the classification and imputation results from [8] still hold for the first-order case. For this experiment we randomly generated two sets of 100 poker hands, where we made sure that each class was represented in both datasets. We trained a single theory (without background knowledge) on the first set and evaluated this theory on the second set.

First we used the standard classification setting, where we removed the class label from the examples in the test set. We then assigned a score for each class label based on the number of clauses violated by the example extended with this class label. For most examples, assigning the correct class label did not violate any clauses, while assigning other class labels did. We repeated this experiment five times on different random datasets, producing an overall accuracy of 97.8%.

We also used our algorithm to complete a partial poker hand of four cards and a class label. We used the same theories learned in the previous setting and applied the same scoring mechanism. In this setting, multiple alternatives where predicted for each example. For example, when given the partial interpretation

$$[card(2, clubs), card(9, diamonds), card(9, spades), card(2, hearts), class(full\_house)]$$

the system would output

$$[card(9, hearts), card(9, clubs), card(2, spades), card(2, diamonds)]$$

as possible fifth cards. In all cases the original card was among those predicted.

6   Conclusions and Future Work

We introduced – for the first time – a framework for stream mining in the setting of logical and relational learning. Our approach upgrades the setting of clausal discovery in interpretations, as used by earlier systems such as Claudien [5], Tertius [6], and Logan-H [7], towards mining of streams. We proposed an algorithm based on the Candidate Elimination algorithm of Mitchell [11], and applied it to the problems of classification and completion of partial interpretations.

Future work in this area includes (1) a further optimization and implementation of our algorithm, (2) a thorough experimental verification of our preliminary results and a comparison with other ILP systems, and (3) application of the framework to other stream mining problems such as concept drift detection.

Acknowledgments

This work was partially supported by the GOA project 08/008 on Probabilistic Logic Learning and the European Commission FP7 project BISON.

References

1. Aggarwal, C.C.: Data streams: models and algorithms. Springer, New York (2007)
2. Gaber, M.M., Zaslavsky, A., Krishnaswamy, S.: Mining data streams: a review. SIGMOD Record 34(2), 18–26 (2005)
3. Valiant, L.G.: A theory of the learnable. Communications of the ACM 27(11), 1134–1142 (1984)
4. De Raedt, L., Džeroski, S.: First order jk-clausal theories are PAC-learnable. Artificial Intelligence 70(1-2), 375–392 (1994)
5. De Raedt, L., Dehaspe, L.: Clausal discovery. Machine Learning 26(2-3), 99–146 (1997)
6. Flach, P.A., Lachiche, N.: Confirmation-guided discovery of first-order rules with Tertius. Machine Learning 42(1-2), 61–95 (2001)
7. Arias, M., Khardon, R., Maloberti, J.: Learning horn expressions with LOGAN-H. Journal of Machine Learning Research 8, 549–587 (2007)
8. Dries, A., Nijssen, S., De Raedt, L.: Mining predictive k-CNF expressions. IEEE Transactions on Knowledge and Data Engineering (2009) (in preprint)
9. Dries, A., Rückert, U.: Adaptive concept drift detection. In: SIAM International Conference on Data Mining, May 2009, pp. 233–244. SIAM, Philadelphia (2009)
10. De Raedt, L.: Logical and Relational Learning. Springer, Heidelberg (2008)
11. Mitchell, T.M.: Machine Learning. McGraw-Hill, New York (1997)
12. Blockeel, H., De Raedt, L., Jacobs, N., Demoen, B.: Scaling up inductive logic programming by learning from interpretations. Data Mining and Knowledge Discovery 3, 59–83 (2000)
13. De Raedt, L., Van Laer, W.: Inductive constraint logic. In: Zeugmann, T., Shinohara, T., Jantke, K.P. (eds.) ALT 1995. LNCS, vol. 997, pp. 80–94. Springer, Heidelberg (1995)

On the Relationship between Logical Bayesian Networks and Probabilistic Logic Programming Based on the Distribution Semantics

Daan Fierens

Katholieke Universiteit Leuven, Department of Computer Science, Celestijnenlaan 200A,
3001 Heverlee, Belgium
Daan.Fierens@cs.kuleuven.be

Abstract. A significant part of current research on (inductive) logic programming deals with probabilistic logical models. Over the last decade many logics or languages for representing such models have been introduced. There is currently a great need for insight into the relationships between all these languages. One kind of languages are those that extend probabilistic models with elements of logic, such as the language of Logical Bayesian Networks (LBNs). Some other languages follow the converse strategy of extending logic programs with a probabilistic semantics, often in a way similar to that of Sato's distribution semantics.

In this paper we study the relationship between the language of LBNs and languages based on the distribution semantics. Concretely, we define a mapping from LBNs to theories in the Independent Choice Logic (ICL). We also show how this mapping can be used to learn ICL theories from data.

Keywords: probabilistic (inductive) logic programming, Bayesian networks.

1 Introduction

In the fields of *probabilistic inductive logic programming (PILP)* [3] and *statistical relational learning (SRL)* [8] there is a large interest in probabilistic logical models. Over the last decade many different languages for representing such models have been introduced.

One class of languages deals with probabilistic extensions of logic programs. Syntactically one typically uses logic programs in which facts, clauses, or heads of clauses are annotated with probabilities. Semantically one often relies (explicitly or implicitly) on Sato's *distribution semantics (DS)* [13]. We refer to languages that fit this description as *DS languages*. Examples of DS languages are PRISM [3, Ch.5], the Independent Choice Logic [12], ProbLog [9] and Logic Programs with Annotated Disjunctions [14].

Another popular class of languages deals with extensions of probabilistic graphical models to the relational case. For instance, Markov Logic [8, Ch.12] and Relational Markov Networks [8, Ch.6] are based on undirected models, while many other languages are based on directed models: Relational Bayesian Networks [3, Ch.13], Probabilistic Relational Models [8, Ch.5], Bayesian Logic Programs [8, Ch.10], BLOG [8, Ch.13], CLP($\mathcal{BN}$) [3, Ch.6], *Logical Bayesian Networks (LBNs)* [4,5] and others. In this paper we focus on the language of LBNs, which is strongly related to other languages based on Bayesian networks, especially BLPs and PRMs [4].

L. De Raedt (Ed.): ILP 2009, LNAI 5989, pp. 17–24, 2010.

1.1   Problem Statement, Goal and Contributions

The plethora of languages in SRL and PILP is sometimes referred to as 'alphabet soup' (consisting of the acronyms of the many languages). There is currently a great need for insight into the relationships between all these languages [2]. The goal of this paper is to study the relationship between LBNs and DS languages. For concreteness, we focus in this paper on one particular DS language, namely the *Independent Choice Logic (ICL)* [12], but our discussion largely applies to each of the other DS languages as well.

One tool for obtaining insight into the relationships between various languages is to define translations or mappings between them [3, Ch.12+13][2]. In this paper we show that each LBN can be mapped to an equivalent ICL theory (this is our first contribution). Based on this mapping, we show how the existing learning algorithms for LBNs can serve as a basis for learning ICL theories (second contribution).

1.2   Structure of This Paper

We briefly review ICL in Section 2 and LBNs in Section 3. In Section 4 we explain how to map an LBN to an equivalent ICL theory. In Section 5 we consider the problem of learning ICL theories from data (by means of the algorithms developed for LBNs).

2   Independent Choice Logic (ICL)

In the definitions below we assume the existence of two disjoint sets of atoms: the set of *base atoms* and the set of *derived atoms*.

An *ICL theory* is a pair $(R, \mathcal{A})$ with R an acyclic logic program and $\mathcal{A}$ a set of annotated alternatives. An *annotated alternative* is a finite set of base atoms each annotated with a probability (to be precise, each atom α is annotated with a number $P_0(\alpha) \in [0, 1]$ such that $\sum_\alpha P_0(\alpha) = 1$). The logic program R in an ICL theory is constrained in the sense that the heads of the clauses cannot unify with base atoms (only with derived atoms). The set of annotated alternatives $\mathcal{A}$ is constrained in the sense that no atom in any alternative can unify with any other atom in the same or a different alternative. These constraints on R and $\mathcal{A}$ are needed to allow for an 'independent choice' of a base atom from each (grounded) annotated alternative. In this paper we assume that there is a finite number of annotated alternatives in $\mathcal{A}$ and that R is functor free.

The semantics of an ICL theory is that it defines a *probability distribution over possible worlds*. A *possible world* is an interpretation of all (base and derived) atoms. The distribution over possible worlds is derived from the annotated alternatives $\mathcal{A}$ and the logic program R as follows. A *total choice* is a set of ground base atoms that can be obtained by selecting from each grounding of each annotated alternative in $\mathcal{A}$ exactly one atom α. The probability of a total choice C is defined as $\prod_{\alpha \in C} P_0(\alpha)$ with $P_0(\alpha)$ the probability of α according to the annotated alternative that it was selected from. To each total choice C corresponds one possible world: the world in which an atom is true if and only if it is entailed by C and R, i.e., if it is in the stable model of $C \cup R$.[1] The probability of this world is the same as that of its total choice C.

---

[1] For negation free programs the stable model is equal to the least Herbrand model.

The above definitions are essentially those of Poole [12][3, Ch.8]. For the purpose of this paper, we need one extension to these definitions: the extension of ICL with *aggregates*. Concretely, we allow aggregate literals in the bodies of the clauses of the logic program R (see the example in Section 4.1). For this we need an extension of the stable model semantics towards logic programs with aggregates. We use the extension of Pelov et al. [11]. One potential complication with using this in ICL is that all stable models in ICL are required to be unique and two-valued [12, p.29]. The stable models of Pelov et al. satisfy these requirements for logic programs that are 'aggregate stratified' [11]. Fortunately, all programs that we consider in this paper are indeed stratified.

3   Logical Bayesian Networks (LBNs)

In LBNs [4,5] we assume that there are some predicates that determine the domain of discourse and that there is no uncertainty about these predicates. For example, in a university domain we could have predicates $student/1$, $course/1$ and $takes/2$. The semantics of an LBN is only defined with respect to a given interpretation of these predicates. We refer to such an interpretation as an *input interpretation* for that LBN.

In LBNs, special predicates are used to represent random variables (RVs). We refer to such predicates and the corresponding atoms as *probabilistic predicates/atoms*. A ground probabilistic atom represents a specific RV, while a non-ground probabilistic atom represents a 'parameterized' RV. Each probabilistic predicate p has an associated 'random variable declaration', or briefly *declaration*, which specifies which RVs built from p exist for a certain input interpretation. Each probabilistic predicate also has an associated *range*, which specifies the (finite) set of values that these RVs can take. In our university example, probabilistic predicates could be $grade/2$ with declaration $random(grade(S, C)) \leftarrow student(S), course(C), takes(S, C)$ and range $\{good, ok, bad\}$, and $graduates/1$ with declaration $random(graduates(S)) \leftarrow student(S)$ and range $\{yes, no\}$. When given an input interpretation I (that specifies for instance the predicates $student/1$, $course/1$, and $takes/2$), we can use the declarations to obtain the set of all RVs that are defined for I. We denote this set by $\mathcal{RV}(I)$.

Another concept in LBNs is that of a *first-order logical probability tree* for a probabilistic predicate p. This is a decision tree in which each internal node contains a boolean test, and each leaf node contains a probability distribution on the range of p. The purpose of such a tree is to specify how RVs built from the predicate p depend on the other RVs. An example of a tree for $graduates/1$ is given in Figure 1 (the parameterized RV $graduates(S)$ is called the 'target' of this tree). As this tree shows, two types of tests can be used in internal nodes.

- The first type is a test on the value of a parameterized RV such as $grade(S, C)=bad$. Logical variables that occur in such tests but not in the target of the tree, such as C, are (implicitly) existentially quantified. Hence the test $grade(S, C)=bad$ checks whether there exists a course C for which the given student S has grade 'bad'.
- The second type is an aggregate test such as $mode(grade(S, C2))=good$. Logical variables that occur in such tests but not in the target of the tree, such as $C2$, are aggregated over. Hence the test $mode(grade(S, C2))=good$ checks whether the most frequent grade of S over all courses $C2$ is 'good'.

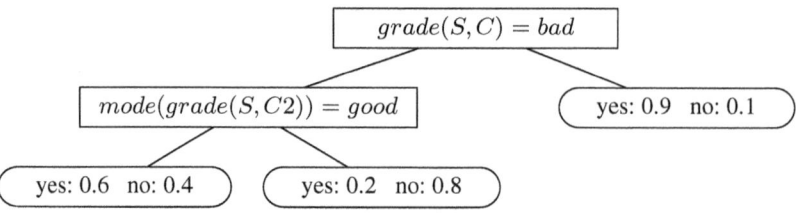

Fig. 1. An example of a first-order logical probability tree for the target *graduates*(*S*). If a test in a node succeeds, the left branch is taken, otherwise the right branch.

An *LBN* consists of one declaration and logical probability tree for each probabilistic predicate. Given an input interpretation I, the trees in an LBN determine a dependency relation between the RVs in $\mathcal{RV}(I)$ (the so-called 'parent relation' [4]). If this dependency relation is acyclic, then we call I a *legal input interpretation* for that LBN.

The semantics of an LBN is that it *maps each legal input interpretation I to a probability distribution over the possible worlds for I*. A possible world is a joint state of the RVs in $\mathcal{RV}(I)$. The probability of a world is defined as a product of conditional probabilities, like in a Bayesian network. The conditional probability distribution for a particular RV given its parents is defined by the corresponding logical probability tree.

4 Mapping an LBN to an ICL Theory

Poole already showed that any discrete Bayesian network can be mapped to an ICL theory that specifies the same probability distribution [3, Ch.8]. In this work we essentially extend this propositional result to the first-order case.

Given an LBN, we want to find an 'equivalent' ICL theory. A technical complication is that an ICL theory directly defines a probability distribution, while an LBN maps input interpretations to probability distributions.[2] Hence we define the mapping problem as follows: given an LBN L, <u>find</u> a logic program R and a set of annotated alternatives $\mathcal{A}$ <u>such that</u> for any input interpretation I that is legal for L, the probability distribution P_{LBN} of L for I is equal to the probability distribution P_{ICL} of the ICL theory $(R \cup I, \mathcal{A})$. We refer to the pair $(R, \mathcal{A})$ as the *equivalent ICL theory* of the LBN (this is a slight abuse of terminology since the actual ICL theories involve not R but $R \cup I$).

It turns out that *each LBN can be mapped to an equivalent ICL theory*. Below we explain the main lines of the mapping. A complete description of the mapping and a proof of correctness can be found in the technical report associated with this paper [6]. In this paper we focus on the mapping from LBNs to ICL, but the mapping to other DS languages (ProbLog, PRISM and Logic Programs with Annotated Disjunctions) is very similar, see the technical report [6].

4.1 The Mapping

To map an LBN to an equivalent ICL theory, we need to map the probability tree (and declaration) of each probabilistic predicate in the LBN. This mapping can be done for each probabilistic predicate separately. In other words, the mapping is local.

---

[2] This is a recurring problem when defining mappings between PILP/SRL languages [3, Ch.13].

To map the probability tree of a probabilistic predicate p, we map each path from the root to a leaf in the tree to an ICL clause $h \leftarrow b$ and a corresponding annotated alternative A. The head h is the target of the considered tree, extended with one extra argument that denotes the value of the target (e.g. if the target is $graduates(S)$ then h is of the form $graduates(S, Val)$). The body b contains literals describing the path to the leaf and also contains a unique base atom. The annotated alternative A specifies the probability distribution for this base atom, which is the same as the distribution in the considered leaf. While the main lines of this mapping are the same as for mapping *propositional* probability trees, some complications arise because we are dealing with *first-order logical* trees here. We now illustrate these issues with an example.

Consider the tree for $graduates(S)$ shown in Figure 1.

1. Let us start with the leftmost leaf of this tree. Note that we end up in this leaf if the tests in both internal nodes succeed. This leaf is mapped to the following clause and annotated alternative in the ICL theory.

$$graduates(S, Val) \leftarrow grade(S, C, bad), mode(G, grade(S, C2, G), good),$$
$$student(S), b_1(S, Val).$$
$$\{\ P_0(b_1(S, yes)) = 0.6\ ,\ P_0(b_1(S, no)) = 0.4\ \}$$

   For the *mode* aggregate, we use syntax similar to that of the *findall/3* predicate in Prolog. The atom $b_1(S, Val)$ is a base atom. It is important to *not* include C as an argument in this atom: if we would include it (by writing $b_1(S, C, Val)$), then we would get a kind of noisy-or effect with as a result that for some S both $graduates(S, yes)$ and $graduates(S, no)$ could become true in the same possible world, and this is unwanted because it is not possible in the original LBN. As another issue, note that the body of the clause contains the atom $student(S)$. This comes from the declaration $random(graduates(S)) \leftarrow student(S)$ in the LBN. While in this particular case the condition $student(S)$ is redundant, in general such conditions need to be included to ensure that atoms in the ICL theory only become true when appropriate (when they are properly typed, etc., see below).

2. The middle leaf in the tree can be mapped in a similar way (note the negated atom).

$$graduates(S, Val) \leftarrow grade(S, C, bad), \neg mode(G, grade(S, C2, G), good),$$
$$student(S), b_2(S, Val).$$
$$\{\ P_0(b_2(S, yes)) = 0.2\ ,\ P_0(b_2(S, no)) = 0.8\ \}$$

3. The rightmost leaf in the tree brings up another issue. We need to express that $grade(S, C, bad)$ fails, but we cannot simply write $\neg grade(S, C, bad)$ since this would cause floundering negation (C is a free variable). The standard solution is to introduce an auxiliary predicate that hides C, and to negate this predicate.

$$hasBadGrade(S) \leftarrow grade(S, C, bad).$$
$$graduates(S, Val) \leftarrow \neg hasBadGrade(S), student(S), b_3(S, Val).$$
$$\{\ P_0(b_3(S, yes)) = 0.9\ ,\ P_0(b_3(S, no)) = 0.1\ \}$$

   Note that the addition of the condition $student(S)$ to the body is really necessary here: if we do not include it, then $graduates(S, Val)$ can become true for some S that is not a student (but for instance a course).

4.2 Discussion

As is required, the ICL theories obtained from the above mapping are always acyclic, or at least 'contingently acyclic' [12, p.30][1]. This is because, for legal input interpretations, the dependency relation specified by an LBN is acyclic as well.

We want to stress that the above mapping does not simply map an LBN to whatever ICL theory that assigns to each possible world the same probability as the LBN. The clauses in the mapped ICL theory in addition also express the same conditional independencies, and even the same *context-specific independencies* [5][3, p.230], as the probability trees in the LBN.

Note that for LBNs that do not contain aggregate tests, the mapped ICL theories do not contain aggregates either. Hence, in such cases we can use the original definition of ICL by Poole [12]. When considering LBNs with aggregate tests, the mapped ICL theories also contain aggregates and we resort to the extension of ICL with aggregates as discussed in Section 2.

5 Learning ICL Theories from Data

We now turn to the problem of learning ICL theories from data. With learning an ICL theory we mean learning not only its *parameters* (the probabilities in the annotated alternatives) but also its *structure* (the set of annotated alternatives itself and the clauses in the logic program). We consider the probabilistic learning from interpretations setting [3, Ch.1]. Concretely, the data that we learn from is an interpretation or set of interpretations of the predicates of interest. The learning task is to find the ICL theory that best fits this data according to a scoring criterion for probabilistic models such as the Bayesian Information Criterion (BIC) [5].

Parameter learning algorithms exist for most DS languages (PRISM [13][3, Ch.5], Logic Programs with Annotated Disjunctions [10], Problog [9], and to some extent ICL [1]). However, to the best of our knowledge, no structure learning algorithms have yet been developed for ICL or any other DS language (with the exception of an algorithm that learns ground Logic Programs with Annotated Disjunctions [10]).

5.1 Learning by Mapping

For LBNs, several structure learning algorithms exist [5]. Hence, the mapping that we described in the previous section directly leads to a simple way of *learning ICL theories*: we first learn an LBN using one of the existing algorithms and then map the LBN to its equivalent ICL theory. The fact that each LBN can be mapped to an equivalent ICL theory (Section 4) guarantees that this approach is always possible.

Poole [3, p.239] recently argued that ICL is a good target language for learning because *"being based on logic programming, it can build on the successes of ILP"*, and *"one of the most successful methods for learning Bayesian networks is to learn a decision tree for each variable [...] these decision trees correspond to a particular form of ICL rules"*. The learning algorithms for LBNs follow exactly this approach.[3]

---

[3] Note that we developed these algorithms in previous work, independently of Poole's remarks.

Moreover, in these algorithms the two suggestions of Poole (ILP + trees) are effectively integrated since LBNs use first-order logical trees. Concretely, each of the learning algorithms for LBNs essentially consists of a search algorithm similar to those used for Bayesian networks (such as ordering-search or structure-search) that repeatedly calls the ILP system TILDE to learn the first-order logical probability trees [5].

As a proof of concept of the above approach for learning ICL theories, and to show what kind of theories can be learned, we applied this approach to the UW-CSE dataset, a popular SRL benchmark [8, Ch.12]. This dataset is about the properties of and relations between the professors, graduate students and courses at a computer science department. The reason why we use this particular dataset is that we have some intuitions and prior knowledge about the dependencies in this domain, which helps in evaluating the learned theories. We applied the ordering-search algorithm for LBNs [5] to this dataset and then mapped the learned LBN to an ICL theory. When investigating this theory we found that many of the learned dependencies are consistent with our intuitions and prior knowledge about the domain. The learned LBN and ICL theory can be found in an online appendix (http://www.cs.kuleuven.be/~dtai/lbn/ilp09).

5.2 Discussion

Our approach is limited to learning certain kinds of ICL theories only, namely ICL theories for which there exists an LBN that maps to them. There are many valid ICL theories for which this is not the case. These are mainly ICL theories that contain dependencies that are not 'tree-structured' (i.e., that do not comply with the structure of a decision tree as is necessary to be representable by LBNs, see the technical report [6, Section 3.3]). Such ICL theories cannot be learned with our approach since they are outside of our hypothesis space.

That we can learn only ICL theories with tree-structured dependencies is a bias of our approach. This bias of course makes the learning process more efficient. Moreover, this bias can also help to learn accurate theories: at least in the context of Bayesian networks it is known that using this bias leads to good results [7].[4] Nevertheless, it is of course useful to try to learn also ICL theories that are not restricted to tree-structured dependencies. Hence, developing a general structure learning algorithm for ICL or other DS languages is an interesting (but challenging) direction for future work.

6 Conclusion

We showed that there is a strong connection between the language of LBNs and ICL: each LBN can be mapped to an equivalent ICL theory (after inclusion of aggregates in the ICL language). Based on this connection we argued that the existing learning algorithms for LBNs can serve as a basis for learning ICL theories. While we focussed on ICL, our discussion applies to a large extent also to other languages based on the distribution semantics like ProbLog, PRISM, and Logic Programs with Annotated Disjunctions [6].

---

[4] In terms of Bayesian networks using this bias means learning conditional probability distributions under the form of propositional probability trees. The fact that this approach performs well and is popular in the Bayesian network community is part of our motivation for using first-order logical probability trees in LBNs.

Acknowledgements. Daan Fierens is supported by the Research Fund K.U.Leuven. This research is also supported by GOA/08/008 'Probabilistic Logic Learning'.

References

1. Carbonetto, P., Kisynski, J., Chiang, M., Poole, D.: Learning a contingently acyclic, probabilistic relational model of a social network. Technical Report TR-2009-08, Department of Computer Science, University of British Columbia (2009)
2. De Raedt, L., Demoen, B., Fierens, D., Gutmann, B., Janssens, G., Kimmig, A., Landwehr, N., Mantadelis, T., Meert, W., Rocha, R., Santos Costa, V., Thon, I., Vennekens, J.: Towards digesting the alphabet-soup of statistical relational learning. In: NIPS Workshop on Probabilistic Programming (2008)
3. De Raedt, L., Frasconi, P., Kersting, K., Muggleton, S.H. (eds.) Probabilistic Inductive Logic Programming. LNCS (LNAI), vol. 4911, pp. 1–27. Springer, Heidelberg (2008)
4. Fierens, D.: Logical Bayesian networks. Chapter 3 of Learning Directed Probabilistic Logical Models from Relational Data. PhD Thesis, Katholieke Universiteit Leuven (2008), http://hdl.handle.net/1979/1833
5. Fierens, D., Ramon, J., Bruynooghe, M., Blockeel, H.: Learning directed probabilistic logical models: Ordering-search versus structure-search. Annals of Mathematics and Artificial Intelligence 54(1-3), 99–133 (2008)
6. Fierens, D.: Mapping logical Bayesian networks to probabilistic logic programs with distribution semantics. Technical Report CW 563, Department of Computer Science, Katholieke Universiteit Leuven (2009), http://www.cs.kuleuven.be/publicaties/rapporten/cw/CW563.abs.html
7. Friedman, N., Goldszmidt, M.: Learning Bayesian networks with local structure. In: Learning and Inference in Graphical Models, pp. 421–459. MIT Press, Cambridge (1999)
8. Getoor, L., Taskar, B.: An Introduction to Statistical Relational Learning. MIT Press, Cambridge (2007)
9. Gutmann, B., Kimmig, A., Kersting, K., De Raedt, L.: Parameter learning in probabilistic databases: A least squares approach. In: Daelemans, W., Goethals, B., Morik, K. (eds.) ECML PKDD 2008, Part I. LNCS (LNAI), vol. 5211, pp. 473–488. Springer, Heidelberg (2008)
10. Meert, W., Struyf, J., Blockeel, H.: Learning ground CP-logic theories by leveraging Bayesian network learning techniques. Fundamenta Informaticae 89(1), 131–160 (2008)
11. Pelov, N., Denecker, M., Bruynooghe, M.: Well-founded and stable semantics of logic programs with aggregates. Theory and Practice of Logic Programming 7(3), 301–353 (2007)
12. Poole, D.: Abducing through negation as failure: Stable models within the independent choice logic. Journal of Logic Programming 44(1-3), 5–35 (2000)
13. Sato, T.: A statistical learning method for logic programs with distribution semantics. In: Proceedings of the 12th International Conference on Logic Programming, pp. 715–729. MIT Press, Cambridge (1995)
14. Vennekens, J., Verbaeten, S., Bruynooghe, M.: Logic programs with annotated disjunctions. In: Demoen, B., Lifschitz, V. (eds.) ICLP 2004. LNCS, vol. 3132, pp. 431–445. Springer, Heidelberg (2004)

Induction of Relational Algebra Expressions

Joris J.M. Gillis and Jan Van den Bussche

Universiteit Hasselt and transnationale Universiteit Limburg

Abstract. We study the problem of inducing relational databases queries expressing quantification. Such queries express interesting multi-relational patterns in a database in a concise manner. Queries on relational databases can be expressed as Datalog programs. Inducing Datalog programs expressing quantification requires both negation and *predicate invention*. Predicate invention has been studied in the ILP literature. However, we propose a radical new approach to inducing quantification. We express queries in the *relational algebra* and we perform a *heuristic search* for the desired expression. A heuristic, which takes the complement operator into account, is proposed. We report some preliminary experimental results of a software prototype implementing our ideas. The results are compared with the results of FOIL and Tilde on the same examples.

1   Introduction

In the theory of database systems [1], a *database query* is defined as a function that maps relational databases to relations. This definition models the situation in practice where one applies an SQL query to a database instance and receives a set of output tuples as the answer to the query on that database. The problem of *relational database query induction* is then naturally stated as follows: we are presented with a finite number of examples, where each example consists of a relational database and an output relation, and we are asked to come up with an expression for a query that agrees with the given examples (and that satisfies the usual requirements placed on induction tasks, most notably generalization).

Of course, we should also specify here which language we are using to express database queries. Whereas SQL is the universal query language used in practice, it is also very complex. A more tractable language to work with is the *relational algebra*, which can be found, together with SQL, in every database textbook, as the relational algebra is the basic underlying query language used for database query processing.

Another common language in which relational database queries can be expressed is Datalog (function-free Prolog). Datalog is also used in Inductive Logic Programming (ILP), and when we restrict attention to nonrecursive programs, Datalog can be translated into the relational algebra [2,3]. As a consequence, relational database query induction could be (and has been [4,5]) considered to be a mere special case of ILP, where the background knowledge consists only of facts (the example databases).

That approach, however, does not cover relational database query induction in its entirety, because not every relational algebra query can be expressed in

L. De Raedt (Ed.): ILP 2009, LNAI 5989, pp. 25–33, 2010.

Datalog. Nonrecursive Datalog without negation can only express the *positive* fragment of the relational algebra, i.e., the fragment without the set-theoretic difference operator. Consider, for example, the following rules over a binary relation $r(A, B)$:

$q(A, B) \leftarrow r(A, C), r(C, B)$
$q(A, B) \leftarrow r(A, C), r(C, D), r(D, B)$

The query q corresponds to the following positive relational algebra expression:

$$\pi_{A,B}(\rho_{B/C}(r) \bowtie \rho_{A/C}(r)) \cup \pi_{A,B}(\rho_{B/C}(r) \bowtie_{B/D} \rho_{A/C}(r) \bowtie \rho_{A/C}(r))$$

Of course, one might use Datalog with negation, and then the difference between two relations r and s can be expressed as follows:

$\textit{diff}\,(A, B) \leftarrow r(A, B), \neg s(A, B)$

If negation can only be placed before input relations, however, this does not suffice to express all relational algebra queries. Notably queries involving universal quantification can still not be expressed. A classical example of such a query is Codd's *relational division* [6] of two relations $r(A, B)$ and $s(B)$, which returns the following unary relation as answer:[1]

$$r \div s = \{A \mid \exists B : r(A, B) \land \forall B : (s(B) \rightarrow r(A, B))\}$$

In relational algebra, division is expressible as $\pi_A(r) \backslash \pi_A((\pi_A(r) \times s) \backslash r)$. Note the nested application of the difference operator $\backslash$. If we want to express such queries in Datalog with negation, we need to introduce auxiliary predicates. In general it is known that with deeper and deeper nested applications of the difference operator (corresponding to alternations of existential and universal quantifiers), more and more queries can be expressed [7], so there is no bound on the number of required auxiliary predicates.

We conclude that Datalog-based approaches to induction of relational algebra queries require a combination of predicate invention and negation. Predicate invention received quite a bit of attention in ILP until the mid 1990s (e.g., [8,9]), and is recently receiving renewed attention [10]. Relatively few ILP systems support negation (e.g., FOIL [11] and TILDE [12]). It is quite conceivable that by combining various of these earlier ILP techniques, relational algebra queries can be induced successfully. In the present paper, however, we explore an approach that directly searches for relational algebra expressions. Nevertheless, we hope that some of our ideas can also be of use in a more classical ILP-based approach. For example, because relational algebra expressions can be translated back into nonrecursive Datalog programs with stratified negation [1], our approach could also be viewed as a form of predicate invention.

---

[1] If r would contain patients with observed symptoms, and s would contain the set of required symptoms to qualify for some specific disease, then $r \div s$ would return all patients that can be diagnosed with the disease.

Related work. We end this introduction with a brief review of some related work, apart from the large body of work in ILP [13]. Acar and Motro [14] focus on the induction of selection queries only (queries without joins) and thus their work essentially boils down to attribute–value learning. In the Clio system [15], mappings between database schemas are inferred in the form of database queries, but the examples given to the induction algorithm are much more fine-grained and consist of explicit value paths between data elements from the source database and the target database. Closest in spirit to our own work is Tupelo [16], a system for inducing data mappings. Tupelo focuses on data restructurings, which are lossless and therefore easier to induce, and is based on a special-purpose algebra with specific restructuring operators (the algebra also lacks the difference operator). Nevertheless, we were influenced by the overall design of Tupelo's search algorithm, which we have copied in the present work.

2   Relational and Cylindric Set Algebra

We assume some familiarity with relational databases and relational algebra, but we fix some terminology and notations.

A relational *database schema* is a finite set of relation names, each with a *relation scheme*, which is a finite set of attribute names. If relation name R has scheme $\{A, B, C\}$, this is often denoted as $R(A, B, C)$. The relational algebra consists of the six operators set union $\cup$; set difference $\setminus$; natural join $\bowtie$; selection $\sigma_{A=B}$; projection π_Z; and renaming $\rho_{A/B}$. Here, A and B stand for attribute names, and Z for a finite set of those. Expressions, of the kind illustrated in the Introduction, are built from relation names using these operators. Each expression has a result relation scheme that can be syntactically determined from the input relation schemes. Semantically, a relational algebra expression over some database schema is evaluated on a database of that schema; the result relation of expression e evaluated on database D is denoted by $e(D)$.

The relational algebra is a typed language: an operator can be applied to expressions only if their result schemes are compatible with the operator. For example, union $e_1 \cup e_2$ can be formed only if e_1 and e_2 have the same result scheme, and renaming $\rho_{A/B}(e)$ can be formed only if the result scheme of e contains attribute A but not attribute B. All these typing restrictions can be something of a nuisance in an induction setting, where we want to systematically combine and generate expressions. Moreover, the subexpressions of an expression can have different result schemes, that can be also different from the final result scheme. In a bottom-up approach where we build up more and more complex expressions, this implies that when testing a hypothesis, we must compare relations of different schemes. Indeed, the given examples are of the final result scheme, whereas intermediate expressions to be tested produce relations of a different scheme.

To avoid these problems, we propose the use of the *cylindric set algebra (CSA)* [17,18,19] as a more suitable alternative to the relational algebra in the context of induction. Fix some finite set U of attribute names. The operators of CSA over

U deal only with relations of scheme U. There are only four operators: union; complementation; selection $\sigma_{A=B}$; and *cylindrification* γ_A, with $A, B \in U$. Union and selection are as in the standard relational algebra. Complementation is the classical set-theoretic operator, but, in order to guarantee finiteness, relativized to the *active domain* $adom(D)$ of the database D on which we are evaluating the expression. The active domain is the set of all values appearing in the relations of D. Then the complement of a relation over U, relative to D, is the set of all tuples over U that take values in $adom(D)$ and that do not belong to r. Finally, the cylindrification $\gamma_A(r)$ of a relation r over U is the set of all tuples over U that agree with some tuple in r on $U \setminus \{A\}$. This intuitively corresponds to existential quantification on the A-column.

As in the relational algebra, given some database schema where all relation names now have scheme U; we call such a schema U we can build up CSA expressions from relation names using the four CSA operators. Unlike the standard relational algebra, there are no restrictions on the formation of expressions, because everything is of the same type U. Formally, let a database schema be given that is U-*uniform* in the sense that every relation name has scheme U. Then each relation name is an expression, and if e_1 and e_2 are expressions, then so are $(e_1 \cup e_2)$, $(e_1)^c$, $\sigma_{A=B}(e)$, and $\gamma_A(e)$ without any restrictions. The result relation of any expression is a relation of scheme U. This makes the CSA very flexible to work with in an induction context.

It can be proven [18] that we do not lose any expressive power by working with CSA instead of the standard relational algebra, as long as we work only with the attributes from U. Difference is expressed using complementation and intersection (a non-primitive operator that we add for convenience); join becomes plain intersection; projection becomes cylindrification; and renaming is simulated using cylindrification and selection.

The choice of U is of course an important parameter. Clearly, we will put in U all attributes from the input database schema and from the output relation scheme. But additional attributes may be needed to be able to express the query. For example, over a binary relation $R(A, B)$ interpreted as the set of edges of a directed graph, the query "output all nodes where a path of length at least two originates" is expressible over $U = \{A, B, C\}$ but not over $U = \{A, B\}$.

3   Heuristic Values Taking into Account Complement

For the induction of an unknown relational database query Q, we are presented with one or more examples of the form (D, r), where D is a database and r is a relation, and the assumption is that $r = Q(D)$. During induction we are searching for an expression e for Q. Thereto we will test hypothesis expressions against the examples to see how well they agree. Thanks to uniformization and the use of CSA, as explained above, both the relation $e(D)$ and the relation $r = Q(D)$ have the same scheme U, which makes comparison quite standard. Indeed, a commonly used metric $d(X, Y)$ on subsets X and Y of some finite universe T uses the symmetric difference: $d(X, Y) = (|X \setminus Y| + |Y \setminus X|)/|V|$. In our case, X and Y are relations of scheme U taking values in the active domain of D, so $|V| = |adom(D)|^{|U|}$.

This simple approach is insufficient, however, for queries whose expression requires set difference (or in CSA, complementation). Indeed, such expressions typically compute a relation that equals the complement of the desired output relation, then perform a final complementation. Consider, for example, the basic *universal quantification* query about two database relations $R(A, B)$ and $S(B)$ that outputs the relation $\{A \mid \exists B : R(A, B) \land \forall B : (R(A, B) \to S(B))\}$ (this query is similar to relational division but the implication is in the other direction). In CSA it is expressible as $\gamma_B (R \cap (R \cap S^c)^c)$. The subexpression $R \cap S^c$ is crucial, but a discovery algorithm based only on direct comparison with the example outputs would discourage further elaboration of this subexpression, as it results in a relation that is extremely different from the final desired result.

We settled on a very direct and simple solution: in order to encourage the exploration of expressions involving complementation, as the heuristic value for an expression e on an example (D, r), we do not use $d(e(D), r)$ but rather $\min\{d(e(D), r), d(e(D)^c, r)\}$. In this way, hypotheses that come close to the example, as well as hypotheses that come close the complement, are favored.

4 Searching for Expressions

As in Tupelo [16], our search space consists of *straight-line programs:* finite sequences of statements. Here a *statement* has the form $R := op(R_1)$ or $R := R_1 \, op \, R_2$, where op is a CSA operator, R is a relation variable, and R_1 and R_2 are either relation names from the database schema or relation variables introduced earlier in the program. Within a program, each relation variable represents an expression, obtained by tracing out its definition. For example, in the program $R_1 := S^c$; $R_2 := \gamma_A(R)$; $R_3 := R_1 \cup R_2$, variable R_3 stands for the expression $S^c \cup \gamma_A(R)$. Important for us are the *top-level* variables in a program, that are those variables that do not occur on the right-hand side of a statement in the program. There can be several top-level variables in a program, and we view a program as representing a set of expressions, one for each top-level variable. Accordingly, as the heuristic value of a program, we use the average heuristic values of its top-level expressions.

The search space can be kept finite because, for each example database D (of which there are a finite number), there are only a finite number of relations of scheme U taking values in $adom(D)$. When exploring the search space by adding a statement to some already encountered program, if that statement defines the same relation[2] on all example databases as some previous statement in the program, the statement will not be added to the program. By imposing an order on the operators and respecting the alphabetical order on attributes in the equivalence operator, we can avoid generating equivalent programs (on the given set of example databases) more than once. If an operator δ uses a variable

---

[2] In practice, such equivalence checks can be implemented efficiently using hashing, at the price of, at least in principle, possible false positives.

X as an operand, X becomes blocked for operators that come before δ in the order. It is easy to show that this restriction does not diminish the expressivity of the search space. We note that, even if U consists of just three attributes, full equivalence of CSA expressions (on all possible databases rather than just the given examples) is undecidable [20]. One can easily imagine other optimizations to trim down the size of the search space. It is, for example unwise to generate expressions containing: $\sigma_{A=B}(\sigma_{A=B}(\ldots))$. The outermost selection operator has no effect whatsoever.

Everything is now in place to find, if it exists, an expression for the unknown query Q given by example. If an U-bounded expression exists that agrees with all examples, it will be found. We perform a best-first search [21] using the heuristic on programs explained above. The initial state is the empty program, which is understood to have as top-level expressions the relation names from the database schema. We generate successor states by appending an extra statement to a program.

5   Experimental Results

We have tested our ideas on a number of typical relational database queries involving universal quantification. We have also tried to induce the queries using the ILP systems FOIL [11] and TILDE [12].[3]

The two obvious testcases were *Codd's relational division* (discussed in the Introduction) and the *universal quantification* query (discussed in Section 3), both about relations $R(A, B)$ and $S(B)$. Both are expressible in CSA over $U = \{A, B\}$, i.e., no additional attribute names are needed, using expressions we have seen earlier. Expressions for these queries were inferred by our search algorithm in a mere 17 (for division) respectively 5 (for universal quantification) steps (where a step means the expansion of a state to all its successor states). Interestingly, whereas for division the obvious expression was found, for universal quantification an equivalent, rather ingenious expression was found that involves more complementation steps but that is shorter overall, namely, $\left(\gamma_B(R \cup S^c)^c\right)^c$.

It must be said, however, that representative examples are important. For division, we presented the algorithm with a small number of databases generated from a small number of constants a_i and b_j which will serve as the nodes of the directed graph R. The b's go in relation S. We generate edges randomly but make make sure to have some nodes with edges to all b's (and also some a's); these nodes will form the output part of the example. We make sure the graph looks sufficiently random otherwise, with edges from a's to a's, from a's to b's, from b's to a's, and from b's to b's. Figure 1 shows a small example database. For universal quantification, we did something similar.

Next we provided FOIL and TILDE with the same examples, to induce both example queries. Both FOIL and TILDE are able to produce Datalog rules with negation. Thus easier queries involving negation, such as the plain difference between two relations, are no problem for both systems. FOIL however was not

---

[3] For TILDE we used the KULeuven ACE data mining system [22].

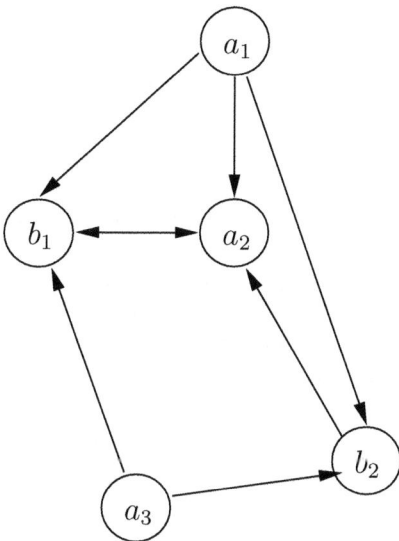

Fig. 1. A small example database for learning Codd's relational division. The database consists of three 'a'-nodes and two 'b'-nodes. Nodes a_1 and a_3 form the output of the relational division.

able to induce the correct expression for the queries with universal quantification. The universal quantification and relational division query are expressible as *First-Order Logical Decision Trees* (FOLDTs). The corresponding trees are shown in table 1. Tilde induced both trees correctly. Again we have to note that representative examples are very important for the induction process.

Obviously, as always, there is no free lunch, and there are queries that take a long time to find using our simple search-based approach. Especially queries that require the renaming operator for their expression seem difficult to find. Renaming is an example of a lossless data transformation of the kind targeted by Tupelo [16]. It might be worthwhile to try to use Tupelo search inside the

Table 1. First-Order Logical Decision Trees expressing the universal quantification query and the relational division query

```
Universal Quantification          Relational Division
q(-A,-B,-C)                       q(-A,-B,-C)
r(A,B,-D) ?                       s(A,-D) ?
+--yes: not_b(A,D) ?             +--yes: not_r(A, B, D) ?
| +--yes: [neg]                  | +--yes: [neg]
| +--no:  [pos]                  | +--no:  [pos]
+--no:  [pos]                    +--no:  [pos]
```

determination of the heuristic value of a candidate expression (search within search), so that pure data transformations are evaluated separately from query operators. This deserves further investigation.

6 Conclusion

Our purpose in this paper was to draw attention to the induction of first-order database queries involving universal quantification, and to the potential of an algebraic approach to ILP. Clearly one can go back and forth between the relational algebra and Datalog, and hopefully some of the ideas we have offered can also be relevant in a more standard ILP setting. Note also that our restriction to relations over a fixed scheme U could be explained as a rudimentary language bias mechanism.

An algebraic approach can sometimes be more convenient. Think, for example, about a genetic programming approach [23] to learning relational algebra expressions. The CSA is convenient in this respect as genetic programming operations on expressions such as mutation and crossover can be freely performed, always resulting in well-formed expressions that can be effectively evaluated.

Acknowledgment. We thank Jan Struyf and Hendrik Blockeel for providing us with the ACE software package and teaching us how to use Tilde.

References

1. Abiteboul, S., Hull, R., Vianu, V.: Foundations of Databases. Addison-Wesley Publishing Company Inc., Reading (1995)
2. Ullman, J.: Principles of Database and Knowledge-Base Systems, vol. 1. Computer Science Publisher, Rockville (1988)
3. Ullman, J.: Principles of Database and Knowledge-Base Systems, vol. 2. Computer Science Publisher, Rockville (1989)
4. Blockeel, H., De Raedt, L.: Relational knowledge discovery in databases (1996)
5. Blockeel, H., De Raedt, L.: Inductive database design (1996)
6. Codd, E.: Relational completeness of data base sublanguages. Database Systems (1972)
7. Chandra, A., Harel, D.: Structure and complexity of relational queries. Journal of Computer and System Sciences 25, 99–128 (1982)
8. Silverstein, G., Pazzani, M.: Relational clichés: Constraining induction during relational learning (1991)
9. Kijsirikul, B., Numao, M., Shimura, M.: Discrimination-based constructive induction of logic programs (1992)
10. Kok, S., Domingos, P.: Statistical predicate invention (2007)
11. Quinlan, J., Cameron-Jones, R.: Induction of logic programs: Foil and related systems. New Generation Computing 13, 287–312 (1995)
12. Blockeel, H., De Raedt, L.: Top-down induction of first-order logical decision trees. Artificial Intelligence 101, 285–297 (1998)
13. De Raedt, L.: Logical and Relational Learning. In: Cognitive Technologies. Springer, Heidelberg (2008)

14. Acar, A., Motro, A.: Intensional encapsulation of database subsets via genetic programming (2005)
15. Miller, R., Haas, L., Hernández, M.: Schema mapping as query discovery (2000)
16. Fletcher, G., Wyss, C.: Data mapping as search. In: Ioannidis, Y., Scholl, M.H., Schmidt, J.W., Matthes, F., Hatzopoulos, M., Böhm, K., Kemper, A., Grust, T., Böhm, C. (eds.) EDBT 2006. LNCS, vol. 3896, pp. 95–111. Springer, Heidelberg (2006)
17. Henkin, L., Monk, J., Tarski, A.: Cylindric Algebras, Part I. North-Holland, Amsterdam (1971)
18. Van den Bussche, J.: Applications of alfred tarski's ideas in database theory (2001)
19. Imielinski, T., Lipski, W.: The relational model of data and cylindric algebras. Journal of Computer and System Sciences 28, 80–102 (1984)
20. Kahr, A., Moore, E., Wang, H.: Entscheidungsproblem recuded to the ∀∃∀ case. Proc. Natl. Acad. Sci. USA 48, 365–377 (1962)
21. Russell, S., Norvig, P.: Artificial Intelligence, A Modern Approach, 2nd edn. Prentice Hall, Englewood Cliffs (2003)
22. Blockeel, H., Dehaspe, L., Demoen, B., Janssens, G., Raons, J., Vandecasteele, H.: Improving the efficiency of inductive logic programming through the use of query packs. Journal of Artificial Intelligence Research 16, 135–166 (2002)
23. Koza, J.: Genetic Programming: On the Programming of Computers by Means of Natural Selection. MIT Press, Cambridge (1992)

A Logic-Based Approach to Relation Extraction from Texts

Tamás Horváth[1,2], Gerhard Paass[2], Frank Reichartz[2], and Stefan Wrobel[2,1]

[1] Dept. of Computer Science III, University of Bonn, Germany
[2] Fraunhofer IAIS, Schloss Birlinghoven, Sankt Augustin, Germany
{tamas.horvath,gerhard.paass}@iais.fraunhofer.de,
{frank.reichartz,stefan.wrobel}@iais.fraunhofer.de

Abstract. In recent years, text mining has moved far beyond the classical problem of text classification with an increased interest in more sophisticated processing of large text corpora, such as, for example, evaluations of complex queries. This and several other tasks are based on the essential step of relation extraction. This problem becomes a typical application of learning logic programs by considering the dependency trees of sentences as relational structures and examples of the target relation as ground atoms of a target predicate. In this way, each example is represented by a definite first-order Horn-clause. We show that an adaptation of Plotkin's least general generalization (LGG) operator can effectively be applied to such clauses and propose a simple and effective divide-and-conquer algorithm for listing a certain set of LGGs. We use these LGGs to generate binary features and compute the hypothesis by applying SVM to the feature vectors obtained. Empirical results on the ACE–2003 benchmark dataset indicate that the performance of our approach is comparable to state-of-the-art kernel methods.

1 Introduction

For a long time, text mining has mostly focused on document classification. In recent years, however, there has been an increased interest in more advanced processing of large text corpora. This problem is motivated by practical applications, e.g., in question answering, information retrieval, ontology learning, and bioinformatics. In case of question answering, for instance, current search engines are not powerful enough for complex queries, such as, for example, "find UN officials born in Africa". Obviously, the internal representation of texts in a search index as sequences of words is insufficient to recover semantics from unstructured texts (e.g., the "born in" relation in the above example). *Relation extraction* is one of the essential steps towards more complex automatic text processing. It is concerned with the problem of detecting and classifying predefined semantic relations among m-tuples (typically between pairs) of entities in unstructured texts.

It is more than evident that relation extraction requires information about the syntactic and semantic structure of the sentences to be processed. Whereas

L. De Raedt (Ed.): ILP 2009, LNAI 5989, pp. 34–48, 2010.

early approaches to relation extraction relied on extracting and exploiting relatively simple patterns (e.g., regular expressions with wildcards), recent directions are based on the utilization of low-level syntactic parse trees, in particular, *dependency trees*, enhancing kernel methods with additional information on the sentences' syntactic structure. While feature-based methods are restricted to a limited number of (structural) features to be used, kernel-based methods offer efficient solutions allowing the exploration of much larger (often exponential, or in some cases, infinite) sets of tree characteristics in polynomial time, without the need of explicit computation of the features. Examples of this approach include, e.g., the dependency tree kernels [6,21] inspired by string kernels.

Similarly to [6], in this work we consider relation extraction as a supervised learning problem and transform the unstructured text to *dependency trees*. We regard dependency trees as relational structures over vocabularies consisting of a single binary predicate corresponding to the edges and a set of unary predicates corresponding to the words associated with the vertices. Accordingly, training examples of a particular m-ary target relation to be learnt are considered as ground atoms of a single m-ary target predicate with constants representing certain distinguished nodes of dependency trees. In this way, relation extraction becomes a typical application of learning logic programs. In contrast to standard problem settings studied in Inductive Logic Programming, we assume a partial order on the set of unary predicates which is defined by the words' concept hierarchy (e.g., the unary predicate `person` is more general than the predicate `physicist`). Such a partial order can automatically be derived from WordNet[7]. WordNet is a lexical database of English words and their relations. The partial order on the words can especially be inferred from the *hypernym* relation (which can be seen as a semantic relation "x is_a y").

Applying a naturally generalized notion of Plotkin's *least general generalization* (LGG) operator [18] to our problem setting, we generate a set of first-order definite non-recursive Horn-clauses satisfying certain frequency and consistency criterion, i.e., all these rules must cover at least a certain number of positive examples while implying at most a certain number of negative examples. In the generation of LGGs, we exploit the specific structure of dependency trees allowing polynomial time rule evaluation defined by the correspondingly generalized θ-subsumption and the fact that the LGG is a closure operator on the power set of the instance space over the target predicate. Using these rules, we generate a binary feature vector for each example and, applying support vector machines to these feature vectors, find a hypothesis separating the positive and negative examples.

Our first experiments on the ACE–2003 benchmark corpus [15] clearly indicate that the above approach compares well to state-of-the-art methods [5,6]. Furthermore, in contrast to other approaches restricted to the special case of binary target relations [5], our approach is applicable to arbitrary arity, i.e., to unary target relations, as well as to arities greater than 2. Preliminary results with various strategies enhancing dependency trees and the generalization operator by semantic features (e.g., by topic model scores for term disambiguation) also suggest that the predictive performance of our approach can further be improved.

The paper is organized as follows. In Section 2 we first overview some state-of-the-art methods and then, in Section 3, we provide a detailed description of the main steps and features of our method including data preprocessing, first-order representation, bottom-up generalization, efficient rule evaluation, and rule generation. In Section 4 we report our empirical results with the ACE–2003 benchmark dataset and finally, in Section 5, we conclude and mention some interesting directions for future research.

2   Related Work

One of the early approaches to relation extraction was based on detecting relatively *simple* patterns expressed as regular expressions with wildcards [10]. The underlying hypothesis of this approach assumes that terms sharing similar linguistic contexts are connected by similar semantic relations. The manual identification of such patterns has inspired other approaches in which an initial set of patterns is used to seed a bootstrapping process that automatically acquires additional patterns for relations [1]. However, the exact meaning of relations identified by such "open information extraction" is often not fixed beforehand. Several authors have proposed related approaches, e.g., using frequent itemset mining to extract characteristic patterns [2] or defining various measures for pattern reliability and filtering incorrect instances using the web (see, e.g., [17]).

More advanced methods analyzing relations in texts take into account additional information about the *sequential structure* of sentences. This can be done with sequential models, e.g., conditional random fields [4]. In this approach, words are annotated with rich features, e.g., part-of-speech information or the location of the verb in the sentence. Each word is assigned to a state indicating whether or not it is part of a specific relation. The model is trained using manually annotated data and can successfully predict semantic relations for instance between diseases and treatments in medical domains [4].

Syntactic parse trees provide an extensive level of structural information and can, for instance, represent relations among subject, verb, and object in a sentence. Due to the potentially huge number of parse tree features traditional feature-based methods are often inapplicable to parse trees; they are able to handle only a limited number of structural features. In contrast, kernel-based methods offer efficient solutions that allow the exploration of much larger sets of tree features in polynomial time, without the need of explicit calculation and representation of the features. Various kernels have been defined for parse trees. The kernels proposed for dependency trees in [6,16,21] are inspired by string kernels. Another kernel introduced in [5] computes similarities between nodes on the shortest path in a dependency tree that connects the entities of a binary relation. In [20], more sophisticated kernels are suggested for dependency parse trees which yield not only improved extraction performance, but also reduced run time. All these kernels are used as input for kernel-based classifiers.

Finally we note that tree kernel-based approaches to relation extraction are not restricted to dependency parse trees. For example, powerful tree kernels can be defined using phrase grammar parse trees [22]. In fact, as shown in a recent work [19], different parse tree types can be combined with each other resulting in improved extraction quality. Furthermore, methods utilizing features associated with words (see, e.g., [9]) have empirically demonstrated the positive effect of the exploitation of the semantic background knowledge present in various sources (e.g., WordNet, FrameNet, PropBank).

3   The Method

In this section we describe the main steps and the most important algorithmic features of the logic-based relation extraction method proposed in this work.

Example 1. In order to demonstrate the main steps of the algorithm, the following running example will be used in the paper. Consider the training sentences

```
Fraunhofer was a German optician.
Heisenberg was a celebrated German physicist and Nobel laureate.
```

and the binary target predicate is_a/2. From the above training sentences, we have the following positive examples of the target predicate:

$$\text{is\_a(Fraunhofer,optician)}$$
$$\text{is\_a(Heisenberg,physicist)} \quad .$$

All other (ordered) pairs of words from the first and from the second sentence are considered to be negative examples (e.g., is_a(physicist,Heisenberg)).

3.1   Data Preprocessing

Given a set of sentences annotated with respect to the target relation, in a preprocessing step we first compute a *dependency tree* for each sentence. Dependency trees are labeled rooted directed trees representing grammatical dependencies among the words in a sentence. They capture a low-level syntactic structure of sentences and have the property that there is a bijective mapping between words in a sentence and nodes in the corresponding dependency tree. Thus one can associate various features of words in a sentence with the corresponding nodes, e.g., part-of-speech type of the word, word sense, or entity type. Since our aim is to detect semantic relationships among entities in a sentence, we merge the nodes of the dependency tree into single artificial nodes that define the same entities. As an example, the nodes corresponding to "Joseph von Fraunhofer" will be merged into a single node labeled by "Joseph_von_Fraunhofer". Thus, in the remainder of this work we assume without loss of generality that all entities are represented by single nodes in the dependency trees.

For the generation of dependency trees we employ the Stanford Parser[1], a state-of-the art natural language parser. The output of the parser generated for the sentences in the running example is given in Figure 1.

---

[1] nlp.stanford.edu/software/lex-parser.shtml

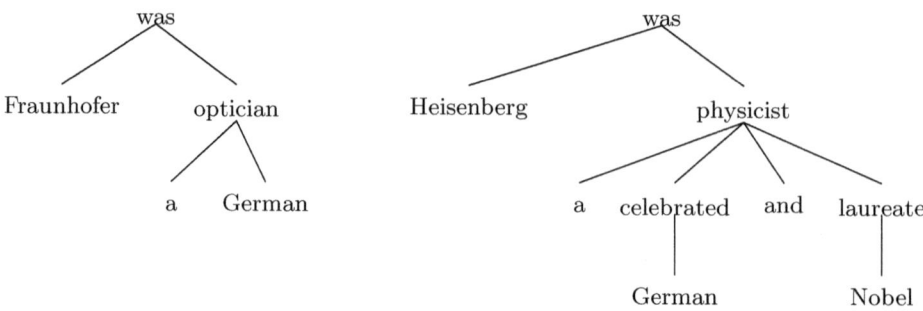

Fig. 1. The dependency trees for the sentences "Fraunhofer was a German optician." (left-hand side) and "Heisenberg was a celebrated German physicist and Nobel laureate." (right-hand side) used in Example 1.

3.2   First-Order Logic Representation

An important feature of the dependency trees obtained in the way described above is that there is an injective mapping from the set of entities in a sentence to the set of nodes in the corresponding dependency tree. Detecting particular instances of a given m-ary target relation on entities thus corresponds to the problem of classifying (ordered) m-tuples of entity nodes in dependency trees. It is important to emphasize that, in contrast to other approaches restricted to the special case of binary target relations, i.e., $m = 2$ (see, e.g., [5]), our approach is applicable to arbitrary arity (also to unary target relation).

We make use of the above feature of dependency trees and the fact that dependency trees can be considered as *relational structures* in the standard natural way; the edges of the trees can be represented by a single binary predicate, say R, while the labels by unary predicates. More precisely, dependency trees are regarded as unordered directed trees where all edges are directed towards the root. For each vertex of a dependency tree a unique constant is introduced. For example, for the dependency trees in Figure 1 we introduce the constants $a_{1,1}, \ldots, a_{1,5}$ for the left-hand side tree and $a_{2,1}, \ldots, a_{2,9}$ for the right-hand side tree, and represent the two trees by the relational structure, i.e., set of facts:

was($a_{1,1}$).   Fraunhofer($a_{1,2}$). optician($a_{1,3}$).   a($a_{1,4}$).
German($a_{1,5}$).

$R(a_{1,2}, a_{1,1})$. $R(a_{1,3}, a_{1,1})$. $R(a_{1,4}, a_{1,3})$. $R(a_{1,5}, a_{1,3})$.

was($a_{2,1}$).   Heisenberg($a_{2,2}$). physicist($a_{2,3}$). a($a_{2,4}$).
celebrated($a_{2,5}$). German($a_{2,6}$).   and($a_{2,7}$).   laureate($a_{2,8}$).
Nobel($a_{2,9}$).

$R(a_{2,2}, a_{2,1})$. $R(a_{2,3}, a_{2,1})$. $R(a_{2,4}, a_{2,3})$. $R(a_{2,5}, a_{2,3})$.
$R(a_{2,6}, a_{2,5})$. $R(a_{2,7}, a_{2,3})$. $R(a_{2,8}, a_{2,3})$. $R(a_{2,9}, a_{2,8})$.

The relational structure corresponding to the dependency trees of *all* sentences to be processed constitutes the (extensional) *background knowledge* of the ILP learning problem to be considered. The binary predicate R introduced for the edges and the unary predicates representing the words will be referred to as *background predicates*.

Using the above representation, the instances of the target relation to be learnt are encoded as m-tuples of vertices of dependency trees. For example, the positive examples

is_a(Fraunhofer,optician) and is_a(Heisenberg,physicist)

in our running example are encoded by

$$\text{is\_a}(a_{1,2}, a_{1,3}) \text{ and } \text{is\_a}(a_{2,2}, a_{2,3}) ,$$

respectively.

This encoding of the dependency trees and the instances of the target relation allows us to represent the training examples (i.e., +/- labeled instance) by ground non-recursive definite Horn-clauses as follows: The head of the Horn-clause is formed by the example and the body is composed of all ground atoms belonging to the background knowledge. For the running Example 1, the positive instance is_a(Fraunhofer,optician) is represented by the ground clause

$$\text{is\_a}(a_{1,2}, a_{1,3}) \longleftarrow R(a_{1,2}, a_{1,1}), \dots, R(a_{1,5}, a_{1,3}),$$
$$R(a_{2,2}, a_{2,1}), \dots, R(a_{2,9}, a_{2,8}),$$
$$\text{was}(a_{1,1}), \dots, \text{German}(a_{1,5}),$$
$$\text{was}(a_{2,1}), \dots, \text{Nobel}(a_{2,9})$$

and the instance is_a(Heisenber,physicist) by the clause consisting of the head is_a$(a_{2,2}, a_{2,3})$ and the same body as above. Thus, in order to calculate the relative LGG, the entire background knowledge is added to the body of the ground clauses corresponding to the examples. This is for formal reasons discussed in the subsequent sections. Practically, it suffices to add only the relational structure representing the dependency tree of the sentence containing the corresponding example.

3.3 Bottom-Up Generalization

Using the first-order representation described in the previous section, we define the inductive generalization of a set of positive examples by applying an adaptation of Plotkin's LGG [18] operator to first-order clauses of the form

$$A_0 \longleftarrow A_1, \dots, A_m , \tag{1}$$

where A_0 is an atom of the target predicate and $A_1, \dots, A_m$ are atoms of the background predicates. The above clause will sometimes be considered as the set

$\{A_0, \neg A_1, \ldots, \neg A_m\}$. Notice that the ground clauses introduced for the training examples in the previous section are also of this form.

For clauses of the form (1), the adapted LGG operator is based on a naturally generalized notion of subsumption, taking into account the partial order on the unary background predicates.[2] More precisely, let C and D be first-order clauses of the form (1) over a vocabulary associated with a partial order $\preceq$ on the set of unary background predicate symbols. Then C *subsumes* D *with respect to* $\preceq$, denoted $C \leq_{\theta, \preceq} D$, if there exists a substitution θ such that

(i) θ maps the head of C to the head of D and
(ii) for every literal $\neg P(t_1, \ldots, t_n) \in C$ there is a literal $\neg P'(t_1', \ldots, P_n') \in D$ satisfying

$$\begin{cases} P' \preceq P \text{ and } t_1 \theta = t_1' & \text{if } n = 1 \\ P = P' \text{ and } (t_1 \theta, \ldots, t_n \theta) = (t_1', \ldots, t_n') & \text{otherwise} . \end{cases}$$

As an example of the above notion, we have

$$(\longleftarrow \texttt{physicist}(X)) \leq_{\theta, \preceq} (\longleftarrow \texttt{optician}(\texttt{Fraunhofer}))$$

with respect to $\texttt{optician} \preceq \texttt{physicist}$. This generalized subsumption will be referred to as $(\theta, \preceq)$-*subsumption*.

The LGG of a set S of clauses with respect to $\leq_{\theta, \preceq}$ can then be defined in the usual way [18], i.e., a clause C is an LGG of S with respect to $\leq_{\theta, \preceq}$ if

- $C \leq_{\theta, \preceq} D$ for every $D \in S$ and
- $C' \leq_{\theta, \preceq} C$ for all C' satisfying $C' \leq_{\theta, \preceq} D$ for every $D \in S$.

One can show that for a set S of clauses of the form (1), a *reduced* LGG with respect to $\leq_{\theta, \preceq}$ always exists and is unique module variable renaming if the partial order $\preceq$ on the unary background predicates is a *join-semilattice* (i.e., for all unary predicates P and Q, the least upper bound of $\{P, Q\}$ exists).

Since the clauses corresponding to the examples are all non-redundant and the bodies of the clauses are all labeled directed trees, the LGG operator with respect to $\leq_{\theta, \preceq}$ is equivalent in our case to the *tensor* (or *weak direct*) *product* of labeled trees [11]. We recall that the tensor product of the directed graphs $G_1 = (V_1, E_1)$ and $G_2 = (V_2, E_2)$ is a directed graph $G = (V, E)$ with $V = V_1 \times V_2$ and $E = \{((u_1, u_2), (v_1, v_2)) \in V \times V : (u_1, v_1) \in E_1 \text{ and } (u_2, v_2) \in E_2\}$. The tensor product of the dependency trees of the two sentences used in the running example is given in Figure 2.

Applying Plotkin's ordinary LGG to labeled graphs G_1 and G_2, we get that a product vertex in G is labeled by λ if and only if its components in G_1 and G_2 are labeled by λ. Otherwise, the product vertex is unlabeled. In contrast to

---

[2] We note that the generalized notions of subsumption and LGG described in this section can easily be generalized further to predicates of arbitrary arity. For the sake of simplicity, however, we restrict the definitions to unary background predicates, as it suffices for the problem setting considered in this work.

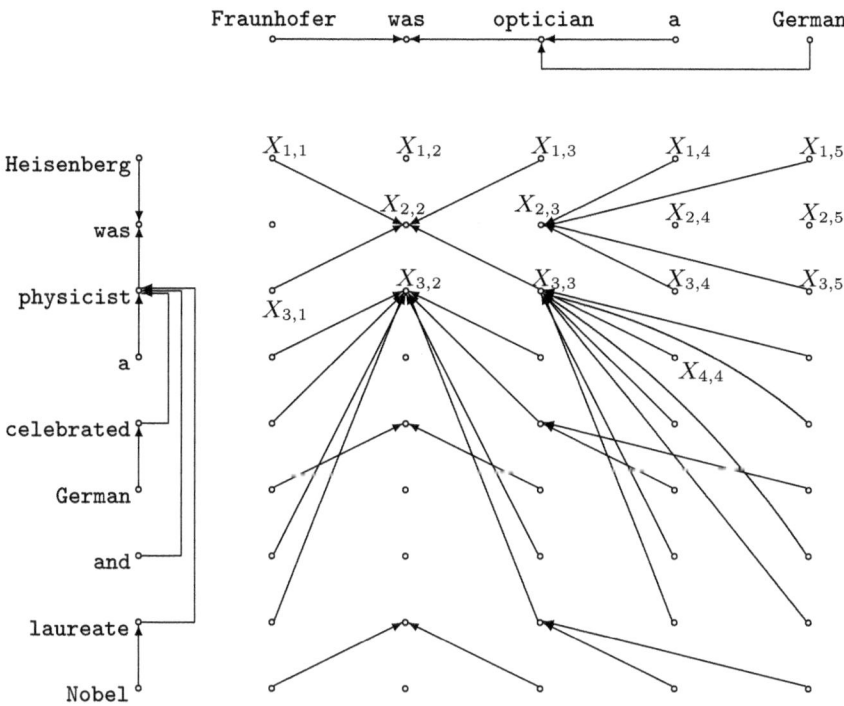

Fig. 2. The tensor product of the dependency trees of the sentences in the running example. For each product vertex in the tensor product, a unique variable is introduced. In particular, the product vertex at the cross of the i-th row and j-th vertex is associated with variable $X_{i,j}$.

this definition, if the labels of the components in a product vertex are different, we generalize them by their least common ancestor in the WordNet. We note that, due to semantic ambiguity, this least common ancestor does not always exist (there can be more than one incomparable minimal common ancestor). This problem can be resolved by disambiguation; we apply the heuristic based on taking the most probable word sense from WordNet. Notice that this generalization of labels (or equivalently, unary predicates) remains equivalent to the ordinary LGG operator by labeling the nodes in the dependency trees by all of their more general categories in the WordNet. In our running example the product vertices have the following labels (see, also, Figure 2):

$$X_{1,1} \mapsto \textit{person entity}$$
$$X_{2,2} \mapsto \texttt{was}$$
$$X_{3,3} \mapsto \texttt{physicist}$$
$$X_{4,4} \mapsto \texttt{a}$$
$$X_{6,5} \mapsto \texttt{German}$$

In case of $X_{3,3}$ this is because an optician could also be a physicist who is specialized on optics. Using this meaning of optician, the least upper bound of

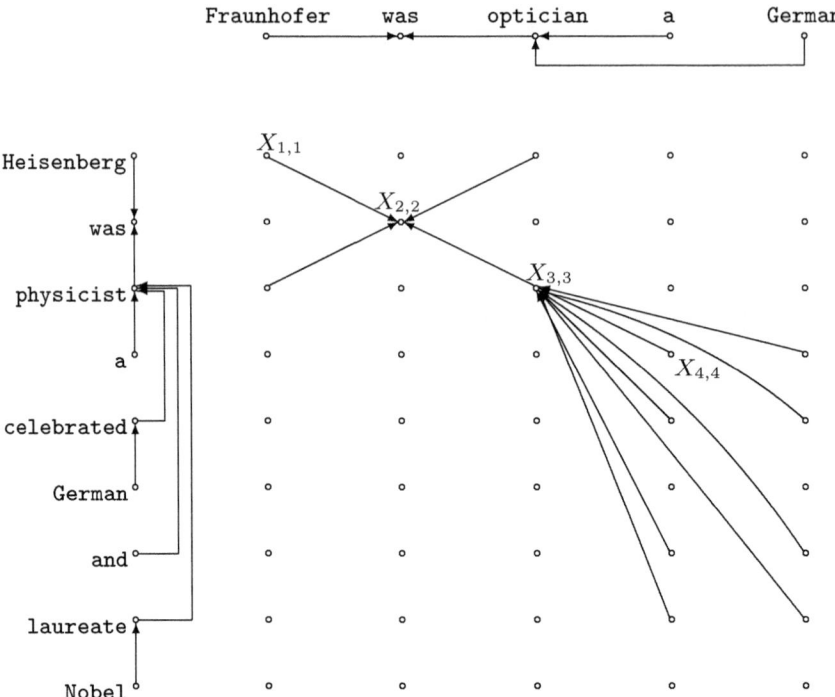

Fig. 3. The tensor product of the dependency trees of the sentences in the running example. For each product vertex in the tensor product, a unique variable is introduced. In particular, the product vertex at the cross of the i-th row and j-th vertex is associated with variable $X_{i,j}$.

optician and physicist is physicist[3]. For all other product variables, the least common ancestor is the most general entry (i.e., the top element) in the Wordnet which subsumes all words. For this reason, all product vertices labeled by the top element of the WordNet are regarded as unlabeled. Notice also that by the generation of the LGGs described above, each product vertex is associated with at most one label. Equivalently, each constant or variable occurs as term in at most one unary background atom.

One can easily see that forests are closed under tensor product implying that the graphs in the bodies of the LGGs are also forests (see, also, Figure 2). Clearly, the LGG of a set of positive examples can be a good generalization only when the product vertices corresponding to the arguments in the head all belong to the same tree in the body's forest. If this is the case, we keep only the tree containing the distinguished product vertices; otherwise, the LGG is considered to be illegal.

---

[3] We note that the least common ancestor of these two words in the current version of Wordnet is person. To demonstrate the proposed method, we still use physicist in our running example.

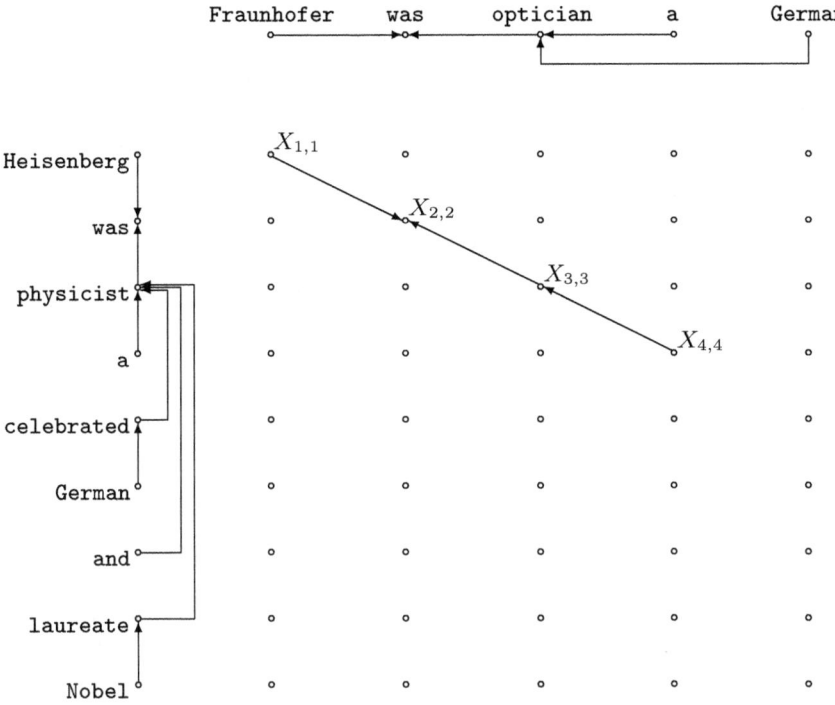

Fig. 4. The tensor product of the dependency trees of the sentences in the running example. For each product vertex in the tensor product, a unique variable is introduced. In particular, the product vertex at the cross of the i-th row and j-th vertex is associated with variable $X_{i,j}$.

Removing the connected components from the product which are disjoint with the connected component containing the distinguished vertices results in an equivalent clause, as all these components can be projected to the background knowledge when evaluating an example. (We recall that the bodies of the examples consist of all atoms from the background knowledge.) This reduction significantly decreases the size of the LGG (see Figure 3). In case of our running example, the product vertices $X_{1,1}$ and $X_{3,3}$ corresponding to the distinguished vertices Heisenberg, Fraunhofer and physicist, optician, respectively belong to the same tree in the tensor product.

The connected component of the distinguished product vertices is then further reduced. In our example on Figure 3, all edges going into $X_{2,2}$ and $X_{3,3}$ are redundant except the edges $(X_{1,1}, X_{2,2})$, $(X_{3,3}, X_{2,2,})$, and $(X_{4,4}, X_{3,3})$; while the other endpoints of the edges going into $X_{3,3}$ are all unlabeled, $X_{4,4}$ is labeled by "was". Removing all these redundant edges we finally get the reduced clause

$$\text{is\_a}(X_1, X_3) \longleftarrow person\_entity(X_1), \text{was}(X_2), \text{physicist}(X_3), \text{a}(X_4),$$
$$R(X_1, X_2), R(X_3, X_2), R(X_4, X_3)$$

(see, also, Figure 4). We note that unary background atoms containing variables that correspond to vertices having become isolated during the reduction must also be removed for the same reason discussed for the deletion of connected components. Though the size of the reduced LGG obtained can still exponentially grow with the number of examples, our experimental results clearly indicate that it remains small in practice.

3.4   Deciding $(\theta, \preceq)$-Subsumption

During the generation of LGGs, there are two different cases where $(\theta, \preceq)$-subsumption must be decided:

Rule reduction, i.e., given a clause C of the form (1) and a binary background literal $L \in C$, decide whether $C \leq_{\theta,\preceq} C \setminus \{L\}$ (see the last part of the previous section).

Query evaluation, i.e., given a ground definite non-recursive Horn-clause D representing an example as defined in Section 3.2 and a clause C of the form (1), decide whether $C \leq_{\theta,\preceq} D$.

For the first case, $(\theta, \preceq)$-subsumption is equivalent to ordinary θ-subsumption. Regarding the second case, we have that $C \leq_{\theta,\preceq} D$ if and only if $C \leq_{\theta} D'$ where D' is defined as follows: For every literal $\neg P(t) \in D$ and for every unary background predicate P' with $P \preceq P'$, add the literal $\neg P'(t)$ to D.

Notice that since the LGGs are of the form (1), they are neither recursive nor self-resolvent. Thus, first-order implication for our case becomes equivalent to θ-subsumption [8], which, in turn, is equivalent to *relational homomorphism* [14]. Although homomorphism between relational structures is computationally intractable, in our case it can be decided in *polynomial time* because homomorphism from labeled trees into graphs can be decided efficiently. In our implementation we applied an adaptation of the bottom-up query evaluation algorithm used for learning acyclic conjunctive queries [12].

3.5   Rule Enumeration and Application

The LGG described in Section 3.3 can also be considered as a function on the power set of the instance space corresponding to the target predicate. More precisely, given a set S of m-tuples of an m-ary target relation, the LGG of the set of clauses representing the m-tuples in S represents a superset of S, i.e., the set of all m-tuples implied by the LGG with respect to the background knowledge (see, also, Section 3.2). One can easily check that this function is extensive, monotone, and idempotent. Thus, the LGG is a *closure operator* on the power set of the instance space (see, e.g., [11] for further details). In the generation of a set of rules, we exploit this fact and, using a simple divide-and-conquer algorithm [3], generate a certain subset of the family of closed sets defined by the LGG as closure operator. More precisely, we generate only a subset of the LGGs that are *frequent* with respect to the positive examples and *infrequent* with respect to the negative examples.

For each closed set we take its reduced LGG representation as described earlier. This set of LGGs is then used as Boolean features for the examples of the target relation. That is, an LGG is true for an m-tuple if and only if it implies the m-tuple with respect to the background knowledge. In this way, we obtain a Boolean feature vector for every training examples and learn a separating hyperplane by using SVM^{light} [13]. For the classification with SVM we need to generate training examples of the target relation from the training sentences. Notice that a sentence may contain more than one instance of the target relation. Thus, for a sentence with n entities and for an m-ary target relation we generate $V_m^n = \frac{n!}{(n-m)!}$ training/test examples.

4   Empirical Results

In this section we empirically compare the performance of our method with the approaches proposed in [5,6]. In our experiments we use the publicly available benchmark dataset ACE–2003[4] [15]. The ACE–2003 corpus was created during the Automatic Content Extraction conference series for the empirical evaluation and comparison of different approaches to information extraction from natural language texts. This corpus consists of 519 natural language text documents arising from different sources. The documents are all news related and consist of newspaper articles, newswire texts, and transcripts of broadcast news. Altogether there are 9256 sentences in the corpus with an average of 18 words per sentence. The named entities and relations among them were annotated by experts. The entities which are annotated belong to one of the following types: *named*, *nominal*, and *pronominal*. Since no algorithm is available to the best of our knowledge that effectively can be used to resolve co-reference for nominal (e.g. "banks") or pronominal (e.g. "ours") entities, we restrict our experiments to named entities. We note, however, that our approach can be applied without any modification to the other two types of entities as well. The texts are annotated according to the following five top level target relations (we also give the number of their occurrences):

relation	#instance
role	732
part	265
near	44
social	55
at	481

The above relations are then further refined into 24 subrelations.

---

[4] Available from LDC (www.ldc.upenn.edu) as corpus LDC2003T11.

Table 1. Table of results on relations between name mentions

Method	Precision	Recall	F_{micro}	F_{macro}
logic-based	68.2%	42.3%	52.2%	45.3%
shortest path kernel [5]	65.5%	53.8%	52.5%	–
subtree kernel [6]	67.1%	35.0%	45.8%	–

Example 2. As an example, we mention the following two sentences from the dataset:

> This was done suddenly without any negotiations discussions or warning said Aziz Tukulj director of domestic sales for Energopetrol.

> In Nigerias joint venture with Shell which produces almost half the country's oil the government takes about 70 percent of the sales price Shell officials said.

The above sentences contain two positive examples of the binary target predicate Role/2:

> Role(Aziz_Tukul, Energopetrol)
> Role(Shell, Nigeria) .

In our experiments we used the five top level target relations only. All multiple word entities (e.g., Aziz Tukul in the above example) were merged and the sentences were processed by the Stanford Parser.

We have performed 5-fold cross-validation and calculated both the averaged macro (per class) and micro (per instance) evaluation measures, as we have a multi-class learning problem. The results of our logic-based method are depicted in Table 1 together with those reported for the shortest path kernel [5] and the subtree kernel [6]. While our method outperforms the subtree kernels both in precision and recall, the shortest path kernel only in precision. It is important to emphasize that the performance of our approach strongly depends, among others, on the semantic correctness and completeness of WordNet (see footnote 3), on the quality of the word sense disambiguation etc. Though the shortest path kernel has better recall, we note that our method is more general than the shortest path kernel, as it is not restricted to binary target relations.

5   Summary

We have proposed a logic-based method for relation extraction from natural language texts. Our method is based on transforming examples into definite Horn-clauses by considering dependency trees as relational structures. The special structure of the clauses obtained enables the effective application of an adaptation of Plotkin's least general generalization operator. Since this operator is a closure operator on the power set of the instance space, we generate them by using a simple divide-and-conquer algorithm. The set of rules enumerated are then used to calculate binary feature vectors for the examples from which the final hypothesis is obtained by applying support vector machines. Empirical results on a popular benchmark dataset indicate that the performance of our method is comparable with the state-of-the-art methods.

It is important to emphasize that, in contrast to several other approaches, our method is not restricted to binary target relations. This is especially important for semantic role labeling consisting of the detection of the semantic arguments associated with the predicate or verb of a sentence. These arguments are associated with semantic roles, e.g. the agent, the object, the temporal specification, or a location. Semantic role labeling is a necessary step towards machine understanding of natural languages.

An important feature of our approach is that, in contrast to other methods based on substructure isomorphism, pattern matching is defined by relational homomorphism. As an example, the shortest path kernel [5] which also exploits, in a broad sense, a part of the tensor product (i.e., the shortest path connecting the two distinguished product vertices) requires the matching function to be injective. One interesting research direction for future work is to compare the performance of these two matching relations on the relation extraction problem. Besides ordinary relational homomorphism, we are going to study constrained homomorphisms as well. For example, it would be interesting to see, whether the performance of our approach can be improved by applying a constrained homomorphism taking into account the word positions in the original sentence.

Acknowledgements

The authors thanks Jan Ramon for his useful comments on an early version of this work. Tamás Horváth, Gerhard Paaß, and Frank Reichartz were partially supported by the German Federal Ministry of Economy and Technology(BMWi) under the Theseus Project.

References

1. Banko, M., Etzioni, O.: The tradeoffs between open and traditional relation extraction. In: Proc. of ACL 2008: HLT, pp. 28–36 (2008)
2. Blohm, S., Cimiano, P.: Scaling up pattern induction for web relation extraction through frequent itemset mining. In: Adrian, B., Neumann, G., Troussov, A., Popov, B. (eds.) Proc. of the KI 2008 Workshop on Ontology-Based Information Extraction Systems (2008)
3. Boley, M., Horváth, T., Poigné, A., Wrobel, S.: Listing closed sets of strongly accessible set systems with applications to data mining (unpublished manuscript)
4. Bundschus, M., Dejori, M., Stetter, M., Tresp, V., Kriegel, H.-P.: Extraction of semantic biomedical relations from text using conditional random fields. BMC Bioinformatics 9 (2008)
5. Bunescu, R.C., Mooney, R.J.: A shortest path dependency kernel for relation extraction. In: Proc. of the conference on Human Language Technology and Empirical Methods in Natural Language Processing, pp. 724–731 (2005)

6. Culotta, A., Sorensen, J.: Dependency tree kernels for relation extraction. In: Proc. of the 42nd Annual Meeting on Association for Computational Linguistics (ACL), pp. 423–429 (2004)
7. Fellbaum: WordNet: An Electronic Lexical Database (Language, Speech, and Communication). The MIT Press, Cambridge (1998)
8. Gottlob, G.: Subsumption and implication. Information Processing Letters 24(2), 109–111 (1987)
9. Harabagiu, S., Bejan, C.A., Morarescu, P.: Shallow semantics for relation extraction. In: Proc. of the Nineteenth International Joint Conference on Artificial Intelligence, IJCAI (2005)
10. Hearst, M.: Automatic acquisition of hyponyms from large text corpora. In: Proc. of the 15th International Conference on Computational Linguistics (COLING), pp. 539–545 (1992)
11. Horváth, T., Turán, G.: Learning logic programs with structured background knowledge. Artificial Intelligence 128(1-2), 31–97 (2000)
12. Horváth, T., Wrobel, S.: Towards discovery of deep and wide first-order structures: A case study in the domain of mutagenicity. In: Jantke, K.P., Shinohara, A. (eds.) DS 2001. LNCS (LNAI), vol. 2226, pp. 100–112. Springer, Heidelberg (2001)
13. Joachims, T.: Making large-scale support vector machine learning practical. In: Schölkopf, B., Burges, C.J.C., Smola, A.J. (eds.) Advances in kernel methods: support vector learning, pp. 169–184. MIT Press, Cambridge (1999)
14. Kolaitis, P., Vardi, M.Y.: Conjunctive-query containment and constraint satisfaction. Journal of Computer and System Sciences 61, 302–332 (2000)
15. Mitchell, A., Strassel, S., Przybocki, M., Davis, J., Doddington, G., Grishman, R., Meyers, A., Brunstein, A., Ferro, L., Sundheim, B.: Ace-2 version 1.0. Linguistic Data Consortium, Philadelphia (2003)
16. Moschitti, A.: Efficient convolution kernels for dependency and constituent syntactic trees. In: Fürnkranz, J., Scheffer, T., Spiliopoulou, M. (eds.) ECML 2006. LNCS (LNAI), vol. 4212, pp. 318–329. Springer, Heidelberg (2006)
17. Pantel, P., Pennacchiotti, M.: Espresso: leveraging generic patterns for automatically harvesting semantic relations. In: Proc. of the 44th annual meeting of the Association for Computational Linguistics (ACL), pp. 113–120 (2006)
18. Plotkin, G.: A note on inductive generalisation. In: Meltzer, B., Michie, D. (eds.) Machine Intelligence, vol. 5, pp. 153–163. Elsevier, North Holland, New York (1970)
19. Reichartz, F., Korte, H., Paass, G.: Composite kernels for relation extraction. In: Proc. of the 47nd Annual Meeting on Association for Computational Linguistics (ACL), pp. 365–368 (2009)
20. Reichartz, F., Korte, H., Paass, G.: Dependency tree kernels for relation extraction from natural language text. In: Buntine, W., Grobelnik, M., Mladenić, D., Shawe-Taylor, J. (eds.) ECML PKDD 2009, Part II. LNCS, vol. 5782, pp. 270–285. Springer, Heidelberg (2009)
21. Zelenko, D., Aone, C., Richardella, A.: Kernel methods for relation extraction. Journal of Machine Learning Research 3, 1083–1106 (2003)
22. Zhang, M., Zhang, J., Su, J.: Exploring syntactic features for relation extraction using a convolution tree kernel. In: Proc. of the main conference on Human Language Technology Conference of the North American Chapter of the Association of Computational Linguistics (NAACL), pp. 288–295 (2006)

Discovering Rules by Meta-level Abduction

Katsumi Inoue[1], Koichi Furukawa[2], Ikuo Kobayashi[2], and Hidetomo Nabeshima[3]

[1] National Institute of Informatics
2-1-2 Hitotsubashi, Chiyoda-ku, Tokyo 101-8430, Japan
ki@nii.ac.jp
[2] SFC Research Institute, Keio University
5322 Endo, Fujisawa 252-8520, Japan
{furukawa,ikuokoba}@sfc.keio.ac.jp
[3] Division of Medicine and Engineering Science, University of Yamanashi
4-3-11 Takeda, Kofu, Yamanashi 400-8511, Japan
nabesima@yamanashi.ac.jp

Abstract. This paper addresses discovery of unknown relations from incomplete network data by abduction. Given a network information such as causal relations and metabolic pathways, we want to infer missing links and nodes in the network to account for observations. To this end, we introduce a framework of meta-level abduction, which performs abduction in the meta level. This is implemented in SOLAR, an automated deduction system for consequence finding, using a first-order representation for algebraic properties of causality and the full-clausal form of network information and constraints. Meta-level abduction by SOLAR is powerful enough to infer missing rules, missing facts, and unknown causes that involve predicate invention in the form of existentially quantified hypotheses. We also show an application of rule abduction to discover certain physical techniques and related integrity constraints within the subject area of Skill Science.

1 Introduction

In many scientific and communication domains, knowledge is often structured in a network or graph form, in which arcs and nodes have important meanings in each application. For example, in biological domains, a sequence of signalings or biochemical reactions constitutes a *pathway*, which specifies a mechanism to explain how genes or cells carry out their functions. However, information of biological networks in public domain databases is generally *incomplete* for several reasons. Incompleteness appears in different forms such as missing (detailed) reactions, missing intermediary metabolites, missing kinetic information of reactions, and unknown reactions. To deal with incompleteness of pathway databases, we need to predict the status of relations which is consistent with the status of nodes [21,22], or augment missing arcs between nodes to explain observations [19,23,11,12]. These goals are often characterized by *abductive* problems called *theory completion* [13] or *graph completion* [19], in which status of nodes or missing arcs are added to account for observations.

Many current abductive methods have several limitations. Sometimes only single missing arcs can be found, but those explanations with more than one missing arc cannot

L. De Raedt (Ed.): ILP 2009, LNAI 5989, pp. 49–64, 2010.

be obtained. In another case, *multiple observations* are often given rather than sequentially, which require abductive systems to infer *multiple missing arcs*. A more difficult case is to abduce *missing nodes* and arcs connecting from/to these unknown nodes.

Similar problems are encountered in analysis of human physical performances. *Skill Science* is a new discipline which explores methods to achieve hard tasks and skills. For example, cello playing requires many skills but humans are often unaware of precise reasons why one performance is better than another. The problem of physical skill discovery is also represented in abduction, which provides a way to suggest players how to well perform hard tasks [9]. However, it is more difficult to explain the principle behind an *empirical rule* such as "one can increase the sound volume if she keeps her arm shut during bowing" [1]. To formulate this explanation, we need to abduce *hidden rules* which explain the empirical rule. Often these abduced rules must be elaborated to explain *multiple goals* simultaneously, and involve *invention of new predicates* [14]. Actually, these requirements give us a strong motivation to conduct this research [1].

In this paper, we provide a simple and powerful method to infer missing nodes as well as missing arcs by abduction. We utilize the deductive procedure SOLAR [15] and realize an advanced form of abduction on top of it. SOLAR is a state-of-the-art inference system based on *SOL resolution* [2], and has been applied to abduction in first-order clausal theories. SOLAR has also been used in *CF-induction* [3], which integrates abductive and inductive inference in theory completion. Here, we propose a different way of SOLAR utilization in abduction. A unique feature in our new abductive system is *rule abduction*, that is, to infer missing rules. Since *explanatory induction* also infers rules that explain observations, we could state that "induction is realized by abduction" (in part) in this work. This is contrasted with the fact that "abduction is realized as a special case of induction" in CF-induction [3]. The merit of our rule abduction lies in the fact that both enumeration of rule-form hypotheses and predicate invention are easily realized by SOLAR in *meta-level abduction*. As an application of the proposed abductive system, we investigate how to discover *knack* in performing skillful tasks [1].

This paper is a full version of an extended abstract [5]. The rest of this paper is organized as follows. Section 2 reviews the logic of hypothesis generation in abduction and its computation by SOLAR. Section 3 introduces the concept of meta-level abduction, and applies it to rule abduction. Section 4 shows that meta-level abduction can also be applied to infer abducible facts and to integrate abduction and induction. Section 5 applies meta-level abduction to problems in Skill Science. Section 6 discusses related work, and Section 7 gives a summary and future work.

2   Generating Abductive Hypotheses

The logical framework of hypothesis generation in abduction can be expressed as follows. A *clausal theory* is a set of clauses. Let B be a clausal theory, which represents the *background theory*, and O be a set of literals, which represents *observations* (or *goals*). Given B and O, the hypothesis-generation problem is to find a formula H called a *hypothesis* such that

$$B \cup H \models O, \text{ and} \tag{1}$$

$$B \cup H \text{ is consistent.} \tag{2}$$

This formalization covers both *abduction* and *(explanatory) induction*. Often, a hypothesis H is constructed with some restricted vocabulary Γ called a (language) *bias*. In abduction, Γ is usually given as a set of literals called *abducibles*, which are candidate assumptions to be added to B for explaining O. Then, H is called an *(abductive) explanation* of O (with respect to B and Γ) if it satisfies (1), (2) and

$$H \text{ is a set of instances of literals from } \Gamma. \qquad (3)$$

Additional conditions on explanations can be introduced such as the maximum number of literals in each explanation. The set of explanations of O with respect to B and Γ is denoted as $Expl(O, B, \Gamma)$. An explanation H of O is *minimal* if no proper subset of H is an explanation of O. An explanation is *ground* if it is a set of ground literals. We identify an explanation H with the conjunction of literals in it.

Later, we will see that an abductive explanation can contain variables which can be assumed existentially quantified at the front. In rule abduction (Section 3), the condition (3) is also generalized to allow rules in an explanation H.

A set H of instances of literals from abducibles Γ is called a *nogood* (with respect to B and Γ) if $B \cup H$ is inconsistent. The set of nogoods with respect to B and Γ is denoted as $Nogood(B, \Gamma)$. A nogood H is *minimal* if no proper subset of H is a nogood. Because of the consistency requirement of explanations (2), any explanation must not include any minimal nogood. Minimal nogoods are also useful to characterize integrity constraints derivable from the background theory B.

2.1   Abduction as Consequence Finding

We use the notion of production fields [2] to represent language biases for hypotheses. A *production field* $\mathcal{P}$ is a pair $\langle \mathbf{L}, Cond \rangle$, where $\mathbf{L}$ is a set of literals and $Cond$ is a certain condition. When $Cond$ is not specified, $\mathcal{P}$ is denoted as $\langle \mathbf{L} \rangle$. A clause C *belongs to* $\mathcal{P} = \langle \mathbf{L}, Cond \rangle$ (and is called a $\mathcal{P}$-*clause*), written $C \in \mathcal{P}$, if every literal in C is an instance of a literal in $\mathbf{L}$ and C satisfies $Cond$. A clause C is a $\mathcal{P}$-*consequence* of a clausal theory Σ if $\Sigma \models C$ and $C \in \mathcal{P}$. The set of all $\mathcal{P}$-consequences of Σ is denoted as $Th_{\mathcal{P}}(\Sigma)$. For example, let $\mathcal{P} = \langle \{ans(\_)\} \rangle$ be a production field where $\{ans(\_)\}$ is the set of atoms with the predicate ans. Then, $Th_{\mathcal{P}}(\Sigma)$ is the set of all positive clauses derivable from Σ such that the predicate of any literal is ans.

The *characteristic clauses* of a clausal theory Σ with respect to a production field $\mathcal{P}$ are

$$Carc(\Sigma, \mathcal{P}) = \mu Th_{\mathcal{P}}(\Sigma) \,,$$

where μT denotes the set of clauses in T that are minimal with respect to subsumption. The *new characteristic clauses* of a clause C with respect to Σ and $\mathcal{P}$ are

$$Newcarc(\Sigma, C, \mathcal{P}) = \mu \left[Th_{\mathcal{P}}(\Sigma \cup \{C\}) \setminus Th_{\mathcal{P}}(\Sigma) \right].$$

Given the observations O, each abductive explanation H of O can be computed by the principle of *inverse entailment* [2], which converts the equation (1) to

$$B \cup \{\neg O\} \models \neg H, \qquad (4)$$

where $\neg O = \bigvee_{L \in O} \neg L$ and $\neg H = \bigvee_{L \in H} \neg L$ are clauses because O and H are sets of literals. Similarly, the equation (2) is equivalent to $B \not\models \neg H$. Hence, for any hypothesis H, its negated form $\neg H$ is deductively obtained as a "new" consequence of $B \cup \{\neg O\}$ which is not an "old" consequence of B alone. Moreover, by (3), any literal in $\neg H$ is an instance of a literal in $\overline{\Gamma} = \{\neg L \mid L \in \Gamma\}$, that is, $\neg H$ is a $\langle \overline{\Gamma} \rangle$-clause. Hence, the set of minimal explanations of O with respect to B and Γ is characterized as follows [2].

$$\mu Expl(O, B, \Gamma) = \{ H \mid \neg H \in Newcarc(B, \neg O, \langle \overline{\Gamma} \rangle) \}. \tag{5}$$

On the other hand, the set of minimal nogoods with respect to B and Γ is equivalent to:

$$\mu Nogood(B, \Gamma) = \{ H \mid \neg H \in Carc(B, \langle \overline{\Gamma} \rangle) \}. \tag{6}$$

2.2 SOLAR

SOLAR (SOL for Advanced Reasoning) [15,16] is a sophisticated deductive reasoning system based on SOL resolution [2], which is complete for finding (new) characteristic clauses with respect to a given production field. Consequence-finding by SOLAR is performed by *skipping* literals belonging to a production field $\mathcal{P}$ instead of resolving them in a *tableau proof*. Those skipped literals are then collected at the end of a proof, which construct a clause as a $\mathcal{P}$-consequence of the axiom set. SOLAR can thus be used to implement an abductive system that is *complete* for finding minimal explanations. Unlike many other top-down resolution-based abductive procedures, which are designed for Horn clauses or normal logic programs, SOLAR is designed for inference in *full clausal theories* containing non-Horn clauses, so can be used in full clausal abduction.

SOLAR avoids producing non-minimal $\mathcal{P}$-consequences as well as redundant computation using state-of-the-art pruning techniques [8,15]. Enumeration of (new) characteristic clauses is also efficiently realized in SOLAR using the *skip-minimality* criterion, which performs subsumption checking of skipped literals against $\mathcal{P}$-consequences obtained so far [16]. A simple way to compute $Newcarc(\Sigma, C, \mathcal{P})$ in SOLAR is: (1) enumerate the subsumption-minimal produced clauses from SOL tableau proofs with the top clause C, and then (2) remove from them those clauses subsumed by some clause in $Carc(\Sigma, \mathcal{P})$ [16]. Consequence enumeration is a strong point of SOLAR as an abductive procedure because many other abductive systems compute just single or a few hypotheses based on their own evaluation methods. Hypothesis enumeration by SOLAR, on the other hand, enables us to compare many different hypotheses and select the most probable ones from them. A statistical method for ranking abductive hypotheses is proposed in [6] based on an EM algorithm working on binary decision diagrams.

A production field $\mathcal{P}$ is *stable* if, for any clauses C and D such that C subsumes D, $D \in \mathcal{P}$ implies $C \in \mathcal{P}$. The stability of a production field is important in efficiently deducing minimal $\mathcal{P}$-consequences [2]. In particular, a stable production field with a condition on a maximum length of a produced clause C is denoted as $\langle \mathbf{L}, |C| \leq k \rangle$, and is practically useful for minimizing search effort because any intermediate proof with more than k skipped literals can be pruned immediately. Note that a production field containing ground literals (or literals containing function symbols or constants) is not stable although such a production field is often used in abduction. In this case, a non-stable production field is automatically converted to a stable production field in SOLAR by way of [18], thereby assuring completeness of ground hypotheses in abduction [16].

3   Meta-level Abduction

This section formalizes rule abduction. To this end, we introduce a framework for *meta-level abduction*, which performs abduction in the meta level. Here, background theory is represented in a network structure called a *causal graph* with the two meta-predicates *caused* and *connected*, which respectively represents causal relations and direct links between facts. This is contrasted with *object-level abduction*, which infers explanations of observed facts using logical formulas representing relations between facts.

3.1   Causal Graphs

Representing background knowledge K, literals in the language represents propositions, events, states, phenomena, or whatever, but we regard each literal in K as a *fact*. K thus represents relations between facts at the object level. Suppose that a fact G is *caused by* a fact S. Here, G is called an *output* (or *goal*) *fact*, and S is called an *input* (or *source*) *fact*. In the object level, this causal relation can be represented intuitively by the rule $G \leftarrow S$, and K can be given as a set of such causal rules.

The problem of the object-level representation is that the meaning of $\leftarrow$ depends on the underlying logic. In classical logic, many formulas other than the relations appearing in K should hold. In particular, $G \leftarrow S$ implies $\neg S \leftarrow \neg G$, but the latter is not always desirable as a causal relation. We thus represent causal relations by formulas at the meta level. Using first-order logic, algebraic properties of causal relations, such as transitivity and non-reflexivity, and constraints can be defined in a natural way. Given the object-level knowledge K, its meta theory is defined using a causal graph as follows.

A *causal graph* is a directed graph representing causal relations, which consists of the sets of nodes and arcs. A *direct causal relation* corresponds to a directed arc, and a *causal chain* is represented by the reachability between two nodes.

$$connected(g, s) \tag{7}$$

$$\leftarrow connected(g, s) \tag{8}$$

$$connected(g, s) \vee connected(h, s) \tag{9}$$

$$connected(g, s) \vee connected(g, t) \tag{10}$$

When there is a direct causal relation from the node s to the node g, we define that $connected(g, s)$ is true (7). Note that $connected(g, s)$ only shows that s is one of possible causes of g, and thus the existence of $connected(g, t)$ ($s \neq t$) means that s and t are alternative causes for g. If we know that there is no direct causal link from s to g, we add an *integrity constraint* of the form (8), which is equivalent to the formula $\neg connected(g, s)$. If a direct causal relation from s has *nondeterministic effects* g and h, it is represented in the disjunction of the form (9). On the other hand, if a direct causal relation to g has *conjunctive causes* s and t, it is represented in the disjunction of the form (10). Any other direct causal relation can be represented in a combination of these components.[1] The logic behind (10) is explained as follows (see Lemma 3.1). The relation that "g is caused by s and t" is intuitively written as $(g \leftarrow s \wedge t)$ in the object level, which is equivalent to $(g \leftarrow s) \vee (g \leftarrow t)$, hence the disjunction $connected(g, s) \vee connected(g, t)$.[2]

When there is a *causal chain* from s to g, we defined that $caused(g, s)$ is true. Then, we have the following formulas as axioms:

$$caused(X, Y) \leftarrow connected(X, Y). \tag{11}$$

$$caused(X, Y) \leftarrow connected(X, Z) \wedge caused(Z, Y). \tag{12}$$

Here, the predicates *connected* and *caused* are both *meta-predicates* for object-level propositions g and s. That is, rules of causal relations in the object level are represented by atoms in the meta level.

We can also allow variables in those object-level expressions like $g(T)$ and $s(T)$. Then, in the meta-level expression $connected(g(T), s(T))$, the predicates g and s are treated as function symbols just in the same way that Prolog can allow higher-order expressions, yet no recursive application of g and s is allowed.

3.2   Rule Abduction

Now we realize *rule abduction* in the meta-level representation. Suppose that a goal fact G is *somehow* caused by an input fact S. Our task of rule abduction is to explain why or how it is caused. The difference from ordinary abduction is that we consider such input-output relations (G, S) as observations rather than simply considering a goal fact G. In this situation, we could add S to K and constructs a hypothesis H such that $K \cup \{S\} \cup H \models G$ in object-level abduction. But we often want to separate S from K as S is usually changeable and is only empirically discovered. Moreover, H in object-level abduction is a set of literals representing facts, so cannot be in the form of rules.

Meta-level abduction is used to infer missing relations (arcs) as well as missing facts (nodes). This process corresponds to filling the gaps in causal graphs. That is, when a causal graph is incomplete, there is no path between a goal fact g and an input fact s. Then, an abductive task infers missing links (and nodes) to complete a path between the two nodes. This is realized by setting the abducibles Γ as the atoms containing *connected* only: $\Gamma = \{connected(\_, \_)\}$. The observation is given in the form of the causal chain $caused(g, s)$, and we usually assume that there is no direct causal relation between them, i.e., $\leftarrow connected(g, s)$, otherwise we do not need abduction.

For example, suppose a goal fact g, an input fact s, and another fact r that causes g, i.e., $connected(g, r)$. Then, suppose that we have observed the causal chain between g and s, i.e., $O = caused(g, s)$. Then, $connected(r, s)$ is a minimal abductive

---

[1] A complex relation of the form $(g \vee h \leftarrow s \wedge t)$ should be handled by decomposing it into two relations, $(s\text{-}t \leftarrow s \wedge t)$ and $(g \vee h \leftarrow s\text{-}t)$, where $s\text{-}t$ represents the intermediate complex.

[2] In the meta level, to construct an abductive proof of g using (10), we need to show abductive proofs of both s and t. Then the meta-level disjunction represents the *case inference* that the whole proof is established if proofs of the both cases are established.

Fig. 1. Hypotheses H_1 (left) and H_2 (right)

explanation of O. Other minimal explanations of O are conceivable if we allow more than one literal in an explanation. For example,

$$\exists X \, (connected(g, X) \wedge connected(X, s)) \tag{13}$$

is a two-literal hypothesis containing an existentially quantified variable X. If X is considered as a new node, then it corresponds to introduction of a *new predicate*. In fact, the formula (13) corresponds to the two rules $\{(g \leftarrow \chi), (\chi \leftarrow s)\}$ where χ is regarded as either a quantified variable or a predicate variable. In this way, *predicate invention* is realized in meta-level abduction.

Abduction by SOLAR enables us to infer such *existentially quantified hypotheses*. In the previous example, the top clause $\neg O$ resolves with the axiom clause (12), which results in the clause $(\neg connected(g, Z) \vee \neg caused(Z, s))$. This can be further resolved with the axiom clause (11), which produces $\neg H_2$. By chaining this inference, any number of variables can be newly introduced. Then we can specify the maximum length of clauses produced by SOLAR as a condition in a production field.

Suppose the *multiple* observations

$$O' : \; caused(g, s) \wedge caused(h, s).$$

Examples of explanations of O' containing two intermediate nodes are:

$$H_1 : \quad \exists X \exists Y (connected(g, X) \wedge connected(h, Y)$$
$$\wedge \, connected(X, s) \wedge connected(Y, s)),$$
$$H_2 : \quad \exists X \exists Y (connected(g, X) \wedge connected(X, Y)$$
$$\wedge \, connected(h, Y) \wedge connected(Y, s)).$$

Fig. 1 shows that H_1 and H_2 have different structures. By this way, we can enumerate different types of network structures that are missing in the original causal graph.

So far, algebraic properties of causal relations are only represented by (11) and (12), but more constraints can be introduced if necessary. For example, non-reflexivity can be assumed for acyclic causal graphs by[3]

$$\leftarrow caused(X, X).$$

Reasoning about properties can be done by associating an attribute with each fact. For example, inheritance of a property p can be represented by

$$p(X) \leftarrow caused(X, Y) \wedge p(Y).$$

We expect that this method can be applied to reasoning about pathways.

---

[3] This relation cannot be used with the axiom (16) for abducibles of fact abduction in Section 4.

3.3   Correctness

We prove the soundness and completeness of rule abduction in meta-level abduction. For the soundness theorem, we suppose a consistent background theory B in the meta level. Let $\lambda(B)$ be the theory obtained by removing (11) and (12) from B and replacing every $connected(g, s)$ appearing in B with the formula $(g \leftarrow s)$.

Lemma 3.1. *If $B \models caused(g, s)$ then $\lambda(B) \models (g \leftarrow s)$.*

Proof. We prove the lemma by induction on the depth of proof trees (SOL tableaux). Since SOL is complete for consequence-finding [2,8], for any unit clause C derived from a consistent axiom set, there is an SOL tableau producing C with a certain depth.

Induction basis. It holds by the meaning of causal graphs that $caused(g, s) \in B$ iff $(g \leftarrow s) \in \lambda(B)$. Moreover, $connected(g, s_1) \vee \cdots \vee connected(g, s_n) \in B$ iff $(g \leftarrow s_i) \vee \cdots \vee (g \leftarrow s_n) \in \lambda(B)$ iff $\lambda(B) \models (g \leftarrow s_1 \wedge \cdots \wedge s_n)$. (†)

Induction hypothesis. Suppose that the lemma holds for any $caused(g, s)$ which is derived from B by an SOL tableau with the depth d such that $d \leq k$.

Induction step. Assume that $caused(g, s)$ can be derived from B by an SOL tableau with the depth $k + 1$. By (12), there is $connected(g, s_1) \vee \cdots \vee connected(g, s_n) \in B$ such that every $caused(s_i, s)$ $(i = 1, \ldots, n)$ is derived from B by an SOL tableau with a depth $d \leq k$. By the induction hypothesis, $\lambda(B) \models (s_i \leftarrow s)$ for every $i = 1, \ldots, n$. Then, by (†), $\lambda(B) \models (g \leftarrow s)$.   □

Theorem 3.1. *Suppose the observation $caused(g, s)$, and let $\Gamma_M = \{connected(\_, \_)\}$. If H is an abductive explanation of $caused(g, s)$ with respect to B and Γ_M then $\lambda(H)$ is a hypothesis satisfying that*

$$\lambda(B) \cup \lambda(H) \models (g \leftarrow s), \ and \tag{14}$$

$$\lambda(B) \cup \lambda(H) \ is \ consistent. \tag{15}$$

Proof. The abductive derivation (14) can be proved by Lemma 3.1. The consistency (15) can be checked by the fact that $B \cup H$ is consistent.   □

The completeness of meta-level abduction is stated as follows.

Theorem 3.2. *Suppose the background knowledge K in the object level, and let $C(K)$ be the meta-theory representing the causal graph associated with K, and define that $\tau(K) = C(K) \cup \{(11), (12)\}$. If a fact g is reached from a fact s in the causal graph of K by augmenting a set E of direct causal relations, then $C(E)$ is an abductive explanation of $caused(g, s)$ with respect to $\tau(K)$ and Γ_M.*

4   Combining Rule Abduction with Fact Abduction

In Section 3, we have seen that meta-level abduction can be used to abduce rules. In this section, meta-level abduction is shown to be applied to ordinary abduction, i.e., abduction of facts. Then the strong point of meta-level abduction is that we can abduce both facts and rules in the framework.

4.1 Fact Abduction

Since $caused(g, s)$ represents the causality that g is caused by s, abduction of a fact g or s in the object level can be formalized as *query answering* in the meta level. Given a query of the form $caused(g, X)$, *abduction of causes* is computed by answer substitutions to the variable X. To enable this inference, we need the new axiom for the causal relations:

$$caused(X, X) \leftarrow abd(X). \tag{16}$$

Here, $abd(X)$ is used to represent that X is abducible. The formula (16) indicates that an abducible can hold by assuming itself. Answer extraction in SOLAR can be realized by giving the top clause of the form:

$$\leftarrow caused(g, X) \wedge abd(X). \tag{17}$$

The variable X is used to collect only abducibles which cause the fact g. Hence, *abd* also plays the role of an *answer predicate* [7]. An integrity constraint that two facts p and q cannot hold at the same time can be represented as:

$$\leftarrow caused(p, X) \wedge caused(q, Y) \wedge abd(X) \wedge abd(Y). \tag{18}$$

This makes any combination of abducibles that causes both p and q a nogood.

On the other hand, a query of the form $caused(X, s)$ is used for *prediction* given a source fact s. If both arguments of *caused* are variables, then *characteristic clauses* with the production field $\langle\{caused(\_, \_)\}\rangle$ can be used for computing *ramification* of the causal theories. All these patterns of inference can be easily realized in SOLAR.

4.2 Abducing Rules and Facts

Recall that abduction in the meta level realizes rule abduction. This is done by giving a query of the form $caused(g, s)$, and abduction in the meta level infers missing causal rules in the object level. On the other hand, fact abduction is realized by giving a query with variables, and answer extraction provides sufficient or necessary facts. Now we combine these two inferences. *Conditional query answering* [7,4] realizes answer extraction under abduced conditions, hence this inference in the meta level realizes abduction of facts and rules in the object level. There are some advanced applications which need to find both missing facts and missing rules [22], and our method will provide a new implementation for such domains.

In conditional query answering, a query of the form $caused(g, X)$ can be used for *abduction of causes and rules*, while a query of the form $caused(X, s)$ is for *conditional prediction*. To avoid complication in these cases, it seems better to restrain predicate invention in meta-level abduction.

Table 1 gives a summary of the correspondence between object-level inference and meta-level inference in SOLAR.

Table 1. Correspondence between object-level inference and meta-level consequence finding

object-level inference	top clause in SOLAR	production field
proving rules	$\neg caused(g, s)$	$\langle \emptyset \rangle$
abducing facts	$\neg caused(g, X) \vee \neg abd(X)$	$\langle \{\neg abd(f_1), \ldots, \neg abd(f_n)\} \rangle$
predicting facts	$\neg caused(X, s) \vee ans(X)$	$\langle \{ans(\_)\} \rangle$
predicting rules	none	$\langle \{caused(\_, \_)\} \rangle$
abducing rules	$\neg caused(g, s)$	$\langle \{\neg connected(\_, \_)\} \rangle$
abducing rules and facts	$\neg caused(g, X) \vee \neg abd(X)$	$\langle \{\neg connected(\_, \_)\} \cup \{\neg abd(f_1), \ldots, \neg abd(f_n)\} \rangle$
predicting conditional facts	$\neg caused(X, s) \vee ans(X)$	$\langle \{\neg connected(\_, \_), ans(\_)\} \rangle$
predicting conditional rules	none	$\langle \{\neg connected(\_, \_), caused(\_, \_)\} \rangle$

5 Application to Skill Science

5.1 Knack Discovery

The goal of rule abduction in a *knack discovery* problem is to find missing rules to explain observed causality. In [1], this is introduced in accomplishing the hard performance task to "increase sound volume" in playing the cello. An observed causality or "knack" for this goal is to "keep the right arm close during bowing", which is empirically proved to achieve the goal. The situation can be represented as follows.

$$connected(\texttt{inc\_sound}, \texttt{bow\_close\_bridge}). \tag{19}$$

$$connected(\texttt{bow\_close\_bridge}, \texttt{stable\_move})$$
$$\vee\, connected(\texttt{bow\_close\_bridge}, \texttt{smooth\_change}). \tag{20}$$

$$connected(\texttt{smooth\_change}, \texttt{flexible\_wrist}). \tag{21}$$

$$\neg connected(\texttt{inc\_sound}, \texttt{keep\_arm\_close}). \tag{22}$$

$$\neg connected(\texttt{stable\_bow\_movement}, \texttt{keep\_arm\_close}). \tag{23}$$

$$\neg connected(\texttt{smooth\_bow\_direction\_change}, \texttt{keep\_arm\_close}). \tag{24}$$

Note that the clause (20) represents the following implication in the object level:

$$\texttt{bow\_close\_bridge} \leftarrow \texttt{stable\_move} \wedge \texttt{smooth\_change}.$$

The constraints (22), (23) and (24) are introduced to prevent direct connections between those facts. Now suppose the observed causal chain:

$$caused(\texttt{inc\_sound}, \texttt{keep\_arm\_close}),$$

which means that one empirically observes that sound volume is increased by keeping one's arm close. The goal of rule abduction is to explain this empirical causality.

One of the explanations of this observation is as follows:

$$\exists X\,(\,connected(\texttt{stable\_bow\_movement}, X) \wedge connected(\texttt{flexible\_wrist}, X)$$
$$\wedge\, connected(X, \texttt{keep\_arm\_close})\,). \tag{25}$$

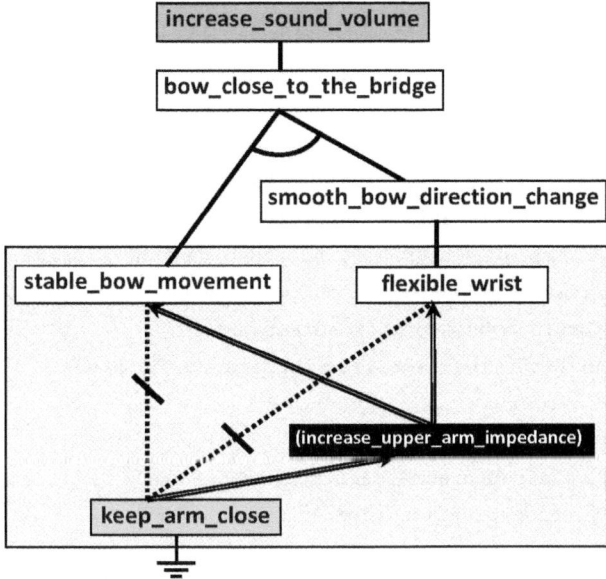

Fig. 2. An explanation for knack discovery

The formula (25) represents three rules with some unknown fact X. By substituting X with `increase_upper_arm_impedance` as a newly introduced predicate, the abduced rules correspond to the three lines connecting with this new node in Fig. 2. In experiments by SOLAR by setting the maximum search depth to 15 and the maximum length of produced clauses to 4, we actually get 52 hypotheses including the hypothesis (25).

The meaning of newly discovered nodes should be given in a domain-dependent manner [1]. In [10], an empirical method to analyze the meaning is proposed for the cello playing domain.

5.2   Skill Discovery

In playing the string instruments, the simultaneous task of achieving both the left hand position change and keeping continuous vibrato is very difficult. Hence we want to find an appropriate way to achieve such difficult tasks using knowledge of body usage including muscle activity patterns and appropriate posture. In the problem, there are several integrity constraints that must be satisfied: neuromuscular constraints, biomechanical constraints and spatio-temporal constraints.

Abducing facts.   We consider the problem of achieving the simultaneous task of the rapid position shift and the continuous vibrato. Here a portion of the background theory in [9] is translated to a meta-level description B_1 in Fig. 3.

The goal is "rapidPositionShift_with_vibrato", and is represented as the clause C_1 labeled by (33). The variable X in (33) extracts abducibles that achieve the goal. Since the goal is a compound task, it is reduced to two subgoals by (26). The first

% Causal graph theory

connected(rapidPositionShift_with_vibrato, rapidPositionShift)
 ∨ connected(rapidPositionShift_with_vibrato, vibrato). (26)
connected(rapidPositionShift, rapidAddAbdOfarm)
 ∨ connected(rapidPositionShift, flexExtOfElbow). (27)
connected(rapidPositionShift, inExCycloOfUpperarm)
 ∨ connected(rapidPositionShift, pronosupinationOfForearm). (28)
connected(vibrato, fixUpperarm) ∨ connected(vibrato, microFlexExtOfElbow). (29)
connected(vibrato, pronosupinationOfForearm). (30)
connected(flexExtOfElbow, activeLeftUpperarmMsclsStrong). (31)

% Integrity constraints

¬caused(activeLeftUpperarmMsclsStrong, X) ∨ ¬abd(X)
 ∨ ¬caused(pronosupinationOfForearm, Y) ∨ ¬abd(Y). (32)

% Top clause : C_1

¬caused(rapidPositionShift_with_vibrato, X) ∨ ¬abd(X). (33)

% Axioms for the causal relation

caused(X, X) ∨ ¬abd(X). (34)
caused(X, Y) ∨ ¬connected(X, Y). (35)
caused(X, Y) ∨ ¬connected(X, Z) ∨ ¬caused(Z, Y). (36)

Fig. 3. Background theory B_1 for skill discovery

subgoal "rapidPositionShift" can be realized in two ways by (27) and (28), while the second subgoal "vibrato" can be solved by (29) and (30), which results in four possible combinations. Then the constraint (32) prunes incompatible sets of abducibles.

In SOLAR, the production field is given as $\mathcal{P}_1 = \langle \overline{Abd_1}, |C| \leq 5 \rangle$, where $\overline{Abd_1}$ is the set of negated abducibles and consists of the 6 literals:

¬abd(rapidAddAbdOfarm), ¬abd(inExCycloOfUpperarm),

¬abd(microFlexExtOfElbow), ¬abd(pronosupinationOfForearm),

¬abd(fixUpperarm), ¬abd(activeLeftUpperarmMsclsStrong).

Note that $\mathcal{P}_1$ is not a stable production field. Then, SOLAR adds a "bridge clause" [18]:

$$abd(X) \lor ans(X)$$

to the back ground theory with a new predicate ans, and replaces $\mathcal{P}_1$ to a stable production field:

$$\mathcal{P}_1' = \langle \{ans(\_)\}, |C| \leq 5 \rangle.$$

In this problem, two different abductive explanations are considerable:

$$H_1 = \{\, \texttt{microFlexExtOfElbow}, \texttt{fixUpperarm},$$
$$\texttt{pronosupinationOfForearm}, \texttt{inExCycloOfUpperarm}\,\},$$
$$H_2 = \{\, \texttt{pronosupinationOfForearm}, \texttt{inExCycloOfUpperarm}\,\}.$$

H_1 is obtained from the choices (28) and (29), while H_2 is from the rules (28) and (30). Since H_1 is subsumed by H_2, the clause corresponding to $\neg H_2$ is obtained by SOLAR as the unique element of $Newcarc(B_1, C_1, \mathcal{P}_1)$:

$$\neg\texttt{abd}(\texttt{pronosupinationOfForearm}) \vee \neg\texttt{abd}(\texttt{inExCycloOfUpperarm}).$$

On the other hand, the choice (27) cannot form a consistent set of abducibles. In fact, the set of negated nogoods is computed by SOLAR as

$$\texttt{abd}(\texttt{fixUpperarm}) \vee \neg\texttt{abd}(\texttt{rapidAddAbdUtarm}),$$
$$\neg\texttt{abd}(\texttt{activeLeftUpperarmMsclsStrong}) \vee \neg\texttt{abd}(\texttt{pronosupinationOfForearm}),$$

and any $\mathcal{P}_1$-consequence derived via (27) is subsumed by either of them.

Abducing rules and facts. Here we suppose the background theory B_2 which is obtained by removing the rule (30) from B_1. Now our goal is to recover this missing rule and to abduce the solutions in the previous problem at the same time.

Put $\overline{\Gamma_M} = \{\neg\texttt{connected}(\_,\_)\}$. The production field $\mathcal{P}_2$ is defined by

$$\mathcal{P}_2 = \langle\, \overline{Abd_1} \cup \overline{\Gamma_M},\ |C| \leq 5 \text{ and } |C \cap \overline{\Gamma_M}| \leq 1 \,\rangle.$$

It is verified that the negation of the intended solution is obtained by SOLAR as one of the 40 clauses in $Newcarc(B_2, C_1, \mathcal{P}_2)$:

$$\neg\texttt{connected}(\texttt{vibrato}, \texttt{pronosupinationOfForearm})$$
$$\vee \neg\texttt{abd}(\texttt{pronosupinationOfForearm}) \vee \neg\texttt{abd}(\texttt{inExCycloOfUpperarm}).$$

6 Discussion

This paper proposes a new method to abduce rules. *Abducible rules* were firstly considered in Theorist [17]. In implementing abducible rules, each rule is given a *name* associated with the free variables in the rule and those names are treated as abducible atoms. This is a convenient method when we know exact patterns of rules as strong biases. However, it is impractical or impossible to prepare all patterns of rules in advance in order to abduce missing rules, and predicate invention is basically impossible by this method. But here, we have realized rule abduction in the meta level and avoid searching in the space of object-level rules.

Abductive reasoning in an earlier version of Robot Scientist [11] has been implemented with a version of SOL resolution restricted to Horn clauses [19]. In our work, SOLAR, SOL for full-clausal theories, is necessary since our formalization allows non-Horn clauses (9) and (10) in background theories. In [19], a *reaction* is given as a pair of sets of compounds representing the substrates and products, and a *metabolic graph* is defined in such a way that each node is given as a set of compounds available by sequences of reactions. In the logical form, a metabolic graph is defined by:

$$edge(X, Y) \leftarrow reaction(A, B) \wedge (A \subseteq X) \wedge (X \cup B = Y),$$
$$path(X, Y) \leftarrow edge(X, Y),$$
$$path(X, Y) \leftarrow edge(X, Z) \wedge path(Z, Y).$$

Here, the last two rules are exactly the same as our definition of *caused* by means of *connected*, but *edge* defined in the first rule is more complicated than *connected*. Then, abduction of reactions produces a combinatorially large number of compound pairs by bidirectional uses of operations member (for $\subseteq$) and append (for $\cup$) in the first rule.

On the other hand, considering classical graphs instead of metabolic graphs, the predicate *caused* could be defined in a similar way to metabolic graphs by associating to a node all sources obtained by a causal chain:

$$caused(X, W) \leftarrow connected(X, [Z_1, ..., Z_k]) \wedge$$
$$caused(Z_1, W_1) \wedge \cdots \wedge caused(Z_k, W_k) \wedge (W = W_1 \cup \cdots \cup W_k).$$

Here, $connected(X, [Z_1, ..., Z_k])$ represents a direct causal link to X from the conjunction of $Z_1, \ldots, Z_k$, viz., $(X \leftarrow Z_1 \wedge \cdots \wedge Z_k)$. In this representation, we need to associate m such rules for $k = 1, \ldots, m$ where m is the maximum number of rule conditions. Again, abductive computation with this method heavily depends on the frequent uses of append ($\cup$) in bidirectional ways, and the number of abducibles with *connected* may grow exponentially. In a related method, Stickel [20] has used the meta-theoretic predicate $fact$ for abduction, and translates an abducible literal L to $fact(L, \{L\})$ and each rule $(X \leftarrow Z_1 \wedge \cdots \wedge Z_k)$ to

$$fact(X, W_1 \cup \cdots \cup W_k) \leftarrow fact(Z_1, W_1) \wedge \cdots \wedge fact(Z_k, W_k),$$

where the second argument of $fact$ represents hypotheses sufficient to prove the literal in the first argument. In [20], this translation is also extended to achieve the goal-directedness in a bottom-up abductive procedure, but rules cannot be abduced.

In this paper, we adopt classical graphs instead of metabolic graphs, but the use of a disjunction of *connected* literals is used to represent a direct multi-causal relationship by (10), which nicely solves these combinatorial problems in representing network structures, and is one of the most important contributions in this paper. Note that a hypothesis of the form (10) can also be computed using SOLAR either by taking a disjunction of explanations of the form $connected(g, \_)$ or by obtaining a *disjunctive answer* [18] for an observation containing free variables like $caused(g, X)$.

In [22], inference about metabolic pathways has been realized using CF-induction [3], in which both abductive and inductive hypotheses are generated to explain observations. Since CF-induction is complete for inverse entailment from full clausal theories, any form of hypotheses can be obtained in principle. CF-induction has also been implemented by calling SOLAR, but requests the user to interact with the system to construct an appropriate hypothesis. On the other hand, SOLAR itself can easily enumerate abductive hypotheses without any user interaction. Moreover, predicate invention by CF-induction could be realized in the framework of *inverse resolution* [14]. Predicate invention within inverse resolution [14] is carried out by the W operator, which constructs three parent clauses from their two resolvents using the lgg (least general generalization) operator, and the literal resolved upon is newly invented. This predicated invention can thus be used in a limited situation, while meta-level abduction by SOLAR naturally enables generation of new predicates in meta-level abduction.

7 Conclusion

In this paper, we have shown a simple and powerful method for rule abduction. In abduction, multiple observations are explained at once, and full clausal theories are allowed for background knowledge. We have proposed the notion of meta-level abduction to realize abduction of both rules and facts. The "AND" connective in causal relations can be captured by disjunctive representation in the meta-level description. Predicate invention is realized as existentially quantified hypotheses. In implementing meta-level abduction, SOLAR can be an effective representation tool to capture this level of complex modeling. The current limitation of our predicate invention is that it is essentially occurred in the ground level, and extension to allow function symbols and recursive expressions is for future work.

Rule abduction has also been applied to Skill Science, in which empirical rules are explained by hidden rules. In the future, we plan to apply rule abduction to Systems Biology. Currently, it is the responsibility of the designer of knowledge bases to guarantee the consistency of the causal theories. However, it is often the case that some property p that holds in the previous state should be changed to $\neg p$, yet introduction of $caused(\neg p, p)$ would be problematic. To precisely describe causality in the domain, the notions of actions and their preconditions and effects could be introduced so that state changes and inertia could be represented in causal theories. This is also future work.

Acknowledgments

This research is supported in part by National Institute of Informatics and by the 2008-2011 JSPS Grant-in-Aid for Scientific Research (A) No. 20240016. We thank Randy Goebel for comments on an earlier draft and Oliver Ray for discussions on this work.

References

1. Furukawa, K., Kobayashi, I., Inoue, K., Suwa, M.: Discovering knack by abductive reasoning. In: SIG-SKL (Skill Science). Japanese Society for Artificial Intelligence (January 2009) (in Japanese)
2. Inoue, K.: Linear resolution for consequence finding. Artificial Intelligence 56, 301–353 (1992)
3. Inoue, K.: Induction as consequence finding. Machine Learning 55, 109–135 (2004)
4. Inoue, K., Iwanuma, K., Nabeshima, H.: Consequence finding and computing answers with defaults. Journal of Intelligent Information Systems 26, 41–58 (2006)

5. Inoue, K., Furukawa, K., Kobayashi, I.: Abducing rules with predicate invention. In: 19th International Conference on Inductive Logic Programming (ILP 2009), Leuven, Belgium (July 2009)
6. Inoue, K., Sato, T., Ishihata, M., Kameya, Y., Nabeshima, H.: Evaluating abductive hypotheses using an EM algorithm on BDDs. In: Proceedings of IJCAI 2009, pp. 810–815 (2009)
7. Iwanuma, K., Inoue, K.: Minimal answer computation and SOL. In: Flesca, S., Greco, S., Leone, N., Ianni, G. (eds.) JELIA 2002. LNCS (LNAI), vol. 2424, pp. 245–257. Springer, Heidelberg (2002)
8. Iwanuma, K., Inoue, K., Satoh, K.: Completeness of pruning methods for consequence finding procedure SOL. In: Proceedings of the 3rd International Workshop on First-Order Theorem Proving, pp. 89–100 (2000)
9. Kobayashi, I., Furukawa, K.: Modeling physical skill discovery and diagnosis by abduction. Information and Media Technologies 3(2), 385–398 (2008)
10. Kobayashi, I., Furukawa, K.: Hypothesis selection using domain theory in rule abductive support for skills. In: SIG-SKL (Skill Science). Japanese Society for Artificial Intelligence (August 2009) (in Japanese)
11. King, R.D., et al.: Functional genomic hypothesis generation and experimentation by a robot scientist. Nature 427, 247–252 (2004)
12. King, R.D., et al.: The automation of science. Science 324, 85–89 (2009)
13. Muggleton, S., Bryant, C.: Theory completion and inverse entailment. In: Cussens, J., Frisch, A.M. (eds.) ILP 2000. LNCS (LNAI), vol. 1866, pp. 130–146. Springer, Heidelberg (2000)
14. Muggleton, S., Buntine, W.: Machine invention of first-order predicate by inverting resolution. In: Proceedings of the 5th International Workshop on Machine Learning, pp. 339–351. Morgan Kaufmann, San Francisco (1988)
15. Nabeshima, H., Iwanuma, K., Inoue, K.: SOLAR: a consequence finding system for advanced reasoning. In: Cialdea Mayer, M., Pirri, F. (eds.) TABLEAUX 2003. LNCS (LNAI), vol. 2796, pp. 257–263. Springer, Heidelberg (2003)
16. Nabeshima, H., Iwanuma, K., Inoue, K., Ray, O.: SOLAR: an automated deduction system for consequence finding. AI Communications, Special Issue on Practical Aspects of Automated Reasoning (2009) (to appear)
17. Poole, D.: A logical framework for default reasoning. Artificial Intelligence 36, 27–47 (1988)
18. Ray, O., Inoue, K.: A consequence finding approach for full clausal abduction. In: Corruble, V., Takeda, M., Suzuki, E. (eds.) DS 2007. LNCS (LNAI), vol. 4755, pp. 173–184. Springer, Heidelberg (2007)
19. Reiser, P.G.K., King, R.D., Kell, D.B., Muggleton, S.H., Bryant, C.H., Oliver, S.G.: Developing a logical model of yeast metabolism. Electronic Transactions in Artificial Intelligence 5-B2(024), 223–244 (2001)
20. Stickel, M.E.: Upside-down meta-interpretation of the model elimination theorem-proving procedure for deduction and abduction. Journal of Automated Reasoning 13(2), 189–210 (1994)
21. Tamaddoni-Nezhad, A., Chaleil, R., Kakas, A., Muggleton, S.: Application of abductive ILP to learning metabolic network inhibition from temporal data. Machine Learning 65, 209–230 (2006)
22. Yamamoto, Y., Inoue, K., Doncescu, A.: Integrating abduction and induction in biological inference using CF-Induction. In: Lodhi, H., Muggleton, S. (eds.) Elements of Computational Systems Biology. John Wiley & Sons, Chichester (2009) (to appear)
23. Zupan, B., Demšar, J., Bratko, I., Juvan, P., Halter, J., Kuspa, A., Shaulsky, G.: GenePath: a system for automated construction of genetic networks from mutant data. Bioinformatics 19(3), 383–389 (2003)

Inductive Generalization of Analytically Learned Goal Hierarchies

Tolga Könik[1], Negin Nejati[1], and Ugur Kuter[2]

[1] Computational Learning Laboratory, Stanford University, Stanford, CA 94305
[2] Department of Computer Science and Institute for Advanced Computer Studies,
University of Maryland, College Park, MD 20742
{konik,nejati}@stanford.edu, ukuter@cs.umd.edu

Abstract. We describe a new approach for learning procedural knowledge represented as teleoreactive logic programs using relational behavior traces as input. This representation organizes task decomposition skills hierarchically and associate explicitly defined goals with them. Our approach integrates analytical learning with inductive generalization in order to learn these skills. The analytical component predicts the goal dependencies in a successful solution and generates a teleoreactive logic program that can solve similar problems by determining the structure of the skill hierarchy and skill applicability conditions (preconditions), which may be overgeneral. The inductive component experiments with these skills on new problems and uses the data collected in this process to refine the preconditions. Our system achieves this by converting the data collected during the problem solving experiments into the positive and negative examples of preconditions that can be learned with a standard Inductive Logic Programming system. We show that this conversion uses one of the main commitments of teleoreactive logic programs: associating all skills with explicitly defined goals. We claim that our approach uses less expert effort compared to a purely inductive approach and performs better compared to a purely analytical approach.

1 Introduction

We describe a new method for *learning by observation* that acquires procedural knowledge from relational behavior traces. Learning from literals that have temporally changing truth values poses new challenges for the relational learning community and may require representations more specialized than general first order logic. In this paper we describe how *teleoreactive logic programs* [1], a first order language for reasoning about actions, goals, and temporally changing states, naturally supports integration of analytical and inductive learning in a temporal relational setting.

One of the challenges of learning by observation is that the purpose for selecting actions is not visible to the learner. For example, in a robotics domain, control actions such as moving and turning may be selected with quite different goals such as avoiding collision with a nearby obstacle or going to a different room. The conditions for selecting those actions often depend on the goals and learning these conditions may be difficult without an explicit goal representation.

Consequently, while some learning by observation systems focus on modeling low level skills (e.g., behavioral cloning [2]), others reason about goals explicitly, usually at

L. De Raedt (Ed.): ILP 2009, LNAI 5989, pp. 65–72, 2010.
© Springer-Verlag Berlin Heidelberg 2010

the cost of more assumed knowledge [3]. RELBO [4], a relational system for behavioral cloning is able to learn complex skills with goals organized hierarchically, but it relies on an expert annotating the behavior traces with goal instances. LIGHT [5] can discover the hierarchical organization of goals automatically using action models and an analysis similar to explanation-based learning, but it may learn over-general skill selection conditions, leading to incorrect performance.

In this paper, we present LIGHTNING (**LIGHT 'N IN**ductive **G**eneralization), a new learning by observation method. Our system first uses LIGHT to build a potentially over-general skill hierarchy by analysing successful traces. Next, it applies those skills on planning problems to generate successful and failed traces. After using these traces to generate positive/negative examples of skill selection conditions, our system refines those conditions using Inductive Logic Programming (ILP), and generates a more accurate skill hierarchy. We claim that LIGHTNING combines the strengths of analytical and inductive learning. We support our claim with experiments showing that LIGHTNING learns more accurate skill hierarchies compared to LIGHT, while it does not need the costly expert annotations RELBO require. We further claim that teleoreactive logic programs' commitment of representing goals explicitly naturally supports LIGHTNING's ability to map the problem of "goal hierarchy refinement using relational traces" to a standard ILP problem of learning Horn clauses.

2   Teleoreactive Logic Programs

Teleoreactive Logic Programs [1] provide a first-order language to reason about actions of physical agents that process relational data over time. In this paper, we use examples from the Escape planning domain (Figure 1) implemented in General Game Playing framework [6]. In this domain, the objective of a learned teleoreactive logic program is to explore an $n \times m$ grid to reach the exit square, picking up and constructing tools to cross barriers along the way. In this example, the explorer agent first must collect the hammer, nails, and logs. It then must combine the nails and the logs using the hammer to build a bridge across the water. Planning scenarios in Escape are challenging since these scenarios involve combinations of puzzle-solving and path-planning capabilities.

Concept instance examples

location(log$_1$, 2, 3)
property(log$_1$, supportsWeight)
inregion(explorer, region$_1$)

Nonprimitive concept definition example

property(ComboItem, ComboProp) ←
    combining(Item$_1$, Item$_2$, ComboItem)
    propertycombine(Prop$_1$, Prop$_2$, ComboProp)
    property(Item$_1$, Prop$_1$)
    property(Item$_1$, Prop$_1$)

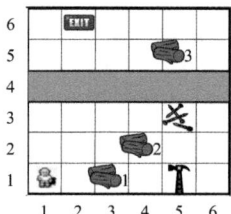

Fig. 1. Concepts and a depiction of a state from the Escape domain. The explorer agent has the goal to reach the exit, which it can achieve by building a bridge over the river using items such as wood, nails, and a hammer.

TLPs consist of two different knowledge bases. (Figure 1) describe situations in an environment, while *skills* (Table 1) describe how to achieve those situations by selecting actions[1]. While conceptual clauses are similar to Horn clauses that are used to make monotonic inferences about a situation, concept instances are ground literals that depict elements of an observed or envisioned situation. Skills are hierarchically organized and describe how to achieve goals by decomposing a goal into subgoals. A skill hierarchy corresponds to a hierarchical task network (HTN)[2] with an additional constraint; all skills are indexed by the goals they achieve.

Even though the same predicate symbols (e.g., *holding* in Table 1) appear in both skills and concepts, there is a clear distinction. A concept p describes what p is and when its instances should be inferred, while a skill p is a *constructive definition* that describes how instances of p can be achieved or constructed through actions. Skills have preconditions that describe (using the conceptual language) when a skill is applicable. We distinguish between nonprimitive skills that decompose a goal into a sequence of subgoals, completion of which may achieve the goal of the skill, and primitive skills that select an action, which can be directly executed in the environment.

Table 1. Skill clauses learned by (**a**) LIGHT (left) and refined by (**b**) LIGHTNING (right)

holding(Type) ← *precond*: combination-type(Type$_1$, Type$_2$, Type) type(Part$_1$, Type$_1$) type(Part$_2$, Type$_2$), *subgoals*: combineable(Part$_1$, Part$_2$) combine(Part$_1$, Part$_2$)	holding(Type) ← *precond*: combination-type (Type$_1$, Type$_2$, Type) type(Par$_1$, Type$_1$) type(Part$_2$, Type$_2$) agent(Explorer) inregion(Explorer, Region) inregion(Part$_1$, Region) inregion(Part$_2$, Region) *subgoals*: combineable(Part$_1$, Part$_2$) combine(Part$_1$, Part$_2$)

For a given main goal and state, a teleoreactive logic program selects an action by decomposing goals into a subgoals using the skill hierarchy until the decomposition terminates with an action, creating a goal path starting from the main goal and ending with the selected action. During this process, the preconditions of all skills in the goal path must match against the state in a mutually unified way. Moreover, each goal at this path must be unsatisfied. If a subgoal of a skill on the path is already satisfied, the path selection moves to the next subgoal of that skill. For example in Table 1(a), when the *combineable* subgoal of the *holding* skill is satisfied, meaning that the required tools to construct a desired object are already collected, the goal path selection moves to the next *combine* subgoal to construct that object. Like clause decomposition in logic programming, goal path selection is a nondeterministic process because there may be multiple skill clauses with the same head and the preconditions can match the state with different instantiations. Each selected action causes the creation of a new world state and the goal path is regenerated based on remaining unachieved goals. As the actions make progress towards the solution, the goals in the selected goal path are achieved and the

---

[1] In this paper, we use Prolog style variables that start with a capital letter.

[2] Over the years, Hierarchical Task Networks have emerged as a prominent framework for representing hierarchically-organized domain knowledge in automated planning [7,8].

selection moves from left to right through the skill hierarchy. Performance terminates when the top level goal is achieved.

Choi and Langley originally proposed teleoreactive logic programs for execution [1]. Instead, in this paper, we focus on automated planning with them. We use a teleoreactive logic program planner [9] that is a modified version of the well-known HTN planner SHOP2 [8]. Planning with TLPs has subtle but important differences with execution. In the latter, a selected action is immediately executed in the world and no backtracking can occur to an earlier decision point, because the actions may irreversibly change the world. In contrast, a planner can backtrack over previously selected actions, since it calculates the world states using an action model instead of execution and perception. By analogy with logic programming, the execution semantics creates cut points that destroy previous choices after actions, while the planning semantics does not do that.

3   Learning Teleoreactive Logic Programs

In this section we present a learning algorithm that generates the skill hierarchy of a teleoreactive logic program using behavior traces and background knowledge. Given a set of *primitive skills* that produce predictable effects under known conditions, a set of *solution traces*, each of which specifying a sequence of state-action pairs, a *goal* that holds in the final state, and *concept definitions* that describe domain knowledge, LIGHT-NING produces a *skill hierarchy* that can achieve goals under similar circumstances.

LIGHTNING has two components. The analytical component, LIGHT [5], analyses the solution trace using the action model and conceptual knowledge, and produces a skill hierarchy, which may be over-general. The inductive component first experiments with those skills using a planner and generates successful and failed traces. Next, it extracts data from those traces to further refine the preconditions of the skills using a standard ILP framework.

3.1   Analytical Learning of Skill Hierarchy Structure

Learning Teleoreactive Logic Programs involves acquiring the hierarchical structure as well as preconditions of the skills. Nejati et al. [5] introduced LIGHT, a system that learns the structure of the skill hierarchy, while generating skill preconditions that may be over-general. Although the skills LIGHT learns show encouraging results in Blocks World, Depots, and Escape domains [5,10], the analytical precondition generation fail to capture patterns in the training examples that the system cannot infer using background knowledge.

LIGHT analyses the expert trace in the context of the background knowledge using goal regression with a deductive approach similar to explanation based learning [11]. Each goal is either explained as an effect of a primitive skill or is decomposed to simpler goals using the conceptual definitions. Unlike most explanation based learning methods, which use only the leaves of an explanation tree to produce a single rule, LIGHT uses the structure of the explanation to determine the structure of the learned skill hierarchy and produces conditions for all skills in the hierarchy. We have discussed the details of the basic LIGHT algorithm at length elsewhere [5]. In the experiments presented in this paper, we use a variant of LIGHT that generates more accurate skill hierarchies [10].

3.2 Inductive Generalization of Conditions for Skill Selection

During skill decomposition, over-general preconditions of skills learned by LIGHT may result in selection of incorrect skill clauses or incorrect instantiations of the subgoal variables. In execution, this results in selection of incorrect actions and ultimately generation of incorrect behavior. In planning, the search time may increase exponentially with the depth of the skill hierarchy, if incorrect selections are available at each level.

Table 1.a shows a clause learned by LIGHT for the skill $holding(Type)$. Here, given a goal $holding(bridge\text{-}type)$, the precondition learned by LIGHT may select two wood pieces (e.g., log_1 and log_2 in Fig. 1) that can be combined to build a bridge. The precondition of this clause is over-general because it can also propose collection of items that are on the other side of the river (e.g., log_3) that is out of reach without a bridge. LIGHTNING fixes this problem by refining the precondition (Table 1(b)) to select only items in the same region as the explorer.

LIGHTNING uses an ILP algorithm to improve the preconditions LIGHT generates. For an over-general skill clause p that is used to decompose a goal at a situation s, the target of learning is a *precondition concept* $precond_p(s, X_h, X_s)$, where X_h is the set of variables in the head of p that are instantiated when p is used for skill decomposition (e.g. $\{Type\}$ in Table 1), and X_s is the set of variables in the subgoals of p that are instantiated by the precondition of p (e.g., $\{Part_1, Part_2\}$ in Table 1). A correct $precond_p$ concept should determine if a skill p should be selected and it should return instantiations for subgoal variables X_s given instantiations for the variables X_h in the skill head.

One of the key contributions in this paper is defining how to convert the solution traces to examples of precondition hypotheses. First, LIGHTNING uses skills created by LIGHT to solve planning problems, generating successful and failed solution traces. Next , it uses those solutions to extract positive and negative examples of precondition concepts. Each state when a skill clause is selected during planning provides an opportunity to extract positive or negative examples of preconditions for that clause. A state s, where a skill clause p is used for decomposition, is used to extract a positive precondition example, if the planner achieves the goal of p at the end of decomposition. On the other hand, if the planner times out or backtracks to s with failure, because it has not found an action sequence that achieves the goal of p following this decomposition, s is used to create a negative example. The examples are simply created by instantiating the variables in target precondition hypothesis with the corresponding variables in the skill clause instance called at s.

For example, suppose s_5 is the state in Fig. 1, where the explorer is decomposing the goal $holding(bridge\text{-}type)$ with the LIGHT skill clause in Table 1.a. Here, the target precondition concept is $precond_p(s, Type, Part_1, Part_2)$ such that $X_h = \{Type\}$, $X_s = \{Part_1, Part_2\}$, meaning that the concept should determine the two items $Part_1$ and $Part_2$ to be collected using the subgoals to construct and object described by $Type$. If during the skill decomposition at s_5, the overgeneral precondition returns the bindings $\{log_1/Part_1, log_2/Part_2\}$ the planner is able to finish the decomposition and returns a plan for building a bridge. From that experience our system and creates the positive example $precondition(s_5, bridge\text{-}type, log_1, log_2)$. On the other hand, if the overgeneral precondition selects the unreachable item log_3 instead of log_2, the planner fails and creates the negative example $precondition(s_5, bridge\text{-}type, log_1, log_3)$.

The preconditions are learned using background knowledge predicates of the form $b(s, X_b)$ where $b(X_b)$ is a concept instance that holds at the state s in a trace generated during planning (e.g., $inregion(s_5, explorer, region_1)$ if $inregion(explorer, region_1)$ holds at the state s_5). LIGHTNING uses a modified version of Aleph [12] to generate hypotheses using the over-general LIGHT precondition as the starting point an adds new literals to this condition using a heuristic search. After learning a target hypothesis with good coverage, our system drops the first state argument from the predicates in the body of the learned hypothesis to obtain the skill preconditions (e.g., $inregion(Explorer, Region)$ in Table 1(b)).

In a hierarchical task network that does not have goals, when a failure occurs, it is difficult to know to which skill in the hierarchy the failure should be attributed. For example, RELBO [4] relies on an expert to annotate the trace for that determination. Because teleoreactive logic programs have a special commitment of associating goals with skills, when a failure occurs, our system attributes the failure to the skill in the lowest level that has not achieved its goal. This assumes that our system first refines the lowest level skills to ensure they can achieve their goals before refining higher level skills that depend on them. Our approach for determining positive and negative examples is similar to the credit assignment approach used by Sleeman, Langley, and Mitchell [13] in that it assigns positive credit to operators that are on a complete solution path and negative credit to the rest, but our approach generalizes that idea to assign credit to abstract operators (skills) and it is not limited to solution paths generated with goal regression.

4   Evaluation

In our evaluation, we generated a random solution for an Escape scenario as input data and ran LIGHT on it to generate a skill hierarchy. From that hierarchy, we selected the skill clause in Table 1(a) for further refinement, which has an over-general condition that may select items from the wrong side of the river. We evaluated the learned skills by using our skill-evaluation system on 160 randomly generated maps and collected successful and failed traces. Next, our system extracted examples from these traces and learned a precondition concept for the selected skill clause using Aleph [12] with best first search setting. We used the precondition LIGHT generated as the starting point of the hypothesis search, since LIGHT generates over-general preconditions. As a result,

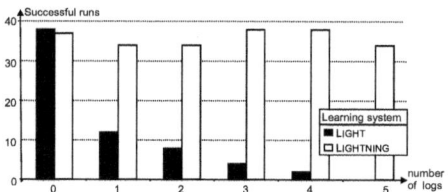

Fig. 2. Number of successful LIGHT and LIGHTING skills versus degree of problem difficulty

LIGHTNING learned the skill in Table 1(b). Finally, we compared the performance of the skill hierarchy learned by LIGHT with the one refined by LIGHTNING on 240 randomly generated test problems. Figure 4 compares the number of problems solved by the two systems as the difficulty of the problem increases (measured by the number of logs on the exit side of the river). While the performance of the over-general LIGHT skill is reduced with difficulty of the problem, the performance of the revised LIGHTNING skill remain uneffected.

5   Related Work

Earlier research has studied learning by observation using both analytical (i.e., macro operator learning [14]) and inductive techniques (i.e., behavior cloning [2]) and few systems integrate those ideas. Mooney and Zelle review [15] systems that integrate EBL and ILP. Estlin and Mooney describe [16] an integrated system that produces planning knowledge to control the search in planning, but their system acquired control rules that specified constraints on desired orderings of actions rather than selecting of high level goals that span multiple actions. X-learn [17] uses EBL in order to generate explanations for action sequences, which is used to identify the relevant predicates in the input planning problem. These predicates are then fed into an ILP component in order to produce decomposition rules. LIGHT[5,10] and HTN-LEARNER[18] use background knowledge to construct the structure and preconditions of methods using analytical techniques and therefore may miss patterns that require induction [19]. Both RELBO [4] and CaMeL++ [19] learn preconditions of hierarchical tasks using ILP and Version Space Candidate Elimination respectively, but both of these techniques require that solution traces are annotated with the tasks. L2ACT [20] learns procedural knowledge that map goals and states into actions, but it does not benefit from an analytical component that reduces the hypothesis space by reasoning about the relations of goals.

6   Conclusion

Our work shows promise in learning procedural knowledge for planning using representations more specialized than first order logic and using a novel combination of analytical and inductive learning. Our current example selection process is based on the assumption that subskills are already learned correctly prior to refinement of a skill. Our future work will use a multi-predicate learning framework to relax this assumption to handle recursive skills. We also plan to tighten the interaction of the analytical and inductive components by using inductively refined skills in improving LIGHT explanations. This approach should also be able to predict the organizational structure of skills inductively.

Acknowledgements. This paper reports research sponsored by DARPA under agreement FA8750-05-2-0283. The U. S. Government may reproduce and distribute reprints for Governmental purposes notwithstanding any copyrights. The authors' views and conclusions should not be interpreted as representing official policies or endorsements, expressed or implied, of DARPA or the Government. We thank Nate Waisbrot, Dongkyu Choi, Pat Langley and Dan Shapiro for their contributions to this paper.

References

1. Choi, D., Langley, P.: Learning teleoreactive logic programs from problem solving. In: ILP 2005. LNCS (LNAI), vol. 3625, pp. 51–68. Springer, Heidelberg (2005)
2. Sammut, C., Hurst, S., Kedzier, D., Michie, D.: Learning to fly. In: Sleeman, D., Edwards, P. (eds.) Proceedings of the 9th International Conference on Machine Learning, pp. 385–393 (1992)
3. Isaac, A., Sammut, C.: Goal-directed learning to fly. In: Fawcett, T., Mishra, N. (eds.) Proceedings of the 20th International Conference on Machine Learning (2003)
4. Könik, T., Laird, J.E.: Learning goal hierarchies from structured observations and expert annotations. Machine Learning 64, 263–287 (2006)
5. Nejati, N., Langley, P., Könik, T.: Learning hierarchical task networks by observation. In: Proceedings of the Twenty-Third International Conference on Machine Learning (2006)
6. Genesereth, M.R., Love, N., Pell, B.: General game playing overview of the aaai competition. AI Magazine 26, 62–72 (2005)
7. Erol, K., Hendler, J., Nau, D.S.: HTN planning: Complexity and expressivity. In: Procceedings of the Nineteenth National Conference on Artificial Intelligence (1994)
8. Nau, D., Au, T.C., Ilghami, O., Kuter, U., Murdock, W., Wu, D., Yaman, F.: SHOP2: An HTN planning system. Journal of Artificial Intelligence Research 20, 379–404 (2003)
9. Waisbrot, N., Kuter, U., Könik, T.: Combining heuristic search with hierarchical task-network planning: A preliminary report. In: Proceedings of the 21st International Florida Artificial Intelligence Research Society Conference (2008)
10. Nejati, N., Könik, T., Kuter, U.: A goal and dependency-driven algorithm for learning hierarchical task networks. In: Procceedings of the Fifth International Conference on Knowledge Capture, pp. 113–120. ACM, New York (2009)
11. Dejong, G., Mooney, R.: Explanation-based learning: An alternative view. Machine Learning 1(2), 145–176 (1986)
12. Srinivasan, A.: The Aleph Manual,
http://www.comlab.ox.ac.uk/activities/
machinelearning/Aleph/aleph.html
13. Sleeman, D., Langley, P., Mitchell, T.M.: Learning from solution paths: An approach to the credit assignment problem. AI Magazine 3(2) (1982)
14. Segre, A.M.: ARMS: Acquiring robotic assembly plans. In: De Jong, G. (ed.) Investigating Explanation-Based Learning. Kluwer, Boston (1993)
15. Mooney, R.J., Zelle, J.M.: Integrating ILP and EBL. SIGART Bulletin 5, 12–21 (1994)
16. Estlin, T.A., Mooney, R.J.: Integrating EBL and ILP to acquire control rules for planning. In: Proceedings of the 3rd International Workshop on Multi-Strategy Learning, pp. 271–279 (1996)
17. Reddy, C., Tadepalli, P.: Learning goal-decomposition rules using exercises. In: Proceedings of the 14th International Conference on Machine Learning (1997)
18. Hogg, C., Muñoz Avila, H., Kuter, U.: HTN-MAKER: Learning HTNs with minimal additional knowledge engineering required. In: Proceedings of the 23rd AAAI Conference on Artificial Intelligence, pp. 950–956. AAAI Press, Menlo Park (2008)
19. Ilghami, O., Muñoz Avila, H., Nau, D., Aha, D.: Learning approximate preconditions for methods in hierarchical plans. In: Proceedings of the 22nd International Conference on Machine Learning, Bonn, Germany (2005)
20. Khardon, R.: Learning action strategies for planning domains. Artificial Intelligence 113, 125–148 (1999)

Ideal Downward Refinement in the $\mathcal{EL}$ Description Logic

Jens Lehmann[1] and Christoph Haase[2]

[1] Universität Leipzig, Department of Computer Science,
Johannisgasse 26, D-04103 Leipzig, Germany
`lehmann@informatik.uni-leipzig.de`
[2] Oxford University Computing Laboratory,
Wolfson Building, Parks Rd, Oxford, OX1 3QD, United Kingdom
`christoph.haase@comlab.ox.ac.uk`

Abstract. With the proliferation of the Semantic Web, there has been a rapidly rising interest in description logics, which form the logical foundation of the W3C standard ontology language OWL. While the number of OWL knowledge bases grows, there is an increasing demand for tools assisting knowledge engineers in building up and maintaining their structure. For this purpose, concept learning algorithms based on refinement operators have been investigated. In this paper, we provide an ideal refinement operator for the description logic $\mathcal{EL}$ and show that it is computationally feasible on large knowledge bases.

1 Introduction

The Semantic Web is steadily growing[1] and contains knowledge from diverse areas such as science, music, people, books, reviews, places, politics, products, software, social networks, as well as upper and general ontologies. The underlying technologies, sometimes called *Semantic Technologies*, are currently starting to create substantial industrial impact in application scenarios on and off the web, including knowledge management, expert systems, web services, e-commerce, e-collaboration, etc. Since 2004, the Web Ontology Language OWL, which is based on description logics (DLs), has been the W3C-recommended standard for Semantic Web knowledge representation and is a key to the growth of the Semantic Web.

However, recent progress in the field faces a lack of well-structured ontologies with large amounts of instance data due to the fact that engineering such ontologies constitutes a considerable investment of resources. Nowadays, knowledge bases often provide large amounts of instance data without sophisticated schemata. Methods for automated schema acquisition and maintenance are therefore being sought (see e.g. [5]). In particular, concept learning methods have attracted interest, see e.g. [2,6,11,13].

---

[1] As a rough size estimate, the semantic index Sindice (http://sindice.com/) lists more than 10 billion entities from more than 100 million web pages.

L. De Raedt (Ed.): ILP 2009, LNAI 5989, pp. 73–87, 2010.

Many concept learning methods borrow ideas from Inductive Logic Programming including the use of *refinement operators*. Properties like ideality, completeness, finiteness, properness, minimality, and non-redundancy are used as theoretical criteria for the suitability of such operators. It has been shown in [12] that no ideal refinement operator for DLs such as $\mathcal{ALC}$, $\mathcal{SHOIN}$, and $\mathcal{SROIQ}$ can exist (the two latter DLs are underlying OWL and OWL 2, respectively). In this article, an important gap in the the analysis of refinement operator properties is closed by showing that ideal refinement operators for the DL $\mathcal{EL}$ do exist, which in turn can lead to an advance in DL concept learning.

$\mathcal{EL}$ is a light-weight DL, but despite its limited expressive power it has proven to be of practical use in many real-world large-scale applications. For example, the Systematized Nomenclature of Medicine Clinical Terms (SNOMED CT) [4] and the GENE ONTOLOGY [18] are based on $\mathcal{EL}$. Since standard reasoning in $\mathcal{EL}$ is polynomial, it is suitable for large ontologies. It should furthermore be mentioned that $\mathcal{EL}^{++}$, an extension of $\mathcal{EL}$, will become one of three profiles in the upcoming standard ontology language OWL 2.

Overall, we make the following contributions in this paper: We

- close a gap in the research of properties of refinement operators in DLs,
- provide an ideal and practically useful refinement operator for $\mathcal{EL}$, and
- show the computational feasibility of the operator.

This paper is structured as follows. Section 2 introduces the preliminaries for our work and the refinement operator is presented in Section 3. There, we prove its ideality and describe how it can be optimised to work efficiently and incorporate background knowledge. We evaluate the operator on real-world knowledge bases in Section 4. Related work is described in Section 5 and conclusions are drawn in Section 6.

2 Preliminaries

In this section, the definitions relevant for the refinement operator in Section 3 are introduced. Besides recalling known facts from the literature, we introduce minimal $\mathcal{EL}$ trees that serve as the basis for the refinement operator.

2.1 The $\mathcal{EL}$ Description Logic

Before we begin to introduce the DL $\mathcal{EL}$, we briefly recall some notions from order theory. Let Q be a set and $\preceq$ a quasi order on Q, i.e., a reflexive and transitive binary relation on Q. Then $(Q, \preceq)$ is called a *quasi ordered space*. The quasi order $\preceq$ induces the equivalence relation $\simeq$ and the *strict quasi order* $\prec$ on Q: $q \simeq q'$ iff $q \preceq q'$ and $q' \preceq q$, and $q \prec q'$ iff $q \preceq q'$ and $q \not\simeq q'$. For $P \subseteq Q$, $max(P) := \{p \in P \mid \text{there is no } p' \in P \text{ with } p \prec p'\}$ defines the *set of maximal elements* of P. We say $(Q, \preceq)$ has a *greatest element* iff there is a $q^* \in Q$ such that $max(Q) := \{q^*\}$.

Table 1. $\mathcal{EL}$ syntax and semantics

Concept constructor	Syntax	Semantics
Top	$\top$	$\Delta^{\mathcal{I}}$
Concept name	A	$A^{\mathcal{I}}$
Conjunction	$C \sqcap D$	$C^{\mathcal{I}} \cap D^{\mathcal{I}}$
Existential restriction	$\exists r.C$	$\{x \in \Delta^{\mathcal{I}} \mid \text{there is } y \in C^{\mathcal{I}} \text{ with } (x,y) \in r^{\mathcal{I}}\}$

Table 2. Knowledge base axioms

Name	Syntax	Restriction on $\mathcal{I}$
Concept inclusion	$A \sqsubseteq B$	$A^{\mathcal{I}} \subseteq B^{\mathcal{I}}$
Role inclusion	$r \sqsubseteq s$	$r^{\mathcal{I}} \subseteq s^{\mathcal{I}}$
Disjointness	$A \sqcap B \sqsubseteq \bot$	$A^{\mathcal{I}} \cap B^{\mathcal{I}} = \emptyset$
Domain	$domain(r) = A$	$x \in A^{\mathcal{I}}$ for all $(x,y) \in r^{\mathcal{I}}$
Range	$range(r) = A$	$y \in A^{\mathcal{I}}$ for all $(x,y) \in r^{\mathcal{I}}$

The expressions in the DL $\mathcal{EL}$ are *concepts*, which are built inductively starting from sets of *concepts names* N_C and *role names* N_R of arbitrary but finite cardinality, and then applying the *concept constructors* shown in Table 1. There and in the following, A, B denote concept names, r, s denote role names, and C, D denote arbitrary $\mathcal{EL}$ concepts. By $\mathcal{C}(\mathcal{EL})$ we refer to the set of all $\mathcal{EL}$ concepts. The size of an $\mathcal{EL}$ concept C is denoted by $|C|$ and is is just the number of symbols used to write it down. When proving properties of $\mathcal{EL}$ concepts, the *role depth* of a concept C is a useful induction argument. It is defined by structural induction as $rdepth(A) = rdepth(\top) := 0$, $rdepth(C \sqcap D) := max(rdepth(C), rdepth(D))$ and $rdepth(\exists r.C) := rdepth(C) + 1$.

The semantics of an $\mathcal{EL}$ concept C is given in terms of an *interpretation* $\mathcal{I} = (\Delta^{\mathcal{I}}, \cdot^{\mathcal{I}})$, where $\Delta^{\mathcal{I}}$ is a set called the *interpretation domain* and $\cdot^{\mathcal{I}}$ is the *interpretation function*. The interpretation function maps each $A \in N_C$ to a subset of $\Delta^{\mathcal{I}}$, and each $r \in N_R$ to a binary relation on $\Delta^{\mathcal{I}}$. It is then inductively extended to arbitrary $\mathcal{EL}$ concepts as shown in Table 1.

In this paper, a *knowledge base* $\mathcal{K}$ is a finite union of knowledge base axioms given in Table 2. An interpretation $\mathcal{I}$ is a *model* of a knowledge base $\mathcal{K}$ iff the conditions on the right-hand side of Table 2 are fulfilled for every knowledge base axiom in $\mathcal{K}$. An $\mathcal{EL}$ concept C is *satisfiable* w.r.t. $\mathcal{K}$ iff there exists a model $\mathcal{I}$ of $\mathcal{K}$ such that $C^{\mathcal{I}} \neq \emptyset$.

A standard reasoning task in DLs is *subsumption*. Given a knowledge base $\mathcal{K}$ and $\mathcal{EL}$ concepts C, D, we say C is *subsumed* by D w.r.t. $\mathcal{K}$ ($C \sqsubseteq_{\mathcal{K}} D$) iff $C^{\mathcal{I}} \subseteq D^{\mathcal{I}}$ for all models $\mathcal{I}$ of $\mathcal{K}$. Intuitively, this states that the concept C is a specialisation of the concept D w.r.t. $\mathcal{K}$. In the remainder of this paper, we always assume a knowledge base to be implicitly present, and we therefore just write $C \sqsubseteq D$. Obviously, $(\mathcal{C}(\mathcal{EL}), \sqsubseteq)$ forms a quasi ordered space, from which we can accordingly derive the relations $\equiv$ (*equivalence*) and $\sqsubset$ (*strict subsumption*).

Example 1. From the following knowledge base $\mathcal{K}$ we can infer $C \sqsubseteq D$.

$$N_C = \{\texttt{Human}, \texttt{Animal}, \texttt{Bird}, \texttt{Cat}\}$$

$$N_R = \{\texttt{has}, \texttt{has\_child}, \texttt{has\_pet}\}$$

$$\mathcal{K} = \{\texttt{has\_pet} \sqsubseteq \texttt{has}, \texttt{has\_child} \sqsubseteq \texttt{has}, \texttt{Bird} \sqsubseteq \texttt{Animal}, \texttt{Cat} \sqsubseteq \texttt{Animal},$$
$$domain(\texttt{has\_pet}) = \texttt{Animal}\}$$

$$C = \texttt{Human} \sqcap \exists\texttt{has\_pet}.\top \sqcap \exists.\texttt{has\_child}.\top$$

$$D = \texttt{Human} \sqcap \exists\texttt{has}.\texttt{Animal}$$

In practice, knowledge bases can be derived from arbitrary ontologies, which may be formulated in DLs other than $\mathcal{EL}$. Concept and role inclusion axioms can be extracted by computing a classification of the respective ontology, and the remaining axioms can be handled in a similar fashion. The presented $\mathcal{EL}$ refinement operator can therefore be used with all DL knowledge bases.

For a role name $r \in N_R$, we define the set of role names that are strictly below r in the subsumption hierarchy as $sh_{\downarrow}(r) := max\{s \mid s \sqsubset r\}$. Given finite sets of concept names $\mathcal{A}, \mathcal{B} \subseteq N_C$, we write $\mathcal{A} \sqsubseteq \mathcal{B}$ iff for every $B \in \mathcal{B}$ there is some $A \in \mathcal{A}$ such that $A \sqsubseteq B$. We sometimes abuse notation and write $\mathcal{A} \sqsubseteq B$ instead of $\mathcal{A} \sqsubseteq \{B\}$. We call $\mathcal{A} \subseteq N_C$ *reduced* if there does not exist $\mathcal{B} \subseteq N_C$ with $|\mathcal{B}| < |\mathcal{A}|$ and $\mathcal{A} \equiv \mathcal{B}$.

2.2 Downward Refinement Operators

Refinement operators are used to structure a search process for concepts. Intuitively, downward refinement operators construct specialisations of hypotheses. This idea is well-known in Inductive Logic Programming [15].

Let $(Q, \preceq)$ be a quasi ordered space and denote by $\mathcal{P}(Q)$ the powerset of Q. A mapping $\rho : Q \to \mathcal{P}(Q)$ is a *downward refinement operator* on $(Q, \preceq)$ iff $q' \in \rho(q)$ implies $q' \preceq q$. In the remainder of this paper, we will call downward refinement operators just refinement operators. We write $q \rightsquigarrow_\rho q'$ for $q' \in \rho(q)$ and drop the index ρ if the refinement operator is clear from the context. A *refinement chain of length* n of a refinement operator ρ that starts in q_1 and ends in q_n is a sequence $q_1 \rightsquigarrow \ldots \rightsquigarrow q_n$ such that $q_i \rightsquigarrow q_{i+1}$ for $1 \leq i < n$. We say that the chain *goes through* q iff $q \in \{q_1, \ldots, q_n\}$. Moreover, $q \rightsquigarrow^* q'$ iff there exists a refinement chain of length n starting from q and ending in q' for some $n \in \mathbb{N}$.

Refinement operators can be classified by means of their properties. Let $(Q, \preceq)$ be a quasi ordered space with a greatest element, and let $q, q', q'' \in Q$. A refinement operator ρ is *finite* iff $\rho(q)$ is finite for all q. It is *proper* iff $q \rightsquigarrow q'$ implies $q \not\equiv q'$. We call ρ *complete* iff $q' \prec q$ implies $q \rightsquigarrow^* q''$ for some $q'' \equiv q'$. Let q^* be the greatest element in $(Q, \preceq)$, ρ is *weakly complete* iff for any $q' \prec q^*$, $q^* \rightsquigarrow^* q''$ with $q'' \equiv q'$. We say ρ is *redundant* iff $q^* \rightsquigarrow^* q'$ via two refinement chains, where one goes through an element q'' and the other one does not go through q''. Finally, ρ is *ideal* iff it is finite, proper and complete.

2.3   Minimal $\mathcal{EL}$ Concepts

An important observation is that $\mathcal{EL}$ concepts can be viewed as directed labeled trees, see e.g. [1]. This allows for deciding subsumption between concepts in terms of the existence of a simulation relation between the nodes of their corresponding trees. Moreover, the graph approach to $\mathcal{EL}$ concepts allows for a canonical representation of $\mathcal{EL}$ concepts as minimal $\mathcal{EL}$ trees. The latter generalise similar approaches found in the literature, namely "reduced $\mathcal{EL}$ concept terms" [10] and "minimal XPath tree pattern queries" [16]. Most proofs are omitted in this section and deferred to the full version of this paper, since they are mostly a straight-forward generalisation of the proofs found in [10].

An $\mathcal{EL}$ *graph* is a directed labeled graph $G = (V, E, \ell)$, where V is the finite set of *nodes*, $E \subseteq V \times N_R \times V$ is the set of *edges*, and $\ell : V \to \mathcal{P}(N_C)$ is the *labeling function*. We define $V(G) := V$, $E(G) := E$, $\ell(G) := \ell$ and $|G| := |V| + |E|$. For an edge $(v, r, w) \in E$, we call w an $(r\text{-})successor$ of v, and v an $(r\text{-})predecessor$ of w. Given a node $v \subset V$, a labelling function ℓ and $L \subseteq N_C$, we define $\ell[v \mapsto L]$ as $\ell[v \mapsto L](v) := L$ and $\ell[v \mapsto L](w) := \ell(w)$ for all $w \neq v$. Given G and $v \in V(G)$, we define $G[v \mapsto L] := (V(G), E(G), \ell(G)[v \mapsto L])$. We say $v_1 \xrightarrow{r_1} \cdots \xrightarrow{r_n} v_{n+1}$ is a *path* of length n from v_1 to v_{n+1} in G iff $(v_i, r_i, v_{i+1}) \in E$ for $1 \leq i \leq n$. A graph G contains a cycle iff there is a path $v \xrightarrow{r_1} \cdots \xrightarrow{r_n} v$ in G.

An $\mathcal{EL}$ concept is represented by an $\mathcal{EL}$ *concept tree*, which is a connected finite $\mathcal{EL}$ graph t that does not contain any cycle, has a distinguished node called the *root* of t that has no predecessor, and every other node has exactly one predecessor along exactly one edge. The set of $\mathcal{EL}$ concept trees is denoted by T. In the following, we call an $\mathcal{EL}$ concept tree just a tree. Figure 1 illustrates two examples of such trees. Given a tree t, we denote by $root(t)$ its root. The tree t corresponding to a concept C is defined by induction on $n = rdepth(C)$. For $n = 0$, t consists of a single node that is labelled with all concepts names occurring in C. For $n > 0$, the root of t is labelled with all concept names occurring on the top-level of C. Furthermore, for each existential restriction $\exists r.D$ on the top-level of C, it has an r-labelled edge to the root of a subtree of t' which corresponds to D. As an example, the tree t corresponding to $A_1 \sqcap \exists r.A_2$ is $t = (\{v_1, v_2\}, \{(v_1, r, v_2)\}, \ell)$ where ℓ maps v_1 to $\{A_1\}$ and v_2 to $\{A_2\}$. By $t_\top$ we denote the tree corresponding to $\top$. Obviously, the transformation from a concept to a tree can be performed in linear time w.r.t. the size of the concept. Similarly, any tree has a corresponding concept[2], and the transformation can be performed in linear time, too.

Let t, t' be trees, $v \in V(t)$ and assume w.l.o.g. that $V(t) \cap V(t') = \emptyset$. Denote by $t[v \leftarrow (r, t')]$ the tree obtained from plugging t' via an r-edge into the node v of t, i.e. the tree $(V(t) \cup V(t'), E(t) \cup E(t') \cup \{(v, r, root(t'))\}, \ell \cup \ell')$, where $\ell \cup \ell'$ is the obvious join of the labeling functions of t and t'. By $t(v)$ we denote the subtree at v. Let C be a concept and t the tree corresponding to C. We define $depth(t) := rdepth(C)$, and for $v \in V(t)$, $level(v) := depth(t) - depth(t(v))$. Moreover, $onlevel(t, n)$ is the set of nodes $\{v \mid level(v) = n\}$ that appear on level n.

---

[2] Strictly speaking, t has a set of corresponding concepts, which are all equivalent up to commutativity.

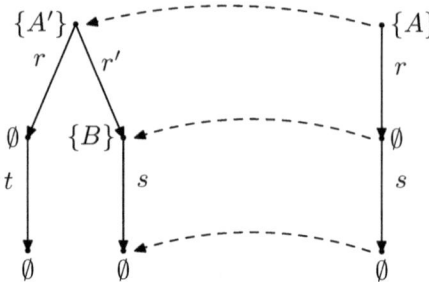

Fig. 1. A (non-maximal) simulation relation w.r.t. the knowledge base $\mathcal{K} = \{A' \sqsubseteq A, r' \sqsubseteq r\}$ from the tree corresponding to $A \sqcap \exists r.\exists s.\top$ to the tree corresponding to $A' \sqcap \exists r.\exists t.\top \sqcap \exists r'.(B \sqcap \exists s.\top)$.

Definition 1. *Let $t = (V, E, \ell), t' = (V', E', \ell')$ be trees. A simulation relation from t' to t is a binary relation $\mathcal{S} \subseteq V \times V'$ such that if $(v, v') \in \mathcal{S}$ then the following* simulation conditions *are fulfilled:*

> *(SC1) $\ell(v) \sqsubseteq \ell'(v')$*
> *(SC2) for every $(v', r, w') \in E'$ there is $(v, r, w) \in E_1$ such that $r \sqsubseteq r'$ and $(w, w') \in \mathcal{S}$*

We write $t \preceq t'$ if there exists a simulation relation $\mathcal{S}$ from t' to t such that $(root(t), root(t')) \in \mathcal{S}$. It is easily checked that $(T, \preceq)$ forms a quasi ordered space, and we derive the relations $\simeq$ and $\prec$ accordingly. A simulation $\mathcal{S}$ from t' to t is *maximal* if for every simulation $\mathcal{S}'$ from t' to t, $\mathcal{S}' \subseteq \mathcal{S}$. It is not hard to check that $\mathcal{S}$ is unique. Using a dynamic programming approach, the maximal simulation can be computed in $\mathcal{O}(|t| \cdot |t'|)$. Figure 1 shows an example of a simulation.

The following lemma is proven by induction on $rdepth(D)$. It allows us to decide subsumption between concepts C, D in terms of the existence of a simulation between their corresponding trees t, t', and moreover to interchange concepts and their corresponding trees. For that reason, the $\mathcal{EL}$ refinement operator presented in the next section will work on trees rather than concepts.

Lemma 1. *Let C, D be concept with their corresponding trees t, t'. Then $C \sqsubseteq D$ iff $t \preceq t'$.*

We can now introduce minimal $\mathcal{EL}$ trees which serve as a canonical representation of equivalent $\mathcal{EL}$ concepts.

Definition 2. *Let $t = (V, E, \ell)$ be a tree. We call t label reduced if for all $v \in V$, $\ell(v)$ is reduced, i.e. no concept name can be removed from the label without resulting in an inequivalent tree. Moreover, t contains redundant subtrees if there are $(v, r, w), (v, r', w') \in E$ with $w \neq w'$, $r \sqsubseteq r'$ and $t(w) \preceq t(w')$. We call t minimal if t is label reduced and does not contain redundant subtrees.*

It follows that the minimality of a tree t can be checked in $\mathcal{O}(|t|^2)$ by computing the maximal simulation from t to t and then checking for each $v \in V(t)$ whether v is label reduced and, using $\mathcal{S}$, whether v is not the root of redundant subtrees. The set of minimal $\mathcal{EL}$ trees is denoted by T_{min}.

We close this section with a small lemma that will be helpful in the next section.

Lemma 2. *Let T_n be the set of minimal $\mathcal{EL}$ trees up to depth $n \geq 0$, and let t, t' be $\mathcal{EL}$ trees with $depth(t) < depth(t')$. Then the following holds:*

1. *$|T_n|$ is finite*
2. *$t \npreceq t'$*

3 An Ideal $\mathcal{EL}$ Refinement Operator

In this section, we define an ideal refinement operator. In the first part, we are more concerned with a description of the operator on an abstract level, which allows us to prove its properties. The next part addresses optimisations of the operator that improve its performance in practice.

3.1 Definition of the Operator

For simplicity, we subsequently assume the knowledge base to only contain concept and role inclusion axioms. We will sketch in the next section how the remaining restriction axioms can be incorporated in the refinement operator.

The refinement operator ρ, to be defined below, is a function that maps a tree $t \in T_{min}$ to a subset of T_{min}. It can be divided into the three base operations *label extension*, *label refinement* and *edge refinement*. Building up on that, the complex operation *attach subtree* is defined. Each such operation takes a tree $t \in T_{min}$ and a node $v \in V(t)$ as input and returns a set of trees that are refined at node v. Figure 2 provides an example.

The base operations are as follows: the operation $el(t, v)$ returns the set of those minimal trees that are derived from t by extending the label of v. Likewise, $rl(t, v)$ is the set of minimal trees obtained from t by refining the label of v. Last, $re(t, v)$ is obtained from t by refining one of the outgoing edges at v. Formally,

- $el(t, v)$: $t' \in el(t, v)$ iff $t' \in T_{min}$ and $t' = t[v \mapsto (\ell(v) \cup \{A\})]$, where $A \in max\{B \in N_C \mid \ell(v) \not\sqsubseteq B\}$
- $rl(t, v)$: $t \in rl(t, v)$ iff $t' \in T_{min}$ and $t' = t[v \mapsto (\ell(v) \cup \{A\}) \setminus \{B\})]$, where $B \in \ell(v)$, $A \in max\{A' \in N_C \mid A' \sqsubset B\}$ and there is no $B' \in \ell(v)$ with $B \neq B'$ and $A \sqsubset B$
- $re(t, v)$: $t' \in re(t, v)$ iff $t' \in T_{min}$ and $t' = (V, E', \ell)$, where $E' = E \setminus \{(v, r, w)\} \cup \{(v, r', w)\}$ for some $(v, r, w) \in E$ and $r' \in sh_\downarrow(r)$

The crucial part of the refinement operator is the attach subtree operation, which is defined by Algorithm 1. The set $as(t, v)$ consists of minimal trees obtained from t that have an extra subtree attached to v. It recursively calls the refinement operator ρ and we therefore give its definition before we explain $as(t, v)$ in more detail.

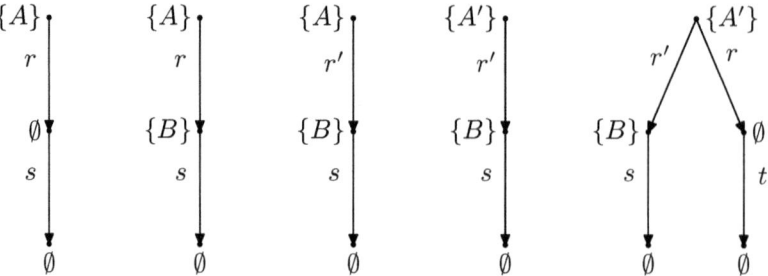

Fig. 2. The tree on the left is refined stepwise to the tree on the right, where we assume a knowledge base $\mathcal{K} = \{A' \sqsubseteq A, r' \sqsubseteq r\}$. The operator performs four different kinds of operations (from left to right): 1. label extension (B added), 2. edge refinement (r replaced by r'), 3. label refinement (A replaced by A'), 4. attaching a subtree ($\exists r.\exists t.\top$ added).

Definition 3. *The refinement operator* $\rho : \mathcal{T}_{min} \rightarrow \mathcal{P}(\mathcal{T}_{min})$ *is defined as:*

$$\rho(t) := \bigcup_{v \in V(t)} (el(t,v) \cup rl(t,v) \cup re(t,v) \cup as(t,v))$$

For $t \in \mathcal{T}_{min}$ and $v \in V$, Algorithm 1 keeps a set of output trees $\mathcal{T}$ and a set $\mathcal{M}$ of candidates which are tuples consisting of a minimal $\mathcal{EL}$ tree and a set of role names. Within the first while loop, an element $(t', \mathcal{R})$ is removed from $\mathcal{M}$. The set $\mathcal{R}'$ is initialized to contain the greatest elements of $\mathcal{R}$, and $\mathcal{R}''$ is initially empty and will later contain role names that need further inspection. In the second while loop, the algorithm iterates over all role names r in $\mathcal{R}'$. First, the tree t'' is constructed from t by attaching the subtree (v, r, w) to v, where w is the root of t'. It is then checked whether t'' is minimal. If this is the case, t'' is a refinement of t and is added to $\mathcal{T}$. Otherwise there are two reasons why t'' is not minimal: Either the newly attached subtree is subsumed by some other subtree of t, or the newly attached subtree subsumes some other subtree of t. The latter case is checked in Line 11, and if it applies the algorithm skips the loop. This prevents the algorithm from running into an infinite loop, since we would not be able to refine t' until t'' becomes a minimal tree. Otherwise in the former case, we proceed in two directions. First, $sh_{\downarrow}(r)$ is added to $\mathcal{R}'$, so it can be checked in the next round of the second while loop whether t' attached via some $r' \in sh_{\downarrow}(r) \cap \mathcal{R}$ to v yields a refinement. Second, we add r to $\mathcal{R}''$, which can be seen as "remembering" that r did not yield a refinement in connection with t'. Finally, once $\mathcal{R}'$ is empty, in Line 19 we add all tupels $(t^*, \mathcal{R}'')$ to $\mathcal{M}$, where t^* is obtained by recursively calling ρ on t'.

Example 2. Let $\mathcal{K}$ be the knowledge base from Example 1 and let $\mathcal{K}' = \mathcal{K} \setminus \{domain(\texttt{has\_pet}) = \texttt{Animal}\}$. Figure 3 depicts the set of all trees in $\rho(\texttt{Human} \sqcap \exists\texttt{has.Animal})$ w.r.t. $\mathcal{K}'$.

Algorithm 1. Computation of the set $as(t, v)$

```
 1: T := ∅; M := {(t_T, N_R)};
 2: while M ≠ ∅ do
 3:    choose and remove (t', R) ∈ M;
 4:    R' := max(R); R'' := ∅;
 5:    while R' ≠ ∅ do
 6:       choose and remove r ∈ R';
 7:       t'' := t[v ← (r, t')]; w := root(t');
 8:       if t'' is minimal then
 9:          T := T ∪ {t''};
10:       else
11:          for all (v, r', w') ∈ E(t'') with w ≠ w' and r ⊑ r' do
12:             if t''(w) ⪯ t''(w') then
13:                nextwhile;
14:             end if
15:          end for
16:          R' := R' ∪ (sh↓(r) ∩ R); R'' := R'' ∪ {r};
17:       end if
18:    end while
19:    M := M ∪ {(t*, R'') | t* ∈ ρ(t'), R'' ≠ ∅};
20: end while
21: return T;
```

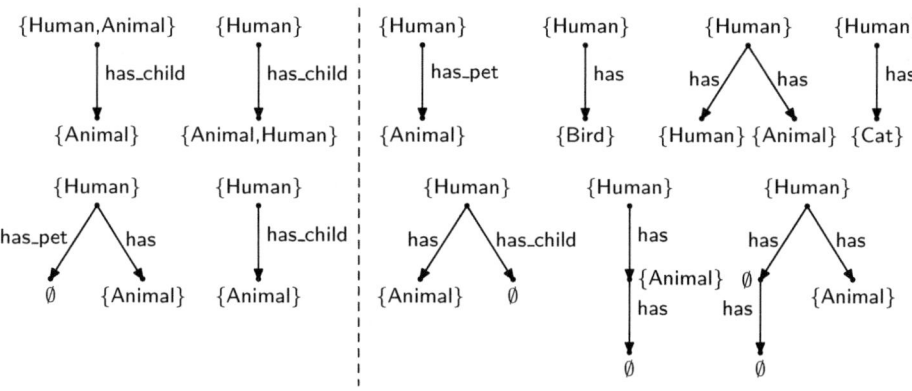

Fig. 3. The set $\rho(\texttt{Human} \sqcap \exists\texttt{has}.\texttt{Animal})$ of minimal trees w.r.t. the knowledge base $\mathcal{K}'$ from Example 2

Proposition 1. ρ is a is a finite, proper and weakly complete downward refinement operator on $(T_{min}, \preceq)$.

Proof. In the following, let $t \in T_{min}$ and $v \in V(t)$.

First, it is easily seen that ρ is a downward refinement operator. Every operation of ρ adds a label or a subtree to a node v, or replaces a label or edge-label by a refined label or edge respectively. Hence, $t' \preceq t$ for all $t' \in \rho(t)$.

Regarding finiteness of ρ, the first part of Lemma 2 guarantees that there is only a finite number of minimal $\mathcal{EL}$ trees up to a fixed depth. It then follows from the second part of Lemma 2 that for a given tree t, $\rho(t)$ only consists of trees of depth at most $depth(t) + 1$. Hence, $\rho(t)$ is finite.

In order to prove properness of ρ, it is sufficient to show $t \not\preceq t'$ for $t' \in \rho(t)$. To the contrary, assume $t \preceq t'$ and that t has been refined at v. Let $\mathcal{S}$ be a simulation from t' to t. Since v has been refined, it follows that $(v, v) \notin \mathcal{S}$. We have that $\mathcal{S}$ is a simulation, so there must be some $v' \in V(t)$ with $level(v') = level(v)$ such that $(v', v) \in \mathcal{S}$. This implies that there is a simulation $\mathcal{S}'$ on t' with $\{(v', v), (v, v)\} \subseteq \mathcal{S}'$. It follows that t' contains a redundant subtree at the predecessor of v, contradicting to the minimality of t'.

Regarding weakly completeness, let $depth(t) \leq n$. We show that t is reachable from $t_\top$ by nested induction on n and $m := |\{(root(t), r, w) \in E(t)\}|$. For the induction base case $n = 0, m = 0$, t is just a single node labeled with some concept names. It is easily seen that by repeatedly applying $e\ell(t, v)$ and $r\ell(t, v)$ to this node we eventually reach t. For the induction step, let $n > 0, m > 0$. Hence, the root of t is has m successor nodes $w_1, \ldots, w_m$ attached along edges $r_1, \ldots, r_m$ to t. By the induction hypothesis, the tree t_{m-1}, which is obtained from t by removing the subtree $t(w_1)$ from t, is reachable from $t_\top$. Also, there is a refinement chain θ from $t_\top$ to $t(w_1)$ such that an intermediate tree t'_{w_1} occurs in θ and $t' = t_{m-1}[root(t) \leftarrow (r'_1, t'_{w_1})] \in as(t_{m-1}, root(t))$ for some r'_1 with $r_1 \sqsubseteq r'_1$. Hence, we can first reach t' from $t_\top$ and then, by applying the remaining refinement steps from θ to t' and refining r'_1 to r_1, eventually reach t.

Still, ρ is not ideal, since it is not complete. It is however easy to derive a complete operator ρ^* from ρ:

$$\rho^*(t) := max\{t' \mid t_\top \rightsquigarrow^*_\rho t', t' \prec t \text{ and}$$
$$depth(t') \leq depth(t) + 1\}.$$

This construction is needed, because we would for example not be able to reach $\exists r.(A_1 \sqcap A_2)$ starting from $\exists r.A_1 \sqcap \exists r.A_2$ with ρ.

Theorem 1. *The $\mathcal{EL}$ downward refinement operator ρ^* is ideal.*

Remark 1. In [12] it has been shown that for languages other than $\mathcal{EL}$ complete and non-redundant refinement operators do not exist (under a mild assumption). The same result carries over to our setting:

Proposition 2. *Let $\psi : T_{min} \rightarrow \mathcal{P}(T_{min})$ be a complete refinement operator. Then ψ is redundant.*

Proof. We assume $\mathcal{K} = \emptyset$ and N_C contains A_1 and A_2. Since ψ is complete and its refinements are minimal, we have $\top \rightsquigarrow^* A_1$. Similarly, $\top \rightsquigarrow^* A_1$, $A_1 \rightsquigarrow^* A_1 \sqcap A_2$, and $A_2 \rightsquigarrow^* A_1 \sqcap A_2$. We have $A_1 \not\sqsubseteq A_2$ and $A_2 \not\sqsubseteq A_1$, which means that $A_1 \not\rightsquigarrow^* A_2$ and $A_2 \not\rightsquigarrow^* A_1$. Hence, $A_1 \sqcap A_2$ can be reached from $\top$ via a refinement chain going through A_1 and a different refinement chain not going through A_1, i.e. ψ is redundant.

3.2 Optimisations

We used two different kinds of optimisations: The first is concerned with the performance of minimality tests and the second reduces the number of trees returned by ρ by incorporating more background knowledge.

Recall from Section 2.3 that checking for minimality of a tree t involves computing a maximal simulation $\mathcal{S}$ on $V(t)$ and is in $\mathcal{O}(|t|^2)$. In order to avoid expensive re-computations of $\mathcal{S}$ after each refinement step, the data-structure of t is extended such that sets $\mathcal{C}_1^{\leftarrow}(v)$, $\mathcal{C}_1^{\rightarrow}(v)$, $\mathcal{C}_2^{\leftarrow}(v)$ and $\mathcal{C}_2^{\rightarrow}(v)$ are attached to every node $v \in V(t)$. Here, the set $\mathcal{C}_1^{\leftarrow}(v)$ contains those nodes w such that (SC1) holds for (v, w) according to Definition 1. Likewise, $\mathcal{C}_2^{\rightarrow}(v)$ is the set of those nodes w such that (SC2) holds for (w, v), and $\mathcal{C}_1^{\leftarrow}(v)$ and $\mathcal{C}_2^{\rightarrow}(w)$ are defined accordingly. When checking for minimality, it is moreover sufficient that each such set is restricted to only consist of nodes from $onlevel(v)$ excluding v itself. This fragmentation of $\mathcal{S}$ allows us to perform local updates instead of re-computation of $\mathcal{S}$ after an operation is performed on v. For example, when the label of v is extended, we only need to recompute $\mathcal{C}_1^{\leftarrow}(v)$, update $\mathcal{C}_1^{\rightarrow}(w)$ for every $w \in \mathcal{C}_1^{\leftarrow}(v)$, and then repeatedly update $\mathcal{C}_2^{\rightarrow}(v')$ and $\mathcal{C}_2^{\leftarrow}(v')$ for every predecessor node v' of an updated node until we reach the root of t. This method saves a considerable amount of computation, since the number of nodes affected by an operation is empirically relatively small.

In order to keep $|\rho(t)|$ small, we use role domains and ranges as well as disjoint concepts inferred from $\mathcal{K}$. The domain restriction axioms can be used to reduce the set of role names considered when adding a subtree or refining an edge: For instance, let w be a node, (v, r, w) the edge pointing to w, and $range(r) = A$. When adding an edge (w, s, u), we ensure that $range(r) \sqcap domain(s)$ is satisfiable. This ensures that only compatible roles are combined. Similar effects are achieved by mode declarations in ILP tools. However, in OWL ontologies role domains and ranges are usually already present and do not need to be added manually. Similar optimisations can be applied to edge refinement. In $as(t, v)$, we furthermore use range restrictions to automatically label a new node with the corresponding role range. For example, if the edge has label r and $range(r) = A$, then the new node w is assigned label $\ell(w) = \{A\}$ (instead of $\ell(w) = \emptyset$).

We now address the optimisation of extending node labels in the implementation of the function $e\ell$. Let A be a concept name for which we want to know whether or not we can add it to $\ell(v)$. We first check $A \sqsubseteq \ell(v)$. If yes, we discard A since we could reach an equivalent concept by refining a concept in $\ell(v)$, i.e. we perform redundancy reduction. Let (u, r, v) be the edge pointing to v and $range(r) = B$. We verify that $A \sqcap B$ is satisfiable and discard A otherwise. Additionally as before, we test whether $\ell(v) \sqsubseteq A$. If yes, then A is also discarded, because adding it would not result in a proper refinement. Performing the last step in a top down manner, i.e. start with the most general concepts A in the class hierarchy, ensures that we compute the maximum of eligible concepts, which can be added to $\ell(v)$. In summary, we make sure that the tree we obtain is label reduced, and perform an on-the-fly test for the satisfiability of its corresponding concept. Applying similar ideas to the case of label refinement is straight forward.

In practice, the techniques briefly described in this section narrow the set of trees returned in a refinement step significantly by ruling out concepts, which are unsatisfiable w.r.t. $\mathcal{K}$ or which can also be reached via other refinement chains. This is is illustrated by the following example.

Example 3. Let $\mathcal{K}$ be as in Example 1 and define $\mathcal{K}' := \mathcal{K} \cup \{domain(\texttt{has\_child}) =$ Human, $range(\texttt{has\_child}) =$ Human, Human $\sqcap$ Animal $\equiv \bot\}$. By incorporating the additional axioms, ρ(Human $\sqcap$ ∃has.Animal) only contains the trees on the right-hand side of the dashed line in Figure 2, except for Human $\sqcap$ ∃has_child.⊤ $\sqcap$ ∃has.Animal, which becomes Human $\sqcap$ ∃has_child.Human $\sqcap$ ∃has.Animal due to the range of has_child.

4   Evaluation of the Operator

In order to evaluate the operator, we computed *random refinement chains* of ρ. A random refinement chain is obtained by applying ρ to ⊤, choosing one of the refinements uniformly at random, then applying ρ to this refinement, etc.

Table 3. Benchmark results on ontologies from the TONES repository. The results show that ρ works well even on large knowledge bases. The time needed to compute a refinement is below one millisecond and does not show large variations.

Name	Logical axioms	Classes	Roles	ρ av. time (in ms)	ρ per ref. (in ms)	Reasoning time (%)	Refinements (av. and max.)		Ref. size (av. and max.)	
GENES	42656	26225	4	167.2	0.14	68.4	1161.5	2317	5.0	8
CTON	33203	17033	43	76.2	0.08	5.1	220.2	28761	5.8	24
GALEN	4940	2748	413	3.5	0.21	37.1	17.0	346	4.9	16
PROCESS	2578	1537	102	193.6	0.16	27.2	986.5	23012	5.7	22
TRANSPORT	1157	445	89	164.4	0.09	5.9	985.2	22651	5.7	24
EARTHREALM	931	559	81	407.4	0.17	23.2	1710.3	27163	5.7	19
TAMBIS	595	395	100	141.6	0.09	1.5	642.4	26685	5.8	23

In order to asses the performance of the operator, we tested it on real ontologies chosen from the TONES repository[3], including some of the most complex OWL ontologies. We generated 100 random refinement chains of length 8 and measured the results. We found experimentally that this allows us to evaluate the refinement operator on a diverse set of concept trees. The tests were run on an Athlon XP 4200+ (dual core 2.2 GHz) with 4 GB RAM. As a reasoner we

---

[3] http://owl.cs.manchester.ac.uk/repository/

used Pellet 1.5. The benchmarks do not include the time to load the ontology into the reasoner and classify it.

The results are shown in Table 3. The first four columns contain the name and relevant statistics of the ontology considered. The next column shows the average time the operator needed on each input concept. In the following column this value is divided by the number of refinements of the input concept. The subsequent column shows how much time is spend on reasoning during the computation of refinements. The two last columns contain the number of refinements obtained and their size. Here, we measure size as the number of nodes in a concept tree plus the sum of the cardinality of all node labels.

The most interesting insight from Table 3 is that despite the different size and complexity of the ontologies, the time needed to compute a refinement is low and does not show large variations (between 0.09 and 0.21 ms). This indicates that the operator scales well to large knowledge bases. It can also be observed that the number of refinements can be very high in certain cases, which is due to the large number of classes and properties in many ontologies and the absence of explicit or implicit disjointness between classes. We want to note that when the operator is used to learn concepts from instances (standard learning task), one can use the optimisations in Section 3.2 and consider classes without common instances instead of class disjointness. In this case, the number of refinements of a given concept will usually be much lower, since no explicit disjointness axioms are required. In all experiments we also note that the time the reasoner requires differs a lot (from 1.5% to 68.4%). However, since the number of reasoner requests is finite and the results are cached, this ratio will decrease with more calls to the refinement operator. Summing up, the results show that efficient ideal refinement on large ontologies can be achieved in $\mathcal{EL}$, which in turn is promising for $\mathcal{EL}$ concept learning algorithms.

5   Related Work

In the area of Inductive Logic Programming considerable efforts have been made to analyse the properties of refinement operators (for a comprehensive treatment, see e.g. [15]). The investigated operators are usually based on horn clauses. In general, applying such operators to DL problems is considered not be a good choice [3]. However, some of the theoretical foundations of refinement operators in Horn logics also apply to description logics, which is why we want to mention work in this area here.

In Shapiro's Model Inference System [17], he describes how refinement operators can be used to adapt a hypothesis to a sequence of examples. In the following years, refinement operators became widely used. [19] found some general properties of refinement operators in quasi-ordered spaces. Nonexistence conditions for ideal refinement operators relating to infinite ascending and descending refinement chains and covers have been developed. The results have been used to show the non-existence of ideal refinement operators for clauses ordered by θ-subsumption. Later, refinement operators have been extended to theories (clause sets) [8].

Within the last decade, several refinement operators for DLs have been investigated. The most fundamental work is [12], which shows for many description languages the maximal sets of properties which can be combined. Among other things, a non-ideality result for the languages $\mathcal{ALC}$, $\mathcal{SHOIN}$, and $\mathcal{SROIQ}$ is shown. We extend this work by providing an ideality result for $\mathcal{EL}$. Refinement operators for $\mathcal{ALER}$ [3], $\mathcal{ALN}$ [7], $\mathcal{ALC}$ [13,9] have been created and used in learning algorithms. It has been stated in [6] and [7] that further research into refinement operator properties is required for building the theoretical foundations of learning in DLs. Finally, [14] provides ideal refinement in $\mathcal{AL}$-log, a hybrid language merging Datalog and $\mathcal{ALC}$, but naturally a different order than DL subsumption was used.

6   Conclusions and Future Work

In summary, we have provided an efficient ideal $\mathcal{EL}$ refinement operator, thereby closing a gap in refinement operator research. We have shown that the operator can be applied to very large ontologies and makes profound use of background knowledge. In future work, we want to incorporate the refinement operator in learning algorithms, and investigate whether certain extensions of $\mathcal{EL}$ may be supported by the operator without losing ideality.

References

1. Baader, F., Molitor, R., Tobies, S.: Tractable and decidable fragments of conceptual graphs. In: Tepfenhart, W.M. (ed.) ICCS 1999. LNCS, vol. 1640, pp. 480–493. Springer, Heidelberg (1999)
2. Baader, F., Sertkaya, B., Turhan, A.-Y.: Computing the least common subsumer w.r.t. a background terminology. J. Applied Logic 5(3), 392–420 (2007)
3. Badea, L., Nienhuys-Cheng, S.-H.: A refinement operator for description logics. In: Cussens, J., Frisch, A. (eds.) ILP 2000. LNCS (LNAI), vol. 1866, pp. 40–59. Springer, Heidelberg (2000)
4. Bodenreider, O., Smith, B., Kumar, A., Burgun, A.: Investigating subsumption in SNOMED CT: An exploration into large description logic-based biomedical terminologies. Artificial Intelligence in Medicine 39(3), 183–195 (2007)
5. Buitelaar, P., Cimiano, P., Magnini, B. (eds.): Ontology Learning from Text: Methods, Evaluation and Applications. Frontiers in Artificial Intelligence, vol. 123. IOS Press, Amsterdam (2007)
6. Esposito, F., Fanizzi, N., Iannone, L., Palmisano, I., Semeraro, G.: Knowledge-intensive induction of terminologies from metadata. In: McIlraith, S.A., Plexousakis, D., van Harmelen, F. (eds.) ISWC 2004. LNCS, vol. 3298, pp. 441–455. Springer, Heidelberg (2004)
7. Fanizzi, N., Ferilli, S., Iannone, L., Palmisano, I., Semeraro, G.: Downward refinement in the ALN description logic. In: HIS, pp. 68–73. IEEE Computer Society, Los Alamitos (2004)
8. Fanizzi, N., Ferilli, S., Di Mauro, N., Altomare Basile, T.M.: Spaces of theories with ideal refinement operators. In: Gottlob, G., Walsh, T. (eds.) Proc. of 18th Int. Joint Conf. on Artificial Intelligence, pp. 527–532. Morgan Kaufmann, San Francisco (2003)

9. Iannone, L., Palmisano, I., Fanizzi, N.: An algorithm based on counterfactuals for concept learning in the semantic web. Applied Intelligence 26(2), 139–159 (2007)
10. Küsters, R.: Non-standard inferences in description logics. Springer, New York (2001)
11. Lehmann, J.: Hybrid learning of ontology classes. In: Perner, P. (ed.) MLDM 2007. LNCS (LNAI), vol. 4571, pp. 883–898. Springer, Heidelberg (2007)
12. Lehmann, J., Hitzler, P.: Foundations of refinement operators for description logics. In: Blockeel, H., Ramon, J., Shavlik, J., Tadepalli, P. (eds.) ILP 2007. LNCS (LNAI), vol. 4894, pp. 161–174. Springer, Heidelberg (2008) (Best Student Paper)
13. Lehmann, J., Hitzler, P.: A refinement operator based learning algorithm for the ALC description logic. In: Blockeel, H., Ramon, J., Shavlik, J., Tadepalli, P. (eds.) ILP 2007. LNCS (LNAI), vol. 4894, pp. 147–160. Springer, Heidelberg (2008) (Best Student Paper)
14. Lisi, F.A., Malerba, D.: Ideal refinement of descriptions in AL-log. In: Horváth, T., Yamamoto, A. (eds.) ILP 2003. LNCS (LNAI), vol. 2835, pp. 215–232. Springer, Heidelberg (2003)
15. Nienhuys-Cheng, S.-H., de Wolf, R. (eds.): Foundations of Inductive Logic Programming. LNCS, vol. 1228. Springer, Heidelberg (1997)
16. Ramanan, P.: Efficient algorithms for minimizing tree pattern queries. In: SIGMOD 2002: Proc. of the 2002 ACM SIGMOD Int. Conf. on Management of data, pp. 299–309. ACM, New York (2002)
17. Shapiro, E.Y.: Inductive inference of theories from facts. In: Lassez, J.L., Plotkin, G.D. (eds.) Computational Logic: Essays in Honor of Alan Robinson, pp. 199–255. The MIT Press, Cambridge (1991)
18. The Gene Ontology Consortium. Gene ontology: tool for the unification of biology. Nature Genetics 25(1), 25–29 (May 2000)
19. van der Laag, P.R.J., Nienhuys-Cheng, S.-H.: Existence and nonexistence of complete refinement operators. In: Bergadano, F., De Raedt, L. (eds.) ECML 1994. LNCS (LNAI), vol. 784, pp. 307–322. Springer, Heidelberg (1994)

Nonmonotonic Onto-Relational Learning

Francesca A. Lisi and Floriana Esposito

Dipartimento di Informatica, Università degli Studi di Bari
Via E. Orabona 4, 70125 Bari, Italy
{lisi,esposito}@di.uniba.it

Abstract. In this paper we carry on the work on Onto-Relational Learning by investigating the impact of having disjunctive DATALOG with default negation either in the language of hypotheses or in the language for the background theory. The inclusion of nonmonotonic features strengthens the ability of our ILP framework to deal with incomplete knowledge. One such ability can turn out to be useful in application domains, such as the Semantic Web. As a showcase we face the problem of inducing an integrity theory for a relational database whose instance is given and whose schema encompasses an ontology and a set of rules linking the database to the ontology.

1 Motivation

An increasing amount of conceptual knowledge is being made available in the form of *ontologies* [8] mostly specified with languages based on *Description Logics* (DLs) [1]. The problem of adding rules to DLs is currently a hot research topic in Knowledge Representation (KR), due to the interest of Semantic Web applications towards the integration of rule systems with ontologies testified by the activity of the W3C Rule Interchange Format (RIF) working group[1] and of the 'Web Reasoning and Rule Systems' conference series[2]. Practically all the approaches in this field concern the study of DL knowledge bases (KBs) augmented with rules expressed in DATALOG [4] and its nonmonotonic (NM) extensions such as disjunctive DATALOG with default negation (DATALOG$^{\neg\vee}$) [6]. Many technical problems arise in this kind of KR systems. In particular, the full interaction between a DL KB and a DATALOG program easily leads to semantic and computational problems related to the simultaneous presence of knowledge interpreted under the *Open World Assumption* (OWA) and knowledge interpreted under the *Closed World Assumption* (CWA) [12]. The KR framework $\mathcal{DL}$+LOG$^{\neg\vee}$ allows for the *tight* integration of DLs and DATALOG$^{\neg\vee}$, through a *weak* safeness condition for variables in rules [13][3].

In [10] we have laid the foundations of an extension of Relational Learning, called Onto-Relational Learning, to account for ontologies. In that work we have

---

[1] http://www.w3.org/2005/rules/wiki/RIF_Working_Group
[2] http://www.rr-conference.org/
[3] We prefer to use the name $\mathcal{DL}$+LOG$^{\neg\vee}$ instead of the original one $\mathcal{DL}$+LOG in order to emphasize the DATALOG$^{\neg\vee}$ component of the framework.

L. De Raedt (Ed.): ILP 2009, LNAI 5989, pp. 88–95, 2010.

proposed to adapt generalized subsumption [3] to a decidable instantiation of $\mathcal{DL}+\text{LOG}^{\neg\vee}$ obtained by integrating DATALOG with the DL underlying the ontology language OWL for the Semantic Web, i.e. the DL $\mathcal{SHIQ}$ [9]. The resulting hypothesis space can be searched by means of refinement operators either top-down or bottom-up. In order to define a coverage relation we have assumed the ILP setting of learning from interpretations. Both the coverage relation and the generality relation boil down to query answering in $\mathcal{DL}+\text{LOG}^{\neg\vee}$. These ingredients for Onto-Relational Learning do not depend on the scope of induction and are still valid for any other decidable instantiation of $\mathcal{DL}+\text{LOG}^{\neg\vee}$, provided that positive DATALOG is still considered. In this paper we carry on the work initiated in [10] by investigating the impact of having DATALOG$^{\neg\vee}$ either in the language of hypotheses or in the language for the background theory. The inclusion of the NM features of $\mathcal{DL}+\text{LOG}^{\neg\vee}$ *full* will strengthen the ability of our ILP framework to deal with incomplete knowledge. One such ability can turn out to be useful in application domains, such as the Semantic Web. As a showcase we face the problem of inducing an integrity theory for a relational database whose instance is given and whose schema encompasses an ontology and a set of rules linking the database to the ontology.

The paper is organized as follows. Section 2 introduces the KR framework of $\mathcal{DL}+\text{LOG}^{\neg\vee}$. Section 3 sketches an ILP solution for learning $\mathcal{DL}+\text{LOG}^{\neg\vee}$ theories. Section 4 concludes the paper with final remarks.

2 $\mathcal{DL}$+log: Integrating DLs and Disjunctive Datalog

Description Logics (DLs) are a family of decidable First Order Logic (FOL) fragments that allow for the specification of knowledge in terms of classes (*concepts*), binary relations between classes (*roles*), and instances (*individuals*) [2]. Complex concepts can be defined from atomic concepts and roles by means of constructors such as atomic negation ($\neg$), concept conjunction ($\sqcap$), value restriction ($\forall$), and limited existential restriction ($\exists$) - just to mention the basic ones. A DL KB can state both is-a relations between concepts (*axioms*) and instance-of relations between individuals (resp. couples of individuals) and concepts (resp. roles) (*assertions*). Concepts and axioms form the so-called TBox whereas individuals and assertions form the so-called ABox[4]. A $\mathcal{SHIQ}$ KB encompasses also a RBox, i.e. axioms defining hierarchies over roles. The semantics of DLs can be defined through a mapping to FOL. Thus, coherently with the OWA that holds in FOL semantics, a DL KB represents all its models. The main reasoning task for a DL KB is the *consistency check* that is performed by applying decision procedures based on tableau calculus.

Disjunctive DATALOG (DATALOG$^{\neg\vee}$) is a variant of DATALOG that admits disjunction in the rules' heads and default negation [6]. The presence of disjunction in the rules' heads because it makes DATALOG$^{\neg\vee}$ inherently nonmonotonic,

---

[4] When a DL-based ontology language is adopted, an ontology is nothing else than a TBox eventually coupled with a RBox. If the ontology is populated, it corresponds to a whole DL KB, i.e. encompassing also an ABox.

i.e. new information can invalidate previous conclusions. Among the many alternatives, one widely accepted semantics for DATALOG$^{\neg\vee}$ is the extension to the disjunctive case of the *stable model semantics* originally conceived for normal logic programs (i.e. logic programs with default negation) [7]. According to this semantics, a DATALOG$^{\neg\vee}$ program may have several alternative models (but possibly none), each corresponding to a possible view of the reality.

The hybrid KR framework of $\mathcal{DL}$+LOG$^{\neg\vee}$ allows for the tight integration of DLs and DATALOG$^{\neg\vee}$ [13]. More precisely, it allows a DL KB to be extended with DATALOG$^{\neg\vee}$ rules of the form:

$$p_1(\boldsymbol{X_1}) \vee \ldots \vee p_n(\boldsymbol{X_n}) \leftarrow$$
$$r_1(\boldsymbol{Y_1}), \ldots, r_m(\boldsymbol{Y_m}), s_1(\boldsymbol{Z_1}), \ldots, s_k(\boldsymbol{Z_k}), not\ u_1(\boldsymbol{W_1}), \ldots, not\ u_h(\boldsymbol{W_h})\ (1)$$

where $n, m, k, h \geq 0$, each $p_i(\boldsymbol{X_i})$, $r_j(\boldsymbol{Y_j})$, $s_l(\boldsymbol{Z_l})$, $u_k(\boldsymbol{W_k})$ is an atom and each p_i is either a DL-predicate or a DATALOG predicate, each r_j, u_k is a DATALOG predicate, each s_l is a DL-predicate. Peculiar to $\mathcal{DL}$+LOG$^{\neg\vee}$ is the condition of *weak safeness*: Every head variable of a rule must appear in at least one of the atoms $r_1(\boldsymbol{Y_1}), \ldots, r_m(\boldsymbol{Y_m})$. It allows to overcome the main representational limits of the approaches based on the DL-safeness condition, e.g. the possibility of expressing conjunctive queries (CQ) and unions of conjunctive queries (UCQ)[5], by keeping the integration scheme still decidable. For $\mathcal{DL}$+LOG$^{\neg\vee}$ a FOL semantics and a NM semantics have been defined. The FOL semantics does not distinguish between head atoms and negated body atoms. Thus, the form (1) is equivalent to:

$$p_1(\boldsymbol{X_1}) \vee \ldots \vee p_n(\boldsymbol{X_n}) \vee u_1(\boldsymbol{W_1}), \ldots, u_h(\boldsymbol{W_h}) \leftarrow$$
$$r_1(\boldsymbol{Y_1}), \ldots, r_m(\boldsymbol{Y_m}), s_1(\boldsymbol{Z_1}), \ldots, s_k(\boldsymbol{Z_k})\ (2)$$

The NM semantics is based on the stable model semantics of DATALOG$^{\neg\vee}$. According to it, DL-predicates are still interpreted under OWA, while DATALOG predicates are interpreted under CWA. Notice that, under both semantics, entailment can be reduced to satisfiability and, analogously, that CQ answering can be reduced to satisfiability. The NMSAT-$\mathcal{DL}$+LOG algorithm has been provided for checking only the NM-satisfiability of finite $\mathcal{DL}$+LOG$^{\neg\vee}$ KBs because FOL-satisfiability can always be reduced (in linear time) to NM-satisfiability by rewriting rules from the form (1) to the form (2). It is shown that in $\mathcal{DL}$+LOG$^{\neg\vee}$ the decidability of reasoning and consequently of ground query answering depends on the decidability of the Boolean CQ/UCQ containment problem in $\mathcal{DL}$.

3   Discovering $\mathcal{DL}$+log$^{\neg\vee}$ Theories

We face the problem of inducing an integrity theory $\mathcal{H}$ for a database Π whose instance Π_F is given and whose schema $\mathcal{K}$ encompasses an ontology Σ and a

---

[5] A *Boolean UCQ* over a predicate alphabet P is a FOL sentence of the form $\exists \boldsymbol{X}.conj_1(\boldsymbol{X}) \vee \ldots \vee conj_n(\boldsymbol{X})$, where $\boldsymbol{X}$ is a tuple of variable or constant symbols and each $conj_i(\boldsymbol{X})$ is a set of atoms whose predicates and arguments are in P and $\boldsymbol{X}$ respectively. A *Boolean CQ* corresponds to a Boolean UCQ for $n = 1$.

set Π_R of rules linking the database to the ontology. We assume that Π and Σ shares a common set of constants so that they can constitute a $\mathcal{DL}+\text{LOG}^{\neg\vee}$ KB.

Example 1. From now on we refer to a database about students in the form of the following $\text{DATALOG}^{\neg\vee}$ program Π:

```
boy(Paul)
girl(Mary)
enrolled(Paul,c1)
enrolled(Mary,c1)
enrolled(Mary,c2)
enrolled(Bob,c3)
```

containing also the rules

```
FEMALE(X) ← girl(X)
MALE(X) ← boy(X)
```

linking the database to an ontology about persons expressed as the following $\mathcal{DL}$ KB Σ:

```
PERSON ⊑ ∃ FATHER⁻.MALE
MALE ⊑ PERSON
FEMALE ⊑ PERSON
FEMALE ⊑ ¬MALE
MALE(Bob)
PERSON(Mary)
PERSON(Paul)
```

Note that Π and Σ can be integrated into a $\mathcal{DL}+\text{LOG}^{\neg\vee}$ KB.

The integrity theory $\mathcal{H}$ we would like to discover is a set of $\mathcal{DL}+\text{LOG}^{\neg\vee}$ rules. It must be induced by taking the background theory $\mathcal{K} = \Sigma \cup \Pi_R$ into account so that $\mathcal{B} = (\Sigma, \Pi \cup \mathcal{H})$ is a NM-satisfiable $\mathcal{DL}+\text{LOG}^{\neg\vee}$ KB. Since the scope of induction is description and a $\mathcal{DL}+\text{LOG}^{\neg\vee}$ KB may be incomplete, this learning task can be considered as a case of characteristic induction from entailment. In the following we sketch the ingredients for an ILP system, named NMDISC-$\mathcal{DL}+\text{LOG}$, able to discover such integrity theories on the basis of NMSAT-$\mathcal{DL}+\text{LOG}$.

3.1   The Hypothesis Space

The language $\mathcal{L}$ of hypotheses allows for generating $\mathcal{DL}+\text{LOG}^{\neg\vee}$ rules starting from three disjoint alphabets $P_{\mathcal{C}}(\mathcal{L})$, $P_{\mathcal{R}}(\mathcal{L})$, and $P_D(\mathcal{L})$ containing names of $\mathcal{DL}$ concepts and roles, and DATALOG predicates, respectively, taken from $\Pi \cup \Sigma$. Also we distinguish between $P_D^+(\mathcal{L})$ and $P_D^-(\mathcal{L})$ in order to specify which DATALOG predicates can occur in positive and negative literals, respectively. Note that the conditions of linkedness and connectedness usually assumed in ILP are guaranteed by the conditions of DATALOG safeness and weak $\mathcal{DL}$ safeness valid in $\mathcal{DL}+\text{LOG}^{\neg\vee}$.

Example 2. The following $\mathcal{DL}+\text{LOG}^{\neg\vee}$ rules:

```
PERSON(X) ← enrolled(X,c1)
boy(X)∨ girl(X) ← enrolled(X,c1)
← enrolled(X,c2), MALE(X)
← enrolled(X,c2), not girl(X)
MALE(X) ← enrolled(X,c3)
```

belong to the language $\mathcal{L}$ built upon the alphabets $P_{\mathcal{C}}(\mathcal{L}) = \{\text{PERSON}(\_), \text{MALE}(\_)\}$, $P_{\mathrm{D}}^{+}(\mathcal{L}) = \{\text{boy}(\_), \text{girl}(\_), \text{enrolled}(\_,\text{c1}), \text{enrolled}(\_,\text{c2}), \text{enrolled}(\_,\text{c3})\}$, and $P_{\mathrm{D}}^{-}(\mathcal{L}) = \{\text{boy}(\_), \text{girl}(\_)\}$.

The order of *relative subsumption* [11] is suitable for extension to $\mathcal{DL}+\text{LOG}^{\neg\vee}$ rules because it can cope with arbitrary clauses and admit an arbitrary finite set of clauses as the background theory.

Definition 1. *Let R_1 and R_2 be $\mathcal{DL}+\text{LOG}^{\neg\vee}$ rules, and $\mathcal{K}$ a $\mathcal{DL}+\text{LOG}^{\neg\vee}$ background theory. We say that R_1 subsumes R_2 relative to $\mathcal{K}$, denoted by $R_1 \succeq_{\mathcal{K}} R_2$, if there exists a substitution θ such that $\mathcal{K} \models \forall(R_1\theta \implies R_2)$.*

Example 3. Let us consider the following rules:

$$R_1 \equiv \text{boy(X)} \leftarrow \text{enrolled(X,c1)}$$
$$R_2 \equiv \text{boy(X)} \vee \text{girl(X)} \leftarrow \text{enrolled(X,c1)}$$
$$R_3 \equiv \text{MALE(X)} \leftarrow \text{enrolled(X,c1)}$$
$$R_4 \equiv \text{PERSON(X)} \leftarrow \text{enrolled(X,c1)}$$

It can be proved that $R_1 \succeq_{\mathcal{K}} R_2$ and $R_3 \succeq_{\mathcal{K}} R_4$.

3.2   The Refinement Operator

A refinement operator for $(\mathcal{L}, \succeq_{\mathcal{K}})$ should generate $\mathcal{DL}+\text{LOG}^{\neg\vee}$ rules good at expressing integrity constraints. Since we assume the database Π and the ontology Σ to be correct, a rule R must be modified to make it satisfiable by $\Pi \cup \Sigma$ by either (i) strenghtening $body(R)$ or (ii) weakening $head(R)$.

Definition 2. *Let $\mathcal{L}$ be a $\mathcal{DL}+\text{LOG}^{\neg\vee}$ language of hypotheses built out of the three finite and disjoint alphabets $P_{\mathcal{C}}(\mathcal{L})$, $P_{\mathcal{R}}(\mathcal{L})$, and $P_{\mathrm{D}}^{+}(\mathcal{L}) \cup P_{\mathrm{D}}^{-}(\mathcal{L})$, and*

$$p_1(\boldsymbol{X_1}) \vee \ldots \vee p_n(\boldsymbol{X_n}) \leftarrow$$
$$r_1(\boldsymbol{Y_1}), \ldots, r_m(\boldsymbol{Y_m}), s_1(\boldsymbol{Z_1}), \ldots, s_k(\boldsymbol{Z_k}), not\ u_1(\boldsymbol{W_1}), \ldots, not\ u_h(\boldsymbol{W_h})$$

be a rule R belonging to $\mathcal{L}$. A refinement operator $\rho^{\neg\vee}$ for $(\mathcal{L}, \succeq_{\mathcal{K}})$ is defined such that the set $\rho^{\neg\vee}(R)$ contains all $R' \in \mathcal{L}$ that can be obtained from R by applying one of the following refinement rules:

$\langle AddDataLit\_B^{+}\rangle$ $body(R') = body(R) \cup \{r_{m+1}(\boldsymbol{Y_{m+1}})\}$ *if*
  1. $r_{m+1} \in P_{\mathrm{D}}^{+}(\mathcal{L})$
  2. $r_{m+1}(\boldsymbol{Y_{m+1}}) \notin body(R)$

$\langle AddDataLit\_B^- \rangle\ body(R') = body(R) \cup \{not\ u_{m+1}(\boldsymbol{W_{h+1}})\}\ if$
    1. $u_{h+1} \in P_{\mathrm{D}}^-(\mathcal{L})$
    2. $u_{h+1}(\boldsymbol{W_{h+1}}) \notin body(R)$

$\langle AddOntoLit\_B \rangle\ body(R') = body(R) \cup \{s_{k+1}(\boldsymbol{Z_{k+1}})\}\ if$
    1. $s_{k+1} \in P_{\mathcal{C}}(\mathcal{L}) \cup P_{\mathcal{R}}(\mathcal{L})$
    2. it does not exist any $s_l \in body(H)$ such that $s_{k+1} \sqsubseteq s_l$

$\langle SpecOntoLit\_B \rangle\ body(R') = (body(R) \setminus \{s_l(\boldsymbol{Z_l})\}) \cup s_l'(\boldsymbol{Z_l})\ if$
    1. $s_l' \in P_{\mathcal{C}}(\mathcal{L}) \cup P_{\mathcal{R}}(\mathcal{L})$
    2. $s_l' \sqsubseteq s_l$

$\langle AddDataLit\_H \rangle\ head(R') = head(R) \cup \{p_{n+1}(\boldsymbol{X_{n+1}})\}\ if$
    1. $p_{n+1} \in P_{\mathrm{D}}^+(\mathcal{L})$
    2. $p_{n+1}(\boldsymbol{X_{n+1}}) \notin head(R)$

$\langle AddOntoLit\_H \rangle\ head(R') = head(R) \cup \{p_{n+1}(\boldsymbol{X_{n+1}})\}\ if$
    1. $p_{n+1} \in P_{\mathcal{C}}(\mathcal{L}) \cup P_{\mathcal{R}}(\mathcal{L})$
    2. it does not exist any $p_i \in head(R)$ such that $p_{n+1} \sqsubseteq p_i$

$\langle GenOntoLit\_H \rangle\ head(R') = (head(R) \setminus \{p_i(\boldsymbol{X_i})\}) \cup p_i'(\boldsymbol{X_i})\ if$
    1. $p_i' \in P_{\mathcal{C}}(\mathcal{L}) \cup P_{\mathcal{R}}(\mathcal{L})$
    2. $p_i \sqsubseteq p_i'$

Note that, since we are working under NM-semantics, two distinct rules, $\langle AddDataLit\_B^- \rangle$ and $\langle AddDataLit\_H \rangle$, are devised for adding negated DATA-LOG atoms to the body and for adding DATALOG atoms to the head, respectively.

Example 4. From the rule:

    `← enrolled(X,c1)`

belonging to the language specified in Example 2 we obtain the following rules:

```
← enrolled(X,c1), boy(X)
← enrolled(X,c1), girl(X)
← enrolled(X,c1), enrolled(X,c2)
← enrolled(X,c1), enrolled(X,c3)
← enrolled(X,c1), not boy(X)
← enrolled(X,c1), not girl(X)
← enrolled(X,c1), PERSON(X)
← enrolled(X,c1), MALE(X)
boy(X) ← enrolled(X,c1)
girl(X) ← enrolled(X,c1)
enrolled(X,c2) ← enrolled(X,c1)
enrolled(X,c3) ← enrolled(X,c1)
PERSON(X) ← enrolled(X,c1)
MALE(X) ← enrolled(X,c1)
```

by applying the refinement operator $\rho^{\neg\vee}$.

NMDISC-$\mathcal{DL}$+log($\mathcal{L}$, $\mathcal{K}$, Π_F)
1. $\mathcal{H} \leftarrow \emptyset$
2. $\mathcal{Q} \leftarrow \{\,\square\,\}$
3. **while** $\mathcal{Q} \neq \emptyset$ **do**
4. $\mathcal{Q} \leftarrow \mathcal{Q} \setminus \{R\}$;
5.   **if** NMSAT-$\mathcal{DL}$+LOG($\mathcal{K} \cup \Pi_F \cup \mathcal{H} \cup \{R\}$)
6.   **then** $\mathcal{H} \leftarrow \mathcal{H} \cup \{R\}$
7.   **else** $\mathcal{Q} \leftarrow \mathcal{Q} \cup \{R' \in \mathcal{L} | R' \in \rho^{\neg\vee}(R)\}$
8.   **endif**
9. **endwhile**
return $\mathcal{H}$

Fig. 1. Main procedure of NMDISC-$\mathcal{DL}$+LOG

3.3 The Algorithm

The algorithm in Figure 1 defines the main procedure of NMDISC-$\mathcal{DL}$+LOG: it starts from an empty theory $\mathcal{H}$ (1), and a queue $\mathcal{Q}$ containing only the empty clause (2). It then applies a search process (3) where each element R is deleted from the queue $\mathcal{Q}$ (4), and tested for satisfaction w.r.t. the data Π_F by taking into account the background theory $\mathcal{K}$ and the current integrity theory $\mathcal{H}$ (5)[6]. If the rule R is satisfied by the database (6), it is added to the theory (7). If the rule is violated by the database, its refinements according to $\mathcal{L}$ are considered (8). The search process terminates when $\mathcal{Q}$ becomes empty (9). Note that the algorithm does not specify the search strategy. In order to get a minimal theory (i.e., without redundant clauses), a pruning step and a post-processing phase can be added to NMDISC-$\mathcal{DL}$+LOG by further calling NMSAT-$\mathcal{DL}$+LOG[7].

Example 5. With reference to Example 4, the following $\mathcal{DL}$+LOG$^{\neg\vee}$ rule:

    PERSON(X) ← enrolled(X,c1)

is the only one passing the NM-satisfiability test at step (5) of the algorithm NMDISC-$\mathcal{DL}$+LOG. It is added to the integrity theory. Other rules returned by the algorithm are reported in Example 2.

4 Conclusions and Future Work

In this paper we have carried on the work on Onto-Relational Learning by considering the problem of learning $\mathcal{DL}$+LOG$^{\neg\vee}$ rules to be used as integrity theory for a database whose schema is represented also by means of an ontology. The main procedure of NMDISC-$\mathcal{DL}$+LOG is inspired by [5] as for the scope of induction and the algorithm scheme but differs from it in several points, notably

---

[6] The NM-satisfiability test includes also the current induced theory in order to deal with the nonmonotonicity of induction in the normal ILP setting.
[7] Based on the following consequence of the Deduction Theorem in FOL: Given a KB $\mathcal{B}$ and a rule R in $\mathcal{DL}$+LOG$^{\neg\vee}$, we have that $\mathcal{B} \models R$ iff $\mathcal{B} \wedge \neg R$ is unsatisfiable.

the adoption of (i) relative subsumption instead of θ-subsumption, (ii) stable model semantics instead of completion semantics, and (iii) learning from entailment instead of learning from interpretations, to deal properly with the chosen representation formalism for both the background theory and the language of hypotheses. The NM feautures as well as the DL component of $\mathcal{DL}+$LOG$^{\neg\vee}$ allow NMDISC-$\mathcal{DL}+$LOG to induce very expressive integrity theories. In this paper we have therefore addressed an issue that has been brought to the attention of the database community with the advent of the Semantic Web, i.e. the issue of how ontologies (and semantics conveyed by them) can help solving typical database problems, through a better understanding of KR aspects related to databases. In the future we plan to further investigate this issue from the ILP perspective.

References

1. Baader, F., Calvanese, D., McGuinness, D., Nardi, D., Patel-Schneider, P.F. (eds.): The Description Logic Handbook: Theory, Implementation and Applications, 2nd edn. Cambridge University Press, Cambridge (2007)
2. Borgida, A.: On the relative expressiveness of description logics and predicate logics. Artificial Intelligence 82(1-2), 353–367 (1996)
3. Buntine, W.: Generalized subsumption and its application to induction and redundancy. Artificial Intelligence 36(2), 149–176 (1988)
4. Ceri, S., Gottlob, G., Tanca, L.: What you always wanted to know about datalog (and never dared to ask). IEEE Transactions on Knowledge and Data Engineering 1(1), 146–166 (1989)
5. De Raedt, L., Bruynooghe, M.: A theory of clausal discovery. In: IJCAI, pp. 1058–1063 (1993)
6. Eiter, T., Gottlob, G., Mannila, H.: Disjunctive DATALOG. ACM Transactions on Database Systems 22(3), 364–418 (1997)
7. Gelfond, M., Lifschitz, V.: Classical negation in logic programs and disjunctive databases. New Generation Computing 9(3/4), 365–386 (1991)
8. Gruber, T.: A translation approach to portable ontology specifications. Knowledge Acquisition 5, 199–220 (1993)
9. Horrocks, I., Sattler, U., Tobies, S.: Practical reasoning for very expressive description logics. Logic Journal of the IGPL 8(3), 239–263 (2000)
10. Lisi, F.A., Esposito, F.: Foundations of Onto-Relational Learning. In: Železný, F., Lavrač, N. (eds.) ILP 2008. LNCS (LNAI), vol. 5194, pp. 158–175. Springer, Heidelberg (2008)
11. Plotkin, G.D.: A further note on inductive generalization. Machine Intelligence 6, 101–121 (1971)
12. Rosati, R.: Semantic and computational advantages of the safe integration of ontologies and rules. In: Fages, F., Soliman, S. (eds.) PPSWR 2005. LNCS, vol. 3703, pp. 50–64. Springer, Heidelberg (2005)
13. Rosati, R.: $\mathcal{DL}+$log: Tight Integration of Description Logics and Disjunctive Datalog. In: Doherty, P., Mylopoulos, J., Welty, C.A. (eds.) Proc. of 10th Int. Conf. on Principles of Knowledge Representation and Reasoning, pp. 68–78. AAAI Press, Menlo Park (2006)

CP-Logic Theory Inference with Contextual Variable Elimination and Comparison to BDD Based Inference Methods

Wannes Meert, Jan Struyf, and Hendrik Blockeel

Dept. of Computer Science, Katholieke Universiteit Leuven, Belgium
{Wannes.Meert,Jan.Struyf,Hendrik.Blockeel}@cs.kuleuven.be

Abstract. There is a growing interest in languages that combine probabilistic models with logic to represent complex domains involving uncertainty. Causal probabilistic logic (CP-logic), which has been designed to model causal processes, is such a probabilistic logic language. This paper investigates inference algorithms for CP-logic; these are crucial for developing learning algorithms. It proposes a new CP-logic inference method based on contextual variable elimination and compares this method to variable elimination and to methods based on binary decision diagrams.

1 Introduction

Formalisms from the field of *Statistical Relational Learning* (SRL) combine concepts from two fields, uncertainty theory and (first-order) logic, to model complex worlds that cannot be represented by using techniques from one of these fields alone. SRL uses uncertainty theory to express the uncertainty that is present in everyday life, while it employs logic to represent complex relations between the objects in the world.

Fast inference algorithms are key to the success of SRL. Inference speed is not only important while answering probabilistic queries, but it is also crucial for developing fast parameter and structure learning algorithms [1]. Uncertainty theory and logic each have their own specific set of inference algorithms. SRL models can be queried by performing both types of inference sequentially [2,3,4]. However, as the field matured, more efficient solutions have been presented in which the two are more interleaved, that is, logic inference is used while performing probabilistic inference and not only as a preprocessing step [5,6,7].

In this paper, we focus on the SRL language CP-logic [8], which is a language that has been designed to model causal processes. Since CP-logic was introduced, several inference algorithms have been proposed for (a subset of) CP-logic. This paper proposes a new CP-logic inference method, which more closely integrates logic an probabilistic inference. The method is based on Poole and Zhang's Contextual Variable Elimination (CVE) [9]. CVE is a Bayesian network (BN) inference algorithm that extends traditional Variable Elimination (VE) [10] by factorizing each factor in the factor representation of the BN further into a product of so-called confactors. This confactor representation compactly

L. De Raedt (Ed.): ILP 2009, LNAI 5989, pp. 96–109, 2010.
© Springer-Verlag Berlin Heidelberg 2010

represents both the logical and probabilistic component of a (ground) CP-logic theory. By exploiting this compact representation, CVE can be orders of magnitude faster than VE on a given inference task.

This paper is organized as follows. In Section 2, we explain the basic notions behind CP-logic. Section 3 presents an overview of the known CP-logic inference techniques. We introduce CP-logic inference by means of CVE in Section 4. Section 5 presents an experimental comparison among the different inference methods. We end with a conclusion and a number of directions for future work in Section 6.

2   CP-Logic

In many applications, the goal is to model the probability distribution of a set of random variables that are related by a causal process, that is, the variables interact through a sequence of non-deterministic or probabilistic events. Causal probabilistic logic (CP-logic) [8] is a probabilistic logic modeling language that can model such processes. The model takes the form of a CP-logic theory (CP-theory), which is a set of CP-events in which each event is represented as a rule of the following form:

$$(h_1 : \alpha_1) \vee \ldots \vee (h_n : \alpha_n) \leftarrow b_1, \ldots, b_m.$$

with h_i atoms, b_i literals, and α_i *causal* probabilities; $0 < \alpha_i \leq 1$, $\sum \alpha_i \leq 1$. We call the set of all $(h_i : \alpha_i)$ the *head* of the event, and the conjunction of literals b_i the *body*. If the body of a CP-event evaluates to true, then the event will happen and cause at most one of the head atoms to become true; the probability that the event causes h_i is given by α_i (if $\sum \alpha_i < 1$, it is also possible that *nothing* is caused).

It is part of the semantics [8] of CP-logic that each rule independently of all other rules makes one of its head atoms true when triggered. CP-logic is therefore particularly suited for describing models that contain a number of independent stochastic events or causal processes.

Example. The CP-theory

$$
\begin{aligned}
shops(john) &: 0.2 \leftarrow . &\quad (c_1) \\
shops(mary) &: 0.9 \leftarrow . &\quad (c_2) \\
(spaghetti : 0.5) \vee (steak : 0.5) &\leftarrow shops(john). &\quad (c_3) \\
(spaghetti : 0.3) \vee (fish : 0.7) &\leftarrow shops(mary). &\quad (c_4)
\end{aligned}
$$

models the situation that John and his partner Mary may independently decide to go out to buy food for dinner. The causal probability associated with each meal indicates the probability that the fact that John (respectively Mary) goes shopping causes that particular meal to be bought.

It may be tempting to interpret the CP-theory parameters as the conditional probability of the head atom given the body, e.g., based on event c_4, one may conclude that $Pr(spaghetti|shops(mary)) = 0.3$, but this is incorrect. The correct value is $0.3 + 0.2 \cdot 0.5 \cdot 0.7 = 0.37$ since Mary buys spaghetti with probability 0.3, but there is also a probability of $0.2 \cdot 0.5 \cdot 0.7$ that the spaghetti is bought by John $(0.2 \cdot 0.5)$ and not by Mary (0.7). Thus, for head atoms that occur in multiple events, the mathematical relationship between the CP-theory parameters and conditional probabilities is somewhat complex, but it is not unintuitive. The meaning of the causal probabilities in the events is quite simple: they reflect the probability that the body causes the head to become true. This is different from the conditional probability, but among the two, the former is the more natural one to express. Indeed, the former is local knowledge: an expert can estimate the probability that $shops(mary)$ causes $spaghetti$ without considering any other possible causes for spaghetti.

CP-logic is closely related to other probabilistic logics such as ProbLog [11], Independent Choice Logic (ICL) [12], Programming in Statistical Modelling (PRISM) [13], and Bayesian Logic Programs (BLPs) [2]. Meert et al. [1] compares CP-logic to these other formalisms.

3   Known CP-Theory Inference Methods

Probabilistic inference in the context of SRL is the process of computing the conditional probability $Pr(q|e)$, with q the query atom and e a conjunction of literals, which is known as the evidence. The example from the previous section, which illustrated how to compute $Pr(spaghetti|shops(mary))$, is as such an example of probabilistic inference.

A simple general probabilistic inference method for CP-logic is to directly apply the process semantics of CP-logic as defined by Vennekens et al. [8]. This is, however, not a computationally efficient method and faster alternatives have been proposed in the mean time. These alternative methods are discussed in the following sections.

3.1   Variable Elimination

Blockeel and Meert [14,1] propose a transformation that can transform any acyclic CP-theory with a finite Herbrand universe into a particular BN, which they call the Equivalent Bayesian Network (EBN) of the CP-theory. Based on this transformation, CP-theory inference is performed by first transforming the CP-theory into its EBN and by subsequently running a traditional BN inference algorithm, such as Variable Elimination (VE) [10] on the EBN to answer the probabilistic query.

In this section, we recall the transformation procedure. (For the proof of correctness we refer to Blockeel and Meert [14]). We only consider ground CP-theories, so a non-ground CP-theory must be grounded first. The following three steps construct the EBN of a ground CP-theory (Fig. 1 shows the resulting EBN for the "shopping" example):

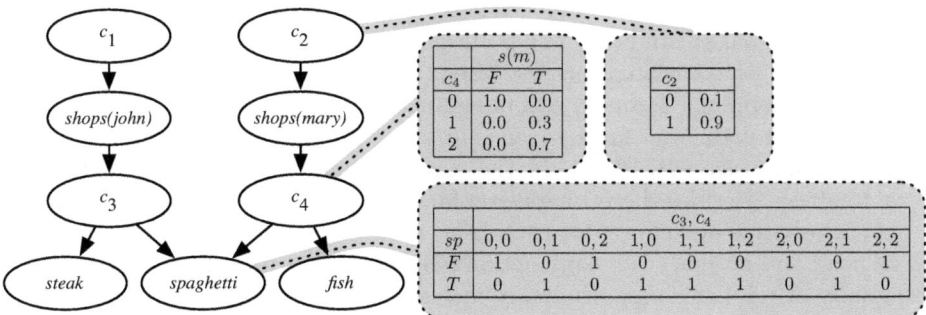

Fig. 1. EBN for the "shopping" example. The CPT for event c_4 represents the probability distribution over its possible outcomes ($c_4 = 0 \ldots 2$) conditioned on its body atom $shops(mary)$. The CPT for the atom node $spaghetti$ represents the deterministic OR $c_3 = 1 \lor c_4 = 1$ (spaghetti is selected by the third event or by the fourth).

1. Create for every atom in the CP-theory a so-called *atom node* in the EBN, and for every event a so-called *choice node*. Next, create for each event in the CP-theory the following edges. Create for every head atom of the event an edge from the event's choice node towards the atom's atom node, and for every literal in the event's body an edge from the literal's atom node towards the event's choice node.
2. A choice node can take $n + 1$ values, where n is the number of head atoms in the CP-event that it represents. It takes the value i if the ith head atom is chosen by the probabilistic process. It takes the value 0 if none of the head atoms is chosen (this can happen if $\sum \alpha_i < 1$). The node's CPT is filled out as follows. Given that the body of the event is true, the probability that the node takes the value $i \neq 0$ is precisely the causal probability α_i given in the event's head. The probability that it takes the value 0 is equal to $1 - \sum \alpha_i$. If the body is not true, then the choice node takes the value 0 with probability 1.0.
3. An atom node is a Boolean node of which the CPT represents the deterministic OR function of the different ways in which the atom can be caused by a CP-event.

3.2 ProbLog

This section explains how to perform CP-theory inference by transforming the CP-theory into ProbLog [5], which is a probabilistic programming language with an efficient inference engine, and by subsequently relying on ProbLog's inference.

A ProbLog program consists of a set of *probabilistic facts* and a set of *Prolog clauses* (Fig. 2, first step). That is, it can be seen as a CP-theory in which all events have precisely one head atom and zero or more positive body literals, and in which the probability of an event's head atom is equal to 1.0 if its body is non-empty.

ProbLog can serve as a target language to which other probabilistic logic modeling languages can be compiled. In particular, acyclic CP-theories without negation can be translated into ProbLog. Fig. 2 illustrates how the CP-theory from the "shopping" example can be translated into ProbLog. The translation introduces probabilistic facts to encode which of the head atoms are chosen (their names start with "c_"). In the example, one probabilistic fact for each event suffices. Besides the probabilistic facts, the ProbLog program includes for each CP-event e one clause for each head atom h of the event. This clause has h as head and its body is a conjunction based on the probabilistic fact for e and the body of e (in this case, the body includes either the probabilistic fact or its negation). For events that have more than two possible outcomes, more than one probabilistic fact is required.

ProbLog's inference engine works as follows. Given a query, it first computes all proofs of the query (using standard Prolog SLD resolution) and collects the probabilistic facts used in these proofs in a DNF formula. The second step of Fig. 2 shows the resulting DNF for the query $Pr(spaghetti)$; each disjunct corresponds to one proof of "spaghetti". Next, it converts this DNF formula to a binary decision diagram (BDD).

A BDD (Fig. 2, right) is a compact representation for a Boolean formula (in this case, the DNF). It is a rooted directed acyclic graph, which consists of decision nodes and two terminal nodes called 0-terminal and 1-terminal. Each decision node is labeled with an atom from the formula and has two outgoing edges: left and right. The left subgraph represents the case that the atom is false and the right subgraph represents the case that it is true. A path from the root to the 1-terminal represents a value assignment that makes the formula true, or in the case of ProbLog it represents a proof of the program.

Relying on this BDD representation, ProbLog computes the query's probability in one bottom-up pass through the BDD (using dynamic programming). To do so, it exploits the fact that the left and right subgraphs of a node represent mutually exclusive cases: the left subgraph represents the cases that the atom is false and the right subgraph the cases that it is true. As a result, the probability of the query can be computed by recursively using the formula $Pr(n_i) = (1 - c_i)Pr(left(n_i)) + c_iPr(right(n_i))$, with $Pr(n)$ the probability computed for node n, c_i the probability listed in the program for probabilistic fact i, and $left(n)$ ($right(n)$) the left (right) child of n. The resulting $Pr(n)$ values are indicated at the top of each node in Fig. 2. Further details can be found in [5].

3.3   cplint

Inspired by ProbLog, Riguzzi [15] proposes cplint, which is a CP-theory inference system that makes use of BDDs in a similar way as ProbLog. There are two differences with the transformation to ProbLog. First, cplint uses a different encoding to represent which head atom is caused by a CP-event. Second, cplint supports negation. When it encounters a negative body literal $\neg a$ in a proof, it computes all proofs of a and includes the negation of the DNF resulting from

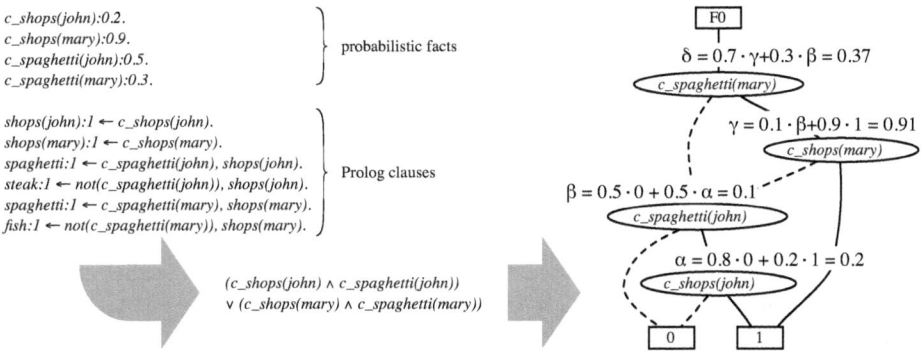

Fig. 2. Steps performed to calculate the probability of the query $Pr(spaghetti)$ for a ProbLog program equivalent to the *shopping* CP-theory

all these proofs into the original DNF. This process is recursive and can be computationally expensive.

4   Inference with Contextual Variable Elimination

As said before, a CP-theory can be transformed to an EBN. However, a CP-theory may contain more structural information than a BN, and this logical structure has to be encoded numerically in the EBN's factors. This may result in factors with redundant information (a factor may have many identical columns) and cause suboptimal inference. This effect can be seen in the factor for *spaghetti* in Fig. 1. To address this problem, we propose to use Contextual Variable Elimination (CVE) [9].

4.1   Contextual Variable Elimination

CVE makes use of a more specific form of conditional independence known as *contextual independence* [16,9].

Definition 1 (Contextual Independence). *Assume that* $\mathbf{x}$, $\mathbf{y}$, $\mathbf{z}$ *and* $\mathbf{c}$ *are sets of variables.* $\mathbf{x}$ *and* $\mathbf{y}$ *are contextually independent given* $\mathbf{z}$ *and context* $\mathbf{c} = \mathbf{C}$, *with* $\mathbf{C} \in \mathrm{dom}(\mathbf{c})$, *iff*

$$Pr(\mathbf{x}|\mathbf{y} = \mathbf{Y_1} \wedge \mathbf{z} = \mathbf{Z_1} \wedge \mathbf{c} = \mathbf{C}) = Pr(\mathbf{x}|\mathbf{y} = \mathbf{Y_2} \wedge \mathbf{z} = \mathbf{Z_1} \wedge \mathbf{c} = \mathbf{C})$$

for all $\mathbf{Y_1}$, $\mathbf{Y_2} \in \mathrm{dom}(\mathbf{y})$ *for all* $\mathbf{Z_1} \in \mathrm{dom}(\mathbf{z})$ *such that* $Pr(\mathbf{y} = \mathbf{Y_1} \wedge \mathbf{z} = \mathbf{Z_1} \wedge \mathbf{c} = \mathbf{C}) > 0$ *and* $Pr(\mathbf{y} = \mathbf{Y_2} \wedge \mathbf{z} = \mathbf{Z_1} \wedge \mathbf{c} = \mathbf{C}) > 0$. *We also say that* $\mathbf{x}$ *is contextually independent* of $\mathbf{y}$ *given* $\mathbf{z}$ *and context* $\mathbf{c} = \mathbf{C}$.

VE represents the joint distribution as a product of factors. CVE factorizes the joint distribution further by replacing each factor by a product of contextual factors or confactors. A confactor r_i consists of two parts: a *context* and a *table*:

$$< \underbrace{v_1 \in V_{1i} \wedge \ldots \wedge v_k \in V_{ki} \wedge \ldots \wedge v_n \in V_{ni}}_{\text{context}}, \quad \underbrace{factor_i(v_k, \ldots, v_m)}_{\text{table}} >$$

The context is a conjunction of set membership tests ($v_j \in V_{ji}, V_{ji} \subseteq domain(v_j)$), which indicates the condition under which the table is applicable. It is used to factorize factors into confactors based on Def. 1. The table stores probabilities for given value assignments for a set of variables ($v_k, \ldots, v_m$). In the original CVE algorithm, the context was limited to equality tests. Our implementation also allows set membership tests, which are required to concisely represent CP-theories (e.g., to represent the inequality tests in the contexts in Fig. 3, left).

The set of confactors that together represent the conditional probability distribution (CPD) of a variable v is mutually exclusive and exhaustive. This means that for each possible value assignment $v = V, \mathbf{pa}(v) = \mathbf{V_p}$, with $\mathbf{pa}(v)$ the parents of v, there is precisely one confactor of which the table includes the parameter $Pr(v = V \mid \mathbf{pa}(v) = \mathbf{V_p})$. These conditions ensure that the product of all confactors is identical to the product of the original factors and thereby equal to the joint distribution.

We describe the CVE algorithm at a high level (the complete algorithm can be found in [9]). Similar to VE, CVE eliminates the non-query non-evidence variables one by one from the joint distribution. To eliminate a variable, it relies on three basic operations: (a) multiplying two confactors with identical contexts; (b) summing out a variable that appears in the table of a confactor; and (c) summing out a variable that appears in the contexts. The operations have as pre-condition that no other compatible confactor exists with a different context. Two contexts are incompatible if there exists a variable that is assigned different values in the contexts; otherwise they are *compatible*. All three operations are only possible if (parts of) the contexts are identical. To satisfy this precondition, CVE uses an operation called *splitting*. Given two compatible confactors, repeated splitting can be used to create two confactors with identical contexts, on which operations (a)-(c) can be performed. These basic operations are repeatedly executed until all non-query non-evidence variables are eliminated; the resulting distribution is the answer to the query.

Confactors may represent CPDs more compactly than tables, but as the previous discussion illustrates, this comes at the cost of more complicated basic operations.

4.2   Converting a CP-Theory to a Set of Confactors

As seen in Section 3, a CP-theory can be represented as an EBN, in which the CPDs are represented by tables. In the previous section, we saw that confactors may lead to a more compact representation. Here, we define the transformation that constructs such a representation. The structure of the EBN is the same as before, but the CPDs are now sets of confactors. The transformation consists of the following two steps (Fig. 3 shows the result for the "shopping" example):

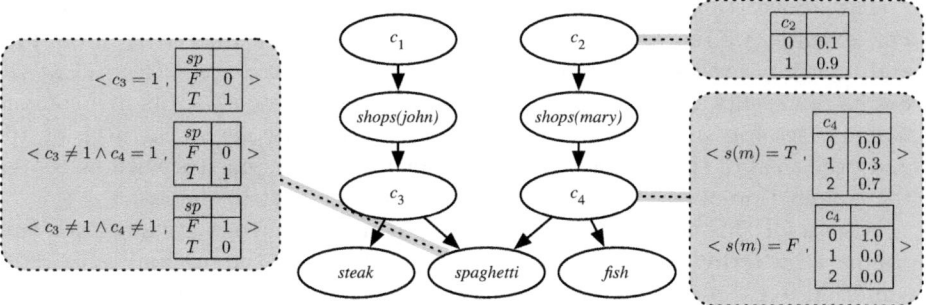

Fig. 3. Confactor representation for node *spaghetti* (left) and c_4 (right)

1. The CPD for a choice node is represented by multiple confactors. The context of one confactor represents the case that the event's body is true. The other confactors constitute the case that the body is false, and make the set of confactors mutually exclusive and exhaustive.

   For example, the first confactor for c_4 (Fig. 3, right) represents the case that the body *shops(mary)* is true and the event chooses to make one of the head atoms true ($c_4 = 1$ for spaghetti, $c_4 = 2$ for fish). The other c_4 confactor corresponds to the case that *shops(mary)* is false; in that case no head atom is caused ($c_4 = 0$).

2. The CPD of an atom node is factorized into multiple confactors that together encode an OR-function (by means of the contexts). If at least one of the events where the atom is in the head has selected the atom, it becomes true; otherwise, it will be false.

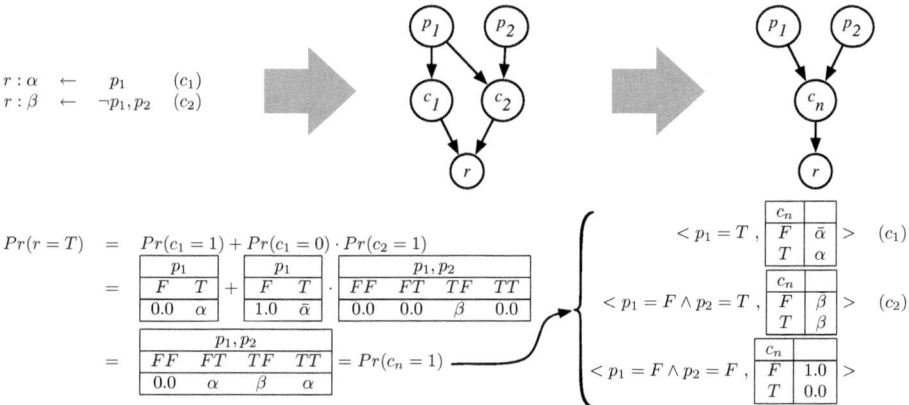

Fig. 4. Example showing how to combine multiple CP-events (c_1 and c_2) with mutually exclusive bodies into a single choice node (c_n). The probability $Pr(r = T)$ can be rewritten such that it depends on c_n. The CPD for c_n combines the distributions for c_1 and c_2 in such a way that the confactors that represent this CPD have the same structure in their contexts as the original set of CP-events has in its bodies ($\bar{\alpha} = 1 - \alpha$).

The above transformation can be extended to improve inference efficiency. For example, we represent a set of CP-events with the same atoms in the head and mutually exclusive bodies by a single choice node. Since such a set of CP-events represents a decision tree this can be encoded by confactors in a compact manner (see Fig. 4). The choice node indicates the outcome of the decision tree and every leaf of the decision tree is converted into a confactor with as context the conjunction of conditions on the path from the root to the leaf.

Also, a CPD is not further factorized if the resulting confactors are not more compact (in terms of the number of parameters) than the original table representation (e.g., in our implementation, c_4's CPD is not split into two confactors as in Fig. 3, but kept as a single factor).

Once the set of confactors that represents the CP-theory is constructed, we use the CVE algorithm [9] to perform CP-theory inference.

5 Experiments

5.1 Setup

We evaluate the inference methods on the task of inferring the marginal distribution of one designated variable in four CP-theories of varying complexity. We always select the variable with the highest inference cost and do not include any evidence (i.e., we consider the most difficult case). The theories are available at http://www.cs.kuleuven.be/~dtai/cplve/ilp09. Fig. 5 presents the results. Graphs (b) and (d) do not include results for ProbLog because they include negation. In the following paragraphs we discuss the used theories.

Growing head. For every atom a_i, $i > 0$ there is a probabilistic fact that causes the atom to become true. Besides these probabilistic facts, there is also a CP-event for every atom a_i in which a_i is in the body and all a_j, $j < i$ are in the head. In other words, atom a_i can cause all other atoms a_j for which $j < i$. This results in an EBN that has a large number of Noisy-OR structures and these are heavily interconnected.

The CP-theory of size 4 is given below as an example (the size parameter indicates the number of atoms in the theory):

$$a0 : 1.0 \leftarrow a1. \qquad a1 : 0.5 \leftarrow .$$
$$a0 : 0.5 \lor a1 : 0.5 \leftarrow a2. \qquad a2 : 0.5 \leftarrow .$$
$$a0 : 0.33 \lor a1 : 0.33 \lor a2 : 0.33 \leftarrow a3. \qquad a3 : 0.5 \leftarrow .$$

Growing body with negation. This CP-theory models a set of decision trees in which the outcome of one tree is used as input by other trees. Each decision tree is represented by a set of CP-events. The theory of size 4 is given below as an example.

$a0 : 0.5 \leftarrow a1.$ $a1 : 0.8 \leftarrow a2.$ $a2 : 0.1 \leftarrow a3.$

$a0 : 0.7 \leftarrow \neg a1, \ a2.$ $a1 : 0.7 \leftarrow \neg a2, \ a3.$

$a0 : 0.2 \leftarrow \neg a1, \neg a2, \ a3.$ $a3 : 0.3 \leftarrow .$

a1

a0:0.5   a2

a0:0.7   a3

a0:0.2   a0:0

a2

a1:0.8   a3

a1:0.7   a1:0

The events that share the same head together form a decision tree because their bodies are mutually exclusive. Inference methods, such as CVE, that exploit that the bodies of such events represent a decision tree can use this knowledge to improve inference efficiency (Section 4.2).

Blood group [2] is a Bayesian network with multi-valued variables and models the probability that a person (p) has a specific blood type (bt) given the chromosomes (pc and mc) that he inherited from his parents (which are again probabilistically dependent on the blood types of their parents). We convert this BN to a CP-theory and perform inference on the resulting theory. We represent each multi-valued variable as a predicate with two arguments: the first argument is an identifier that identifies the person and the second argument is instantiated to one of the possible values of the multi-valued variable. The variable's CPT is encoded by several CP-events. The head of such an event encodes the distribution over the variable's values given a particular value assignment, encoded in the event's body, for the variable's parents in the BN.

A fragment of the theory of size 4 is given below.

$$pc(p, a) : 0.3 \lor pc(p, b) : 0.3 \lor pc(p, nil) : 0.4 \leftarrow .$$
$$mc(p, a) : 0.3 \lor mc(p, b) : 0.3 \lor mc(p, nil) : 0.4 \leftarrow .$$

. . .

$$bt(p, a) : 0.9 \lor bt(p, b) : 0.03 \lor bt(p, ab) : 0.03 \leftarrow pc(p, a), mc(p, a).$$
$$bt(p, a) : 0.03 \lor bt(p, b) : 0.03 \lor bt(p, ab) : 0.9 \leftarrow pc(p, b), mc(p, a).$$

. . .

UW-CSE. This data set [17] records information about the University of Washington's department of Computer Science and Engineering. For this experiment, we start from a model that was learned from this data by means of ordering search [18] and convert this model into a CP-theory. The structure of the resulting theory is complex and contains a combination of the structures of previous examples; it includes both Noisy-OR and decision trees. Note that this theory and also Blood Type are first order theories. We use a simple grounder based on Prolog's SLD resolution to create the relevant grounding for the given probabilistic query.

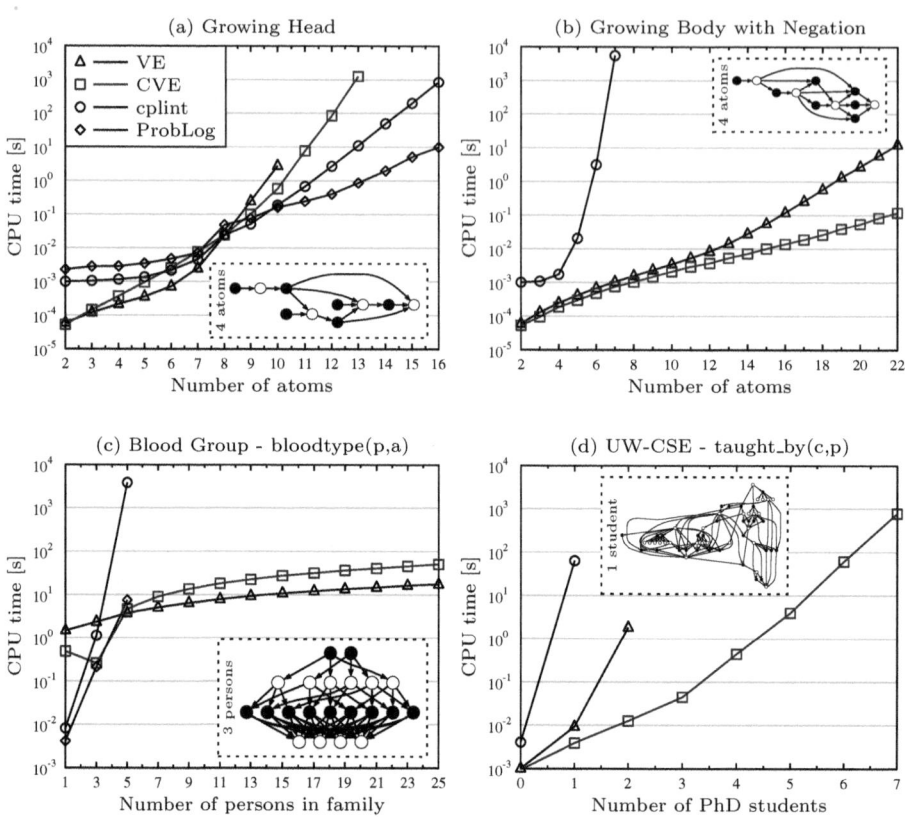

Fig. 5. Experimental results (including example EBNs for small theories)

5.2   Results

For theory (a), the BDD based inference methods (cplint and ProbLog) are faster than CVE and VE for large problem instances. CVE and VE are slower partly because they compute the probability that a variable is true, but also the probability that it is false (separately). It is well known that for Noisy-AND, which occurs in the ProbLog program that theory (a) is transformed into (see Fig. 2), it is more efficient to only compute the probability P_T that its outcome is true and to calculate the probability that it is false as $1 - P_T$. This explains the advantage of the BDD based methods on this theory: they only compute the probability that an atom is true.

The results for theory (b) do not include measurements for ProbLog because ProbLog's inference engine does not support this type of CP-theories. ProbLog only allows negation on predicates that are defined in terms of probabilistic facts; it does not support negation on predicates that are defined by non-trivial clauses. However, the latter type of negation is required to express the decision trees that are modeled by this CP-theory. Supporting full negation in ProbLog is ongoing work (A. Kimmig, personal communication).

Theory (b) contains sets of CP-events that together express decision trees. Since CVE can combine such CP-events compactly into a set of confactors (see Fig. 4) it is more efficient than the other methods. Moreover, CVE and VE outperform the BDD based method cplint due to the complexity of cplint's method for handling negation.

Theory (c) models a Bayesian network in CP-logic. Because CVE has more complex basic operations and tries to represent additional contextual independencies it is slower than VE for large theories in this case. VE and CVE both outperform the BDD based methods. The BDD methods' lower performance is partly due to their inability to represent a BN that has no extra structure beyond conditional independence in its most compact form, which is with tables. Another reason is that the same atom will be encountered multiple times in ProbLog's proofs; if that happens then the DNF formula will contain an identical subexpression for each occurrence. Computing all these subexpressions will require repeatedly proving the same goal and may lead to a larger BDD and to slower inference. Some of these redundant computations can be avoided by 'tabling' proofs [19,20]. Preliminary experiments, however, have shown that tabling does not always improve inference time.

Theory (d) includes a combination of the structures found in theories (a)-(c): it contains Noisy-OR relations, decision trees, and atoms appear repeatedly in proofs. CVE outperforms all other methods in this most general case.

CVE outperforms VE for large problem instances on theories (a), (b), and (d). This is due to the compact representation with confactors instead of factors. VE runs out of memory after size 10 in (a) and size 2 in (d).

6   Conclusion and Future Work

We have proposed a new CP-theory inference method that transforms the given CP-theory to a set of confactors and then performs inference by running contextual variable elimination (CVE) on this representation. CVE outperforms VE with regard to both time and memory consumption for most large problem instances. Depending on the theory, CVE may also be faster than current BDD based methods.

In future work, we plan to incorporate (some of) the above inference methods into CP-theory learning algorithms. Second, we would like to investigate lifted inference for CP-theories. Known lifted-inference methods employ VE; we will try to extend this to CVE. A third item of interest is to investigate inference and learning methods for cyclic CP-theories.

Acknowledgments. Institute for the Promotion of Innovation through Science and Technology in Flanders (IWT-Vlaanderen) to WM. Research Fund K.U.Leuven to JS. GOA/08/008 'Probabilistic Logic Learning'. The authors are grateful to A. Kimmig and D. Fierens for the fruitful discussions and to F. Riguzzi for his suggestions about cplint.

References

1. Meert, W., Struyf, J., Blockeel, H.: Learning ground CP-logic theories by leveraging Bayesian network learning techniques. Fundamenta Informaticae 89, 131–160 (2008)
2. Kersting, K., De Raedt, L.: Bayesian logic programming: Theory and tool. In: Statistical Relational Learning, pp. 291–322. MIT Press, Cambridge (2007)
3. Haddawy, P.: Generating Bayesian networks from probability logic knowledge bases. In: 10th Conference on Uncertainty in Artificial Intelligence (UAI), pp. 262–269 (1994)
4. Wellman, M.P., Breese, J., Goldman, R.: From knowledge bases to decision models. Knowledge Engineering Review 7, 35–53 (1992)
5. Kimmig, A., Santos Costa, V., Rocha, R., Demoen, B., De Raedt, L.: On the efficient execution of ProbLog programs. In: Garcia de la Banda, M., Pontelli, E. (eds.) ICLP 2008. LNCS, vol. 5366, pp. 175–189. Springer, Heidelberg (2008)
6. de Salvo Braz, R., Amir, E., Roth, D.: Lifted first-order probabilistic inference. In: Statistical Relational Learning, pp. 433–452. MIT Press, Cambridge (2007)
7. Singla, P., Domingos, P.: Lifted first-order belief propagation. In: Proceedings of the 23th AAAI Conference on Artificial Intelligence, pp. 1094–1099 (2008)
8. Vennekens, J., Denecker, M., Bruynooghe, M.: Representing causal information about a probabilistic process. In: Fisher, M., van der Hoek, W., Konev, B., Lisitsa, A. (eds.) JELIA 2006. LNCS (LNAI), vol. 4160, pp. 452–464. Springer, Heidelberg (2006)
9. Poole, D., Zhang, N.: Exploiting contextual independence in probabilistic inference. Journal of Artificial Intelligence Research 18, 263–313 (2003)
10. Zhang, N., Poole, D.: A simple approach to bayesian network computations. In: 10th Biennial Canadian Artificial Intelligence Conference, pp. 171–178 (1994)
11. De Raedt, L., Kimmig, A., Toivonen, H.: ProbLog: A probabilistic Prolog and its application in link discovery. In: Proceedings of the 20th International Joint Conference on Artificial Intelligence, IJCAI (2007)
12. Poole, D.: The independent choice logic for modelling multiple agents under uncertainty. Journal of Artificial Intelligence 94, 7–56 (1997)
13. Sato, T., Kameya, Y.: New advances in logic-based probabilistic modeling by PRISM. In: De Raedt, L., Frasconi, P., Kersting, K., Muggleton, S.H. (eds.) Probabilistic Inductive Logic Programming. LNCS (LNAI), vol. 4911, pp. 118–155. Springer, Heidelberg (2008)
14. Blockeel, H., Meert, W.: Towards learning non-recursive LPADs by transforming them into Bayesian networks. In: Blockeel, H., Ramon, J., Shavlik, J., Tadepalli, P. (eds.) ILP 2007. LNCS (LNAI), vol. 4894, pp. 94–108. Springer, Heidelberg (2008)
15. Riguzzi, F.: A top down interpreter for LPAD and CP-logic. In: Proceedings of the 10th Congress of the Italian Association for Artificial Intelligence, AI*IA (2007)
16. Boutilier, C., Friedman, N., Goldszmidt, M., Koller, D.: Context-specific independence in Bayesian networks. In: 12th Conference on Uncertainty in Artificial Intelligence (UAI 1996), pp. 115–123 (1996)
17. Richardson, M., Domingos, P.: Markov logic networks. Machine Learning 62, 107–136 (2006)

18. Ramon, J., Croonenborghs, T., Fierens, D., Blockeel, H., Bruynooghe, M.: Generalized ordering-search for learning directed probabilistic logical models. Machine Learning 70, 169–188 (2008)
19. Riguzzi, F.: The SLGAD procedure for inference on Logic Programs with Annotated Disjunctions. In: Proceedings of the 15th RCRA workshop on Experimental Evaluation of Algorithms for Solving Problems with Combinatorial Explosion (2009)
20. Mantadelis, T., Janssens, G.: Tabling relevant parts of SLD proofs for ground goals in a probabilistic setting. In: International Colloquium on Implementation of Constraint and LOgic Programming Systems, CICLOPS (2009)

Speeding Up Inference in Statistical Relational Learning by Clustering Similar Query Literals

Lilyana Mihalkova[1],[*] and Matthew Richardson[2]

[1] Department of Computer Sciences
The University of Texas at Austin
Austin, TX 78712 USA
lilyanam@cs.utexas.edu
[2] Microsoft Research
One Microsoft Way
Redmond, WA 98052 USA
mattri@microsoft.com

Abstract. Markov logic networks (MLNs) have been successfully applied to several challenging problems by taking a "programming language" approach where a set of formulas is hand-coded and weights are learned from data. Because inference plays an important role in this process, "programming" with an MLN would be significantly facilitated by speeding up inference. We present a new meta-inference algorithm that exploits the repeated structure frequently present in relational domains to speed up existing inference techniques. Our approach first clusters the query literals and then performs full inference for only one representative from each cluster. The clustering step incurs only a one-time up-front cost when weights are learned over a fixed structure.

1 Introduction

Markov logic networks (MLNs) [1] represent knowledge as a set of weighted first-order clauses. Roughly speaking, the higher the weight of a clause, the less likely is a situation in which a grounding of the clause is not satisfied. MLNs have been successfully applied to a variety of challenging tasks, such as information extraction [2], and ontology refinement [3], among others. The general approach followed in these applications, is to treat MLNs as a "programming language" where a human manually codes a set of formulas, for which weights are learned from the data. This strategy takes advantage of the relative strengths of humans and computers: human experts understand the structure of a domain but are known to be poor at estimating probabilities. By having the human experts define the domain, and the computer train the model empirically from data, MLNs can take advantage of both sets of skills.

---

[*] A significant portion of this work was completed while the author was an intern at Microsoft Research in summer 2007.

L. De Raedt (Ed.): ILP 2009, LNAI 5989, pp. 110–122, 2010.
© Springer-Verlag Berlin Heidelberg 2010

Nevertheless, producing an effective set of MLN clauses is not foolproof and involves several trial-and-error steps, such as determining an appropriate data representation and tuning the parameters of the weight learner. Inference features prominently throughout this process. It is used not only to test and use the final model, but also multiple rounds of inference are performed by many popular weight learners [4]. Therefore, just as the availability of fast compilers significantly simplifies software development, "programming" with an MLN would be facilitated by speeding up inference. Speeding up inference would also expand the applicability of MLNs by allowing them to be used for modeling in applications that require efficient test-time, or are too large for training to be done in a reasonable amount of time.

This paper presents a novel meta-inference approach that can speed up any available inference algorithm B by first clustering the query literals based on the evidence that affects their probability of being true. Inference is performed using B for a single representative of each cluster, and the inferred probability of this representative is assigned to all other cluster members. In the restricted case, when clauses in the MLN each contain at most one unknown literal, our approach returns the same probability estimates as performing complete inference using B, modulo random variation of B. Because our new algorithm first breaks down the inference problem to each of the query atoms and then matches, or clusters them, we call it BAM for Break And Match inference.

The remainder of this paper is organized as follows. In Section 2 we introduce necessary background and terminology. We then describe the main contribution of this paper, the BAM algorithm, in Section 3. Section 4 presents our experimental results, showing that BAM maintains the accuracy of the base inference algorithm while achieving large improvements in speed. Finally, we conclude with a discussion of related and future work.

2   Background on MLNs

An MLN [1] consists of a set of weighted first-order clauses. Let $\mathbf{X}$ be the set of all propositions describing a world, $\mathbf{Q}$ be a set of query atoms, and $\mathbf{E}$ be a set of evidence atoms. Without loss of generality, we assume that $\mathbf{E} \cup \mathbf{Q} = \mathbf{X}$. If there are atoms with unknown truth values in whose probabilities we are not interested, they can be added to $\mathbf{Q}$ and then simply ignored in the results. Further, let $\mathcal{F}$ be the set of all clauses in the MLN, w_i be the weight of clause f_i, and $n_i(\mathbf{q}, \mathbf{e})$ be the number of true groundings of f_i on truth assignment $(\mathbf{q}, \mathbf{e})$. The probability that the atoms in $\mathbf{Q}$ have a particular truth assignment, given as evidence the values of atoms in $\mathbf{E}$ is

$$P(\mathbf{Q} = \mathbf{q} | \mathbf{E} = \mathbf{e}) = \frac{1}{Z} \exp \left(\sum_{f_i \in \mathcal{F}} w_i n_i(\mathbf{q}, \mathbf{e}) \right).$$

Ground clauses satisfied by the evidence $\mathbf{E}$ cancel from the numerator and Z and do not affect the probability. Thus, a ground clause G containing one or more atoms from **E** falls in one of two categories: **(A)** G is satisfied by the evidence and can be ignored, or **(B)** all literals from **E** that appear in G are false and G can be simplified by removing these literals; in this case the truth value of G hinges on the assignment made to the remaining literals, which are from the set **Q**.

In its most basic form, inference over an MLN is performed by first grounding it out into a Markov network (MN) [5], as described by Richardson and Domingos [1]. Although several approaches to making this process more efficient have been developed (e.g. [6], which reduces the memory requirement, and [7], which speeds up the process of grounding the MLN), this basic approach is most useful to understanding BAM. Given a set of constants, the ground MN of an MLN is formed by including a node for each ground atom and forming a clique over any set of nodes that appear together in a ground clause. In the presence of evidence, the MN contains nodes only for the unknown query atoms and cliques only for ground clauses that are not made true by the evidence.

Inference over the ground MN is intractable in general, so MCMC approaches have been introduced. We use MC-SAT as the base inference procedure because it has been demonstrated to be faster and more accurate than other methods [8]. However, BAM is independent of the base inference algorithm used and can in principle be applied to speed up any inference method.

3 Speeding Up Inference Using BAM

We first describe BAM in the case where each of the clauses in the MLN contains at most one unknown literal. This case, which we call *restricted*, arises in several applications, such as when modeling the function of independent chemical compounds whose molecules have relational structure [9]. This restricted scenario also arose in work by Wu and Weld [3] on automatic ontology refinement, and, as we will see in Section 4, also in our experimental study.

3.1 BAM in the Restricted Case

In the restricted case, the ground MN constructed from the given MLN consists of a set of disconnected query nodes. Thus the probability of each query literal $Q \in \mathbf{Q}$ being true can be computed independently of the rest and depends only on the number of groundings of each MLN clause that contain Q and fall in category **(B)** described in Sect. 2. This probability is given by:

$$P(Q = q | E = e) = \frac{\exp\left(\sum_{f_i \in \mathcal{F}} w_i \cdot n_{i,Q}(q)\right)}{\exp\left(\sum_{f_i \in \mathcal{F}} w_i \cdot n_{i,Q}(0)\right) + \exp\left(\sum_{f_i \in \mathcal{F}} w_i \cdot n_{i,Q}(1)\right)},$$

where $n_{i,Q}(q)$ is the number of groundings of clause i that contain Q and are true when setting $Q = q$. In the denominator of the expression, we have set q to 0 and 1 in turn. In the restricted case, these counts constitute the *query signature* of a literal, i.e., the signature for a literal Q consists of a set of (C_i, n_i) pairs where, for each clause C_i, n_i is the number of groundings of C_i containing Q that

---

Algorithm 1. Break and Match Inference (BAM)

1: **Q**: set of ground query literals, B: Base inference algorithm
2: **for each** $Q \in \mathbf{Q}$ **do**
3:   SIG_Q = calculateQuerySignature(Q,0)   (Alg. 2 in general case)
4: Partition **Q** into clusters of queries with identical signatures.
5: **for each** Cluster K found above **do**
6:   Pick an arbitrary query literal from K as the representative R
7:   Calculate $P(R = true)$ using B on the portion of the MN used to calculate
     SIG_R.
8:   **for each** $Q \in K$ **do**
9:     Set $P(Q = true) = P(R = true)$

---

are not satisfied by the evidence. Literals with the same query signature have the same probability of being true. We can therefore partition all literals from **Q** into clusters of literals with identical signatures. The probability of only one representative from each cluster is calculated and can be assigned to all other members of the cluster. This is formalized in Alg. 1.

3.2 BAM in the General Case

The algorithm in the general case differs only in the way query signatures are computed. The intuition behind our approach is that the influence a node has on the query node diminishes as we go further away from it. So, by going enough steps away, we can perform a simple approximation to the value of the distant nodes, while only slightly affecting the accuracy of inference. The computation begins by first grounding the given MLN to its corresponding Markov network, simplifying it by removing any evidence nodes and any clauses satisfied by the evidence. All nodes in the resulting Markov network are unknown query nodes from the set **Q**. The signature of each node is computed using a recursive procedure based on the signatures of the nodes adjacent to it.

The probability that a node Q is true depends on the probabilities that its adjacent nodes are true. The adjacent nodes are those with which Q participates in common clauses. Figure 1 provides an illustration. In this figure, we are trying to compute the query signature of the node in the middle that is labeled with q^*. It participates in clauses with four nodes, its immediate neighbors, which are colored in slightly lighter shades in the figure, and thus the probability that is computed for q^* being true will depend on the probabilities computed for these adjacent nodes. The probability that each adjacent node is true, on the other hand, depends on the probabilities that *its* adjacent nodes are true and so on. In this way, BAM expands into the ground MN until it reaches a pre-defined depth $maxDepth$. We used $maxDepth = 2$ in the experiments. At this point, it cuts off the expansion by assigning to the outermost nodes their most likely values found using the MaxWalkSat algorithm [10]. Figure 1 shows these nodes in black, with their most likely assignment (**True** or **False**) written inside them. In this way, q^* is rendered conditionally independent from nodes that are further than depth

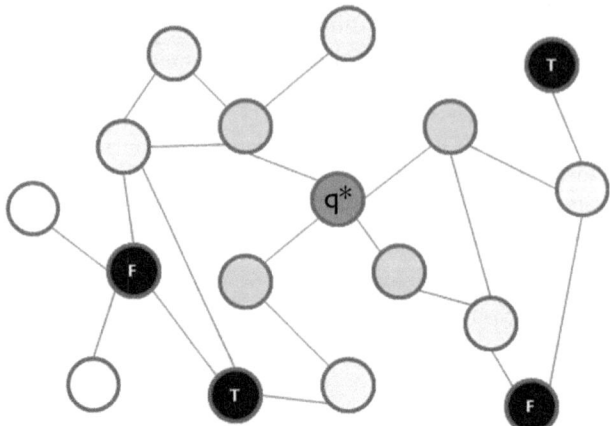

Fig. 1. Illustrating the operation of BAM

$maxDepth$ from it in the Markov network, given the most likely values of nodes that are exactly at depth $maxDepth$.

If Q is selected as a cluster representative in Alg. 1, inference to determine its probability of being true will be carried out only over this portion of the MN (line 7 of Alg. 1). Alg. 2 formalizes the query signature calculation process. Rather than returning the signature itself, Alg. 2 returns a unique identifier associated with each unique signature. In this way, the clustering of nodes occurs alongside the calculation of their signatures, and signatures can be efficiently compared once their identifiers are determined. The identifiers of the outermost nodes whose values are set using MaxWalkSat are 1 (0) for True (False) assignments.

3.3   Using BAM in the Loop of Weight Learning

When we use BAM for weight learning, we would like to compute all signatures up-front, rather than having to re-cluster the nodes for each iteration. For this to be possible, the query signature of a node must not depend on the weights of the clauses. Upon inspection of Alg. 2, we notice that the only aspect of this algorithm that depends on the clause weights is in line 4, where we use MaxWalkSat to set the outer-most nodes to their most likely values. A simple solution is to assign arbitrary values to the outer-most nodes instead of using MaxWalkSat. We experimented with setting the values of all outer-most nodes to False and observed that the accuracy of inference degrades only slightly. A further discussion of this issue is provided in Sect. 4.

3.4   Further Optimizations

The running time of BAM may suffer because inference may be performed several times for the same query literal. This happens because the portion of the MN over

---

Algorithm 2. `calculateQuerySignature(Q, d)` (General case)

1: Input: Q, query literal whose signature is being computed
2: Input: d, depth of literal
3: **if** $d == maxDepth$ **then**
4:    Return value (0 or 1) assigned to Q by MaxWalkSat
5: **for each** Grounding G of a clause in the MLN that contains Q **do**
6:    **for each** Unknown literal U in G whose signature is not yet computed, $U \neq Q$
      do
7:       $SIG_U =$ `calculateQuerySignature(U, d + 1)`
8: **for each** Unground clause C in the MLN **do**
9:    **for each** Distinct way a of assigning signature identifiers to the *other* unknown
      literals in groundings of C that contain Q **do**
10:       Include a triple C, a, n in the signature where n is the number of times the
         particular assignment a was observed
11: Return the unique identifier for the signature

---

which we perform inference in order to compute the probability of the cluster representative R contains additional literals that may themselves be chosen as representatives or may appear in the MNs of multiple representatives. To address this problem, we modified the algorithm to perform inference for more than one representative at a time: suppose we would like to perform inference for literal $L1$ from cluster $C1$, but this would involve inference over literal $L2$ from cluster $C2$. If $C2$ is not yet represented, we include in the MN all the literals up to the desired depth necessary for inference over $L2$ as well. If a cluster is already represented, further representatives are not considered.

4   Experimental Set-Up and Results

The goal of our experiments was to answer the following questions:

1. Does BAM produce correct probability estimates?
2. Are the probability estimates output by BAM close to those output by the base inference algorithm?
3. Does BAM provide an advantage in inference time over the base inference algorithm?
4. What is the effect of setting the outer-most nodes to `false` rather than to their most likely values?

There is a subtle difference between the first two questions. The first is concerned solely with BAM's accuracy of inference, whereas the second one focuses on how well BAM approximates the behavior of the base inference algorithm. To answer these questions, we implemented BAM within Alchemy [11] and used the implementation of MC-SAT provided with it. MC-SAT was run as a stand-alone inference algorithm and as the base inference algorithm of BAM. The same parameter settings were used for all systems: all of Alchemy's defaults were kept, except that the number of sampling steps was set to $10,000$ in order to decrease the amount of variation due to sampling and to better simulate a scenario in which BAM is used in the loop of weight-learning.

To answer the fourth question, we ran two flavors of BAM, which differed in how the values of the outer-most nodes were set. The original algorithm that uses MaxWalkSat to set the outer nodes to their most likely values is called BAM. The version of the algorithm that simply sets the outer nodes to false is called BAM-False. Note that, barring random variation of the base inference algorithm, those two versions of the algorithm are indistinguishable in the restricted case.

We compared the systems in terms of three measures:

- **CLL.** Average conditional log-likelihood, which helps answer the first question above and lends evidence towards answering the second question.
- **MAE.** Mean absolute error, measured as the absolute difference between the probability output by BAM and that output by MC-SAT and averaged over all query atoms. This measure provides a direct comparison between BAM and the base inference algorithm.
- **Time.** Inference time, which helps answer the third question. To ensure a fair comparison for the timings, all the experiments of the two systems within the same run were performed on the same dedicated machine. Furthermore, when performing inference for cluster representatives, BAM executes the same code as that executed by MC-SAT.

We performed experiments on two sets of data: a synthetically generated one that allowed us to control the complexity and size of the models, as well as the UW-CSE domain [1], which contains information on social interactions in an academic setting.

4.1 Synthetic Data Sets

We first describe how the synthetic data was generated and then discuss the experimental results.

Synthetic Data Generation. To generate synthetic MLNs and corresponding datasets, we used a heuristic procedure in which we varied the number of objects and clauses and the complexity of the clauses. This procedure, which we describe next, was designed to model the sparsity typical in relational data. In each case we designated one predicate P_{tar} as the target predicate, all of whose groundings were the query literals. The remaining predicates provided evidence. All predicates were binary. A synthetic dataset and its MLN are specified by three parameters: the number of constants, the number of clauses, and the complexity of the clauses. We considered two levels of clause complexity: restricted (type 1) and non-restricted (type 2). MLNs with restricted complexity contain only clauses that mention P_{tar} once. In non-restricted MLNs, half of the clauses mention P_{tar} once and the other half mention it twice. A dataset/MLN pair was generated as follows. We first randomly assigned truth values to the evidence ground literals. To simulate the fact that relational domains usually contain

Table 1. Possible values characterizing synthetic data

Model Complexity	Num Clauses	Num Objects
Type 1, Type 2	5,10	100, 150, 200

very few true ground literals, for each evidence predicate/constant pair (P, C), we drew n_{true}, the number of true literals of P that have C as the first argument, from a geometric distribution with success probability 0.5. We then randomly selected n_{true} constants to serve as second arguments in the true literals. Each clause was of the form (a conjunction of two literals) $\Rightarrow$ (conclusion), where the conjunction in the antecedents contained two evidence literals different from P_{tar} in models with type 1 complexity, and one evidence literal and one literal of P_{tar} in models with type 2 complexity. The conclusion was always a literal of P_{tar}. Truth values were assigned to the ground literals of P_{tar} by first using the clauses of type 1 and then the clauses of type 2 (if applicable). A grounding of P_{tar} was assigned a value of true 90% of the time a grounding of a clause implied it. Finally, to add more noise 1% of the ground literals of P_{tar} had their truth values flipped. We used the default discriminative weight learning procedure in Alchemy [11] to learn weights for the synthetic MLNs. Table 1 summarizes the possible values for the various characteristics of a dataset/MLN pair. In this way, we obtained $2 \times 2 \times 3 = 12$ distinct dataset/MLN pairs.

Results on Synthetic Data. For each of the 12 dataset/MLN pairs generated above, we performed 5 random runs with each of the systems (BAM and MC-SAT), using the same random seed for both systems within a run.

Table 2 displays the results in terms of CLL. The "Experiment" column of this table describes the experiment as a (number of objects), (number of clauses), (clause complexity type) tuple. Each "Difference" column gives the difference in CLL between the corresponding version of BAM and MC-SAT. A positive (negative) value shows a slight advantage of BAM(MC-SAT). The results presented in this table allow us to draw three conclusions. First, both systems give good accuracy in terms of CLL, as all values are close to 0. Second, the difference in the CLL scores of MC-SAT and BAM is very small; thus, BAM is able to mirror the quality of probability estimates output by MC-SAT. Third, BAM-False leads to only a very slight degradation, as compared to BAM.

Further evidence supporting the conclusion that BAM's probability estimates are very close to those of MC-SAT is presented in Table 3. This table displays the mean absolute differences between the probability output by BAM and that output by MC-SAT, over all test atoms. The fact that these differences are tiny indicates that BAM's performance is very close to that of its base algorithm.

Moreover, BAM runs consistently faster than MC-SAT, and the improvement in speed increases as the number of constants in the domain grows. This can be observed in Fig. 2, which shows only the running times of BAM. The running time of BAM-False is always at most that of BAM because the potentially time-consuming MaxWalkSat step is omitted. On average, BAM performed inference

Table 2. Average CLL of MC-SAT, BAM and BAM-False on synthetic data

Experiment	MC-SAT	BAM	Difference	BAM-False	Difference
100,5,1	-0.067	-0.067	0.000	-0.067	0.000
150,5,1	-0.064	-0.064	-0.000	-0.064	-0.000
200,5,1	-0.061	-0.061	0.000	-0.061	0.000
100,5,2	-0.338	-0.316	-0.021	-0.302	-0.036
150,5,2	-0.344	-0.340	-0.004	-0.333	-0.011
200,5,2	-0.220	-0.207	-0.014	-0.205	-0.016
100,10,1	-0.080	-0.080	0.000	-0.080	0.000
150,10,1	-0.069	-0.069	0.000	-0.069	0.000
200,10,1	-0.068	-0.068	0.000	-0.068	0.000
100,10,2	-0.491	-0.489	-0.001	-0.489	-0.002
150,10,2	-0.598	-0.614	0.015	-0.619	0.020
200,10,2	-0.668	-0.668	0.000	-0.674	0.006

Table 3. Mean Absolute Error (MAE) of BAM and BAM-False compared to MC-SAT on synthetic data

Experiment	BAM	BAM-False
100, 5, 1	0.003	0.003
150, 5, 1	0.002	0.002
200, 5, 1	0.002	0.002
100, 5, 2	0.036	0.041
150, 5, 2	0.020	0.025
200, 5, 2	0.008	0.007
100, 10, 1	0.003	0.003
150, 10, 1	0.003	0.003
200, 10, 1	0.002	0.002
100, 10, 2	0.008	0.009
150, 10, 2	0.021	0.055
200, 10, 2	0.033	0.060

for 37% of the query atoms in domains with 100 constants; 42% in domains with 150 constants; and 31% in domains with 200 constants.

4.2   UW-CSE Domain

We additionally tested BAM on the UW-CSE domain [1][1] using models[2] learned with BUSL, which gave good predictive accuracy on this data set [12]. We used the MLNs from the last point on the learning curve, which were trained on all

---

[1] Available from `http://alchemy.cs.washington.edu/` under "Datasets."
[2] Available from `http://www.cs.utexas.edu/~ml/mlns/` under "BUSL."

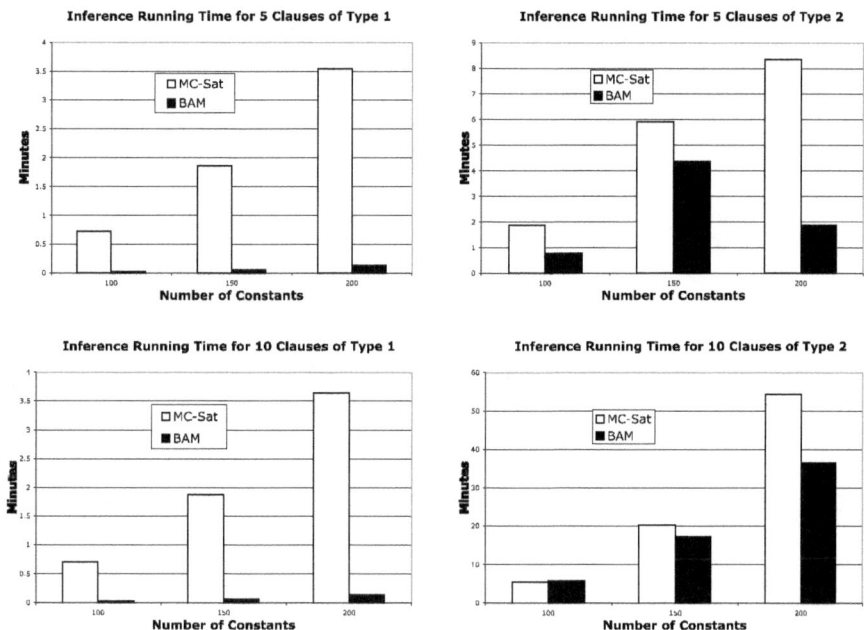

Fig. 2. Average inference running time in minutes

but one of the available examples. Thus, there are five possible models, one for each of the examples left out for testing. The goal of inference was to predict the advisedBy relation. We found that all of these models fall in the restricted case; i.e. they contain at most one mention of the advisedBy predicate per clause. This provides an example of a situation in which the restricted case arose naturally in practice. Thus, on this domain the behavior of BAM is identical to that of BAM-False; hence we report results only for the former.

Table 4 compares the CLL scores of MC-SAT and BAM. The "Difference" column provides the difference, where a positive (negative) value shows a slight advantage of BAM (MC-SAT). As on the synthetic data, we see that both systems have very high scores, and that their scores are of similar quality.

Table 4. The first four columns compare the average CLL of MC-SAT and BAM on each of the test examples in UW-CSE. The last column lists the MAE of BAM compared to MC-SAT.

Example	MC-SAT	BAM	Difference	MAE
1 (AI)	-0.045	-0.045	0.000	0.002
2 (Graphics)	-0.044	-0.052	-0.008	0.004
3 (Language)	-0.060	-0.060	0.000	0.002
4 (Systems)	-0.040	-0.055	-0.015	0.006
5 (Theory)	-0.031	-0.031	0.000	0.001

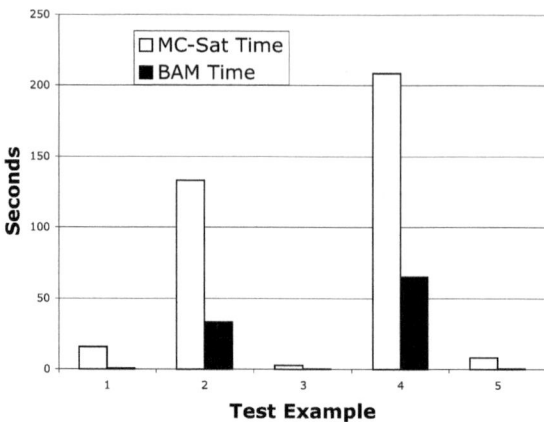

Fig. 3. Average inference time in seconds. All times are plotted, although some are extremely small.

The MAE is shown in the last column of Table 4, and the inference time is shown in Fig. 3. From these experiments, we can conclude that MC-SAT and BAM produce probability estimates of similar quality and that BAM is consistently faster than MC-SAT. On average, BAM was 12.07 times faster than MC-SAT and performed actual inference only for 6% of the unknown query atoms on average over the five test examples.

In both the synthetic and UW-CSE experiments, the results of inference exhibited very little variance across the random runs. Variance is therefore not reported to reduce clutter.

5   Related Work

BAM is related to work on lifted inference in which variable elimination (VE), aided by clever counting and ordering heuristics, is used to eliminate a large number of instantiations of variables at once [13,14,15]. Sen *et al.* [16] introduce an algorithm, based on VE, that constructs a graph whose nodes represent the original and intermediate factors used by VE. By inspecting this graph and carefully computing node labels, factors that carry out identical computations are identified. In a recent extension [17], to allow for approximate inference, the authors exploit the idea that the influence between two nodes diminishes as the distance between them increases, analogous to the idea exploited in the present work.

Jaimovich *et al.* [18] introduce an algorithm based on belief propagation in which inference is performed on the template, i.e. variablized, level. Their approach targets the case when no evidence is present and takes advantage of the fact that in this case, the same inference steps are carried out for all literals in the model. This algorithm has been extended to the case when evidence is present [19]. Furthermore, a recent extension [20] allows for belief propagation to be carried out in an on-demand fashion, grounding out the model as time allows and maintaining bounds on the probability estimates. The approaches based on belief propagation calculate exact probabilities when belief propagation would, but suffer from the same limitations as ordinary belief propagation in the presence of loops.

All above techniques are tied to a particular inference approach and are therefore better viewed as stand-alone inference algorithms, in contrast to BAM, which is a meta-inference technique in that it can be applied to any existing inference algorithm.

6   Future Work

An interesting future extension to BAM is to allow "soft" query signature matching, so that signatures need only be very similar to each other to be placed in the same cluster. It would also be interesting to provide a method for quickly recomputing query signatures as the clauses of the MLN are refined, added, or deleted, allowing efficient use of BAM for structure learning. Another possible improvement includes developing a lazy version of BAM, analogous to the work of Singla and Domingos [6], that would not require the MLN to be grounded to its corresponding Markov network ahead of time.

Acknowledgment

We would like to thank Raymond Mooney for suggesting the use of mean absolute error; Tuyen Huynh for helpful discussions; and the anonymous reviewers for their comments. Some of the experiments were run on the Mastodon Cluster, provided by NSF Grant EIA-0303609, at UT Austin.

References

1. Richardson, M., Domingos, P.: Markov logic networks. Machine Learning 62, 107–136 (2006)
2. Poon, H., Domingos, P.: Joint inference in information extraction. In: Proceedings of the Twenty-Second Conference on Artificial Intelligence (AAAI 2007), Vancouver, Canada (2007)
3. Wu, F., Weld, D.: Automatically refining the Wikipedia infobox ontology. In: Proceedings of the Seventeenth International World Wide Web Conference (WWW 2008), Beijing, China (2008)
4. Lowd, D., Domingos, P.: Efficient weight learning for Markov logic networks. In: Kok, J.N., Koronacki, J., Lopez de Mantaras, R., Matwin, S., Mladenič, D., Skowron, A. (eds.) PKDD 2007. LNCS (LNAI), vol. 4702, pp. 200–211. Springer, Heidelberg (2007)
5. Pearl, J.: Probabilistic Reasoning in Intelligent Systems: Networks of Plausible Inference. Morgan Kaufmann, San Mateo (1988)

6. Singla, P., Domingos, P.: Memory-efficient inference in relational domains. In: Proceedings of the Twenty-First Conference on Artificial Intelligence (AAAI 2006), Boston, MA (2006)
7. Shavlik, J., Natarajan, S.: Speeding up inference in Markov logic networks by preprocessing to reduce the size of the resulting grounded network. In: Proceedings of the Twenty-first International Joint Conference on Artificial Intelligence (IJCAI 2009), Pasadena, CA (2009)
8. Poon, H., Domingos, P.: Sound and efficient inference with probabilistic and deterministic dependencies. In: Proceedings of the Twenty-First Conference on Artificial Intelligence (AAAI 2006), Boston, MA (2006)
9. Frasconi, P., Passerini, A.: Learning with kernels and logical representations. In: De Raedt, L., Frasconi, P., Kersting, K., Muggleton, S.H. (eds.) Probabilistic Inductive Logic Programming. LNCS (LNAI), vol. 4911, pp. 56–91. Springer, Heidelberg (2008)
10. Kautz, H., Selman, B., Jiang, Y.: A general stochastic approach to solving problems with hard and soft constraints. In: Gu, D., Du, J., Pardalos, P. (eds.) DIMACS Series in Discrete Mathematics and Theoretical Computer Science, vol. 35, pp. 573–586. American Mathematical Society, Providence (1997)
11. Kok, S., Singla, P., Richardson, M., Domingos, P.: The Alchemy system for statistical relational AI. Technical report, Department of Computer Science and Engineering, University of Washington (2005),
http://www.cs.washington.edu/ai/alchemy
12. Mihalkova, L., Mooney, R.J.: Bottom-up learning of Markov logic network structure. In: Proceedings of 24th International Conference on Machine Learning (ICML 2007), Corvallis, OR (2007)
13. Poole, D.: First-order probabilistic inference. In: Proceedings of the Eighteenth International Joint Conference on Artificial Intelligence (IJCAI 2003), Acapulco, Mexico (2003)
14. de Salvo Braz, R., Amir, E., Roth, D.: MPE and partial inversion in lifted probabilistic variable elimination. In: Proceedings of the Twenty-First Conference on Artificial Intelligence (AAAI 2006), Boston, Massachusetts (2006)
15. Milch, B., Zettlemoyer, L.S., Kersting, K., Haimes, M., Kaelbling, L.P.: Lifted probabilistic inference with counting formulas. In: Proceedings of the Twenty-Third Conference on Artificial Intelligence, AAAI 2008 (2008)
16. Sen, P., Deshpande, A., Getoor, L.: Exploiting shared correlations in probabilistic databases. In: Proceedings of the 34th International Conference on Very Large Data Bases (VLDB 2008), Auckland, New Zealand (2008)
17. Sen, P., Deshpande, A., Getoor, L.: Bisimulation-based approximate lifted inference. In: Proceedings of 25th Conference on Uncertainty in Artificial Intelligence, UAI 2009 (2009)
18. Jaimovich, A., Meshi, O., Friedman, N.: Template based inference in symmetric relational Markov random fields. In: Proceedings of 23rd Conference on Uncertainty in Artificial Intelligence (UAI 2007), Vancouver, Canada, pp. 191–199 (2007)
19. Singla, P., Domingos, P.: Lifted first-order belief propagation. In: Proceedings of the Twenty-Third Conference on Artificial Intelligence (AAAI 2008), Chicago, IL (2008)
20. de Salvo Braz, R., Natarajan, S., Bui, H., Shavlik, J., Russell, S.: Anytime lifted belief propagation. In: International Workshop on Statistical Relational Learning (SRL 2009), Leuven, Belgium (2009)

Chess Revision: Acquiring the Rules of Chess Variants through FOL Theory Revision from Examples

Stephen Muggleton[1], Aline Paes[1,2],
Vítor Santos Costa[3], and Gerson Zaverucha[2]

[1] Department of Computing, Imperial College London, UK
shm@doc.ic.ac.uk
[2] Department of Systems Eng. and Computer Science, UFRJ, Brazil
{ampaes,gerson}@cos.ufrj.br
[3] CRACS and DCC/FCUP, Universidade do Porto, Portugal
vsc@dcc.fc.up.pt

Abstract. The game of chess has been a major testbed for research in artificial intelligence, since it requires focus on intelligent reasoning. Particularly, several challenges arise to machine learning systems when inducing a model describing legal moves of the chess, including the collection of the examples, the learning of a model correctly representing the official rules of the game, covering all the branches and restrictions of the correct moves, and the comprehensibility of such a model. Besides, the game of chess has inspired the creation of numerous variants, ranging from faster to more challenging or to regional versions of the game. The question arises if it is possible to take advantage of an initial classifier of chess as a starting point to obtain classifiers for the different variants. We approach this problem as an instance of theory revision from examples. The initial classifier of chess is inspired by a FOL theory approved by a chess expert and the examples are defined as sequences of moves within a game. Starting from a standard revision system, we argue that abduction and negation are also required to best address this problem. Experimental results show the effectiveness of our approach.

1 Introduction

Game playing is a fundamental human activity, and has been a major topic of interest in AI communities since the very beginning of the area. Games quite often follow well defined rituals or rules on well defined domains, hence simplifying the task of representing the game as a computer program. On the other hand, good performance in games often requires a significant amount of reasoning, making this area one of the best ways of testing human-like intelligence. Namely, datasets based on games are common testbeds for machine learning systems [7]. Usually, machine learning systems may be required to perform two different kinds of tasks. A first task is to learn a model that can be used to decide whether a move in a game is legal, or not. Having such a model is fundamental

L. De Raedt (Ed.): ILP 2009, LNAI 5989, pp. 123–130, 2010.

for the second task, where one wants to learn a winning strategy. We focus on the the first task in this work.

In order to acquire a meaningful representation of the classifier of the game, one could take advantage of the expressiveness of first-order logic and use *Inductive Logic Programming (ILP)* [11] methods to induce the game's rules written as a logic program, from a set of positive and negative examples and background knowledge (BK). Previous work has demonstrated the feasibility of using ILP to acquire a rule-based description of the rules of chess [8] using structured induction.

Game playing is a dynamic environment where games are always being updated, say, to be more challenging to the player, or to produce an easier and faster variant of the original game. In fact, popular games often have different regional versions. Consider, for example, the game of Chess, arguably, the most widely played board game in the world. It also has been a major game testbed for research on artificial intelligence and it has offered several challenges to the area, since any application involving chess must focus on intelligent reasoning. There are numerous chess variants, where we define *chess variant* as any game that is derived from, related to or inspired by chess, such that the capture of the enemy king is the primary objective [13]. For instance, the *Shogi* game, is the most popular Japanese version of Chess. Although both games have similar rules and goal, they also have essential differences. For example, in Shogi a captured piece may change sides and return to the board [1], which is not allowed in Western Chess.

Knowledge acquisition is a time consuming and error prone task. Ideally, if the rules of a variant of a game have been obtained, we would like to take advantage of them as a starting point to obtain the rules of a variant. However, such rules may need to be modified in order to represent the particular aspects of the variant. In a game such as chess this is a complex task that may require addressing different board sizes, introducing or deleting new promotion and capture rules, and may require redefining the role of specific pieces in the game. Modifying a set of rules so that they could explain a new set of examples is the task of *Theory Revision from Examples* [16]. Besides, transferring a set of rules learnt from a domain to another possibly related domain is studied in the area of *transfer learning* [15,2]. In this work we handle the problem of transfering the rules of chess to some variants of this game as an instance of Theory Revision from Examples. This task is particularly challenging due to the necessity of representing all the official rules of the game, which makes the theory be very large, with several branches and sub-concepts being used to classify a correct move.

We show that we can learn rules between different variants of the game of chess. Starting from the state-of-the-art FORTE revision system [14,5], we contribute with **(i)** a new strategy designed to *simplify* the initial theory by removing rules that will *not* be transferred between variants; **(ii)** support for *abduction*; and **(iii)** support for *negation as failure*. Experimental evaluation on real variants of chess

---

[1] It is suggested that this innovative drop rule was inspired by the practice of 16th century mercenaries who switched loyalties when captured [13].

shows that our technique can transfer between variants with smaller and larger boards, acquire unusual rules, and acquire different pieces.

The paper is organized as follows. We start by discussing the revision system, emphasizing the modifications performed on it in section 2. Next, we briefly describe the examples and the initial knowledge in section 3. We present experimental results in section 4 and finally we conclude the paper and discuss future work in section 5.

2   Chess Revision

Inductive Logic Programming (ILP) systems learn using a set of examples and background knowledge (BK), both described as logic programs and assumed as correct. On the other hand, theory revision from examples consider that the BK could also contain incorrect rules, which, after being modified, would better reflect the dataset. Thus, in theory revision the BK is divided into two sets: one containing the rules subject to modification, named here as *initial theory* and the other composed of rules known to be correct and therefore not modifiable, containing intensional definitions of the fundamental relations used to define a domain, named as *fundamental domain theory (FDT)* [14]. The goal of theory revision is to identify points in the initial theory preventing it from correctly classifying some example, and propose modifications to such points, so that the revised theory together with the FDT is correct. Theory Revision is particularly powerful and challenging because it must deal with the issues arising from revising multiple clauses (theory) and even multiple predicates (target).

Usually, the first step in a revision process is to find the clauses and/or antecedents responsible for the misclassification of some example, the *revision points*. Then, modifications are suggested to these revision points, through *revision operators*, such as rule deletion and antecedent addition, if one wants to specialize, or antecedent deletion or rule addition, if one wants to generalize [16].

In this work we follow the new version of FORTE revision system [14] as modified in [5] to allow the use of bottom clause and modes declarations [9] when searching for literals to be added to the body of clauses. In order to best address the revision of the rules of Chess, we performed three main modifications on the current version of the system, described as follows.

Starting the revision process by deletion of rules. In an attempt to decrease the complexity of the theory and consequently of the whole revision process, we introduced a first step of deletion of rules. This process is performed as a hill-climbing iterative procedure, where at each iteration the facts used in proofs of negative examples are selected, the deletion of each one is scored and the one able of improving the score at most is chosen. This step restarts from the modified theory and finishes when no deletion is able to improve the score. We found this procedure both reduces theory size and noise, namely when the target theory is a specialized version of the initial theory.

Using abduction during the revision process. Abduction is concerned about finding explanations for observed facts, viewed as missing premises in an argument, from available knowledge deriving those facts [6]. Usually, theory revision systems, including FORTE, use abduction when searching for generalization revision points, to locate faults in a theory and suggest repairs to it, determining a set of assumptions that would allow the positive example be proved [14]. We further benefit from abduction in two distinguished moments of the revision process. First, when searching for generalization revision points (points failing on proving positive examples), we assume that faulting abducible predicates are true and continue to search for further revision points possibly depending on them. Note that in this moment the literal is ground. Then, such abducible predicates are included in the theory under revision, up to a maximum number of included abducible predicates. The abducible predicates might be eventually generalized/specialized in the next iterations, in case they become a faulting point in the theory.

The second further moment we use abdcution is when constructing a bottom clause for intermediate predicates. Those are predicates in the head of clauses but the dataset does not contain examples for them, since the examples are of the top-level predicate(s) only. However, to construct the bottom clause it is necessary to start from a positive example with the same predicate as the head of the clause being specialized. Thus, from a positive example proved by the current clause we obtain the required literal using the initial theory and FDT. The procedure takes a top-level goal (an example) and the intermediate predicate and instantiates such a predicate to the first call to that predicate encountered when attempting to prove the goal. The proof starts from the example and return the instantiation of the specified intermediate predicate. Next, we construct the bottom clause from such a literal and use it as search space for adding literals to the clause. In [10] a similar procedure is used to obtain "contra-positive examples" to the intermediate predicates and then construct a bottom clause for them.

Using negated literals in the theory. FORTE was neither able to introduce negated literals in the body of the clause nor revise negated antecedents. Negation is essential to elegantly model the chess problem, since we need to represent concepts such as *the king is* **not** *in check*, among others. In order to add negated literals in the body of the clause, we allow the bottom clause procedure construction to generate negated literals. To do so, one could take advantage of constructive negation techniques [3,4], ensuring that there are no free variables in any not(*Goal*) that might be called. We also introduced a procedure for handling a faulty negated literal during the revision process. Roughly speaking, if the negated literal is responsible for a failed proof of positive examples, it is treated as a specialization revision point. On the other hand, if the negated literal takes part in a proof of a negative example, it is treated as a generalization revision point. This is a preliminary attempt at introducing non-monotonic reasoning in FORTE.

3   The Examples and the Background Knowledge

Examples. In our framework, the legal moves in the chess variant are the positive examples and the illegal moves are the negative examples. The dataset is composed of a set of simulated games, where each game is limited to a specified maximum number of legal moves, considering both players. The moves are within a game, aiming to represent castling and en-passant, which require the history of the game (p.ex., the rook and the king involved in a castling must have not moved yet in the whole game), and promotion, which requires updating the board to represent the promoted piece. We take advantage of FORTE examples representation where the examples are in the format

Ground instances of target predicate ← Conjunction of facts from the context

and we represent each game as a set of legal and illegal moves within the game (they are the ground instances of the target predicate) and the positions of the pieces related to each round (the board of the game) are the facts from the context. Thus, each game has its separate set of legal and illegal moves and set of positions of the pieces during the game, in the format

> *Target Predicate* :
> *Positives* :
> $move(Round, Piece_1, Colour_1, File_1, Rank_1, Piece_2, Colour_2, File_2, Rank_2), ...$
> *Negatives* :
> $move(Round, Piece_1, Colour_1, File_1, Rank_1, Piece_2, Colour_2, File_2, Rank_2), ...$
> *Context* :
> $board(Round, Piece, Colour, File, Rank), ...$
> $out\_board(Round, Piece, Colour, -1, -1), ..., out\_board(Round, Piece, Colour, 0, 0), ...$

The terms of the ground instances of the target predicate are the number of the move, the current and next status of the piece. For example, $move(9, pawn, white, c, 7, rook, white, c, 8)$ states that in the $9th$ move executed in the game a *white pawn* moves from $c, 7$ to $c, 8$ and is promoted to a *rook*. The facts from the context represent the position of the pieces on the board and the pieces removed from the games. Considering the example above, the set of facts would contain $board(10, rook, white, c, 8)$, stating that after the $9th$ move, there is a *white rook* in the position $c, 8$ on the board and $out\_board(10, pawn, white, 0, 0)$, giving the information that a *white pawn* was removed from the board. The board setting is updated according to the legal move(s) performed on the previous round. A move generator procedure is responsible for creating the dataset of simulated games.

Background Knowledge. In the chess revision problem, the initial theory describes the rules of the standard game of chess, which will be revised using the set of examples for its variant. The theory encompass all the possible branches of a move, considering the official rules of chess [1]. For example, *in case* the king is under attack, a piece *must* only move to stop the attack, by *either* capturing the attacking piece *or* moving to the way between the king and the attacking piece. This theory is inspired on the one learned in [8] using hierarchical structured induction

ILP and Progol [9]. There, the BK was composed of 61 clauses and 61 clauses were learned.[2] The major differences between the theory used in the present work and the previous one are (1) the clauses we include to describe castling, en-passant and promotion, since the authors of that work opted to not represent any such special move and (2) the clauses we modified to use constructive negation, to avoid free variables in negated literals.

The FDT contains fundamental rules to the problem of chess in general, such as to calculate the difference between two positions (files or ranks) of pieces on the board, definitions of equality, among others. This knowledge is valid to the standard chess and also to the chess variants. During the revision process, the clauses in FDT are not modified. It is composed of 40 clauses, approximately.

The initial theory describes the rules for achieving a legal move following the rules of chess. The top-level predicate is the same as the target predicate, related to the move of the piece. The sub-concepts such as *simple and double check, pin, attack, promotion* and *basic moves of each piece* are described through intermediate predicates. The pieces, files, ranks and colours used in the game are enumerated by facts. All in all, the initial theory has approximately 100 clauses.

4 Experimental Results

Experimental methodology. To experiment the proposal of the paper, we generated datasets with 5 simulated games where each stage of the game has 1 positive and 5 negative examples and the maximum number of legal moves is 20, for 3 different chess variants. We performed 5-fold cross validation and scored the revisions using f-measure. We proceed with this section by describing the chess variant followed by the automatic revisions performed by the system to obtain its theory. The variants include a smaller version of chess, a version with an unusual rule and a variant with larger board and unusual pieces.

- Using smaller boards: Gardner's Minichess: This is the smallest chess game (5X5) in which all original chess pieces and legal moves are still used, including pawn double move, castling and en-passant [13]. The revisions are as follows.

  1. The delete rule step was able to remove the following clauses from the theory: file(f). file(g). file(h). rank(6). rank(7). rank(8). promotion_zone (pawn,white,7,8).
  2. The add rule generalization operator created the following clause: promotion_zone(pawn,white,4,5) (white pawn promoting in rank 5).

  The final accuracy was 100% and the returned theory matches the rules of Gardner Minichess.

- Unusual rule: Free-capture chess: This variant of chess allows a piece to capture friendly pieces, except for the friendly king, besides the opponent pieces [13].

---

[2] The resulting theory was approved by Professor Donald Michie, who could be considered a chess expert.

1. The chess theory makes sure that a piece when attacking does not land on a friendly piece, requiring that the colours of them both are different. The revision created a new clause by first deleting the predicate requiring the colours are different and then adding a literal restricting the attacked pieces to be different from the king. Note that the original rule is kept on the theory.
2. There is a specific rule in the case of a king attacking, since the restrictions on the king's move must be observed. The revision system deleted the literal requiring the colour of a piece attacked by the king be different from the colour of the king.

The final accuracy was 100% and we can say that the returned theory perfectly corresponds to the target theory.

- Unusual pieces and larger board: Neunerschach: This is a chess variant played on a 9x9 board. There is a piece called as *marschall* replacing the queen and moving like it. The extra piece is the *hausfrau*, which moves as a queen but only two squares [13]. The theory was revised as follows.

1. The delete rule step removed the clause defining the queen as a piece;
2. The abduction procedure included facts defining the *marschall* and *hausfrau* as pieces on the theory;
3. From the rule defining the basic move of the queen, the add rule operator created a rule for the *marschall*.
4. New rules were added to the theory, defining the basic move of *hausfrau* and introducing the facts $file(i)$ and $rank(9)$.

Since in this dataset no promotion moves were generated, due to the size of the board, the revision process failed on correcting the promotion on the last rank. We expect that using games with a larger number of total moves will allow us to represent such a promotion. Nevertheless, the final accuracy was 100%.

5   Conclusions

We presented a framework for applying the knowledge learned to the rules of chess to learn variants of chess through theory revision and a set of generated examples. We described the modifications implemented in the revision system, including the introduction of an initial step for deleting rules, the use of abduction and negation. Three variants of chess were experimented and the system was able to return final theories correctly describing the rules of the variants, except for one case, the promotion in the last rank of a 9X9 board.

In order to decrease the runtime of the revision process we intend to use the stochastic local search algorithms developed in [12]. We would like to try induce the variants of Chess using standard ILP system, such as Progol [9], with the chess theory as BK. Beforehand, it is expected that these systems are not successful in the cases requiring specialization, since they do not perform such operation. If the chess theory is not used as BK, we would not take advantage of the initial knowledge about the domain. Additionally, we want to take a further step towards the acquirement of more complex chess variants, such as the regional chess games *Shogi* and *Xiangqi*.

Acknowledgements

Stephen Muggleton would like to thank the Royal Academy of Engineering and Microsoft Research for funding his Research Chair in Machine Learning. Aline Paes and Gerson Zaverucha are financially supported by Brazilian Research Council (CNPq) and Vítor Santos Costa by Fundação para a Ciência e Tecnologia.

References

1. Burg, D.B., Just, T.: IUS Chess Federation Official Rules of Chess. McKay, New York (1987)
2. Caruana, R.: Multitask Learning. Machine Learning 28(1), 41–75 (1997)
3. Chan, D.: Constructive Negation Based on the Completed Database. In: Proc. of the 5th Int. Conf. and Symp. on Logic Programming, pp. 111–125. The MIT Press, Cambridge (1988)
4. Drabent, W.: What is Failure? An Approach to Constructive Negation. Acta Inf. 32(1), 27–29 (1995)
5. Duboc, A.L., Paes, A., Zaverucha, G.: Using the Bottom Clause and Mode Declarations in FOL Theory Revision from Examples. Machine Learning 76, 73–107 (2009)
6. Flach, P., Kakas, A.: Abduction and Induction: Essays on their Relation and Integration. Kluwer Academic Publishers, Dordrecht (2000)
7. Fürnkranz, J.: Recent Advances in Machine Learning and Game Playing. OGAI-Journal 26(2), 147–161 (2007)
8. Goodacre, J.: Master thesis, Inductive Learning of Chess Rules Using Progol. Programming Research Group, Oxford University (1996)
9. Muggleton, S.: Inverse Entailment and Progol. New Generation Computing 13(3&4), 245–286 (1995)
10. Muggleton, S., Bryant, C.H.: Theory completion using inverse entailment. In: Cussens, J., Frisch, A.M. (eds.) ILP 2000. LNCS (LNAI), vol. 1866, pp. 130–146. Springer, Heidelberg (2000)
11. Muggleton, S., De Raedt, L.: Inductive Logic Programming: Theory and Methods. J. Log. Program. 19/20, 629–679 (1994)
12. Paes, A., Zaverucha, G., Costa, V.S.: Revising FOL Theories from Examples through Stochastic Local Search. In: Blockeel, H., Ramon, J., Shavlik, J., Tadepalli, P. (eds.) ILP 2007. LNCS (LNAI), vol. 4894, pp. 200–210. Springer, Heidelberg (2007)
13. Pritchard, D.B.: The Classified Encyclopedia of Chess Variants. John Beasley (2007)
14. Richards, B.L., Mooney, R.J.: Automated Refinement of First-order Horn-Clause Domain Theories. Machine Learning 19(2), 95–131 (1995)
15. Thrun, S.: Is Learning the nth Thing any Easier than Learning the First? In: Adv. in Neural Inf. Proc. Systems. NIPS, vol. 8, pp. 640–646. MIT Press, Cambridge (1995)
16. Wrobel, S.: First-order theory refinement. In: De Raedt, L. (ed.) Advances in Inductive Logic Programming, pp. 14–33. IOS Press, Amsterdam (1996)

ProGolem: A System Based on Relative Minimal Generalisation

Stephen Muggleton, José Santos, and Alireza Tamaddoni-Nezhad

Department of Computing, Imperial College London
{shm,jcs06,atn}@doc.ic.ac.uk

Abstract. Over the last decade Inductive Logic Programming systems have been dominated by use of top-down refinement search techniques. In this paper we re-examine the use of bottom-up approaches to the construction of logic programs. In particular, we explore variants of Plotkin's Relative Least General Generalisation (RLGG) which are based on subsumption relative to a bottom clause. With Plotkin's RLGG, clause length grows exponentially in the number of examples. By contrast, in the Golem system, the length of ij-determinate RLGG clauses were shown to be polynomially bounded for given values of i and j. However, the determinacy restrictions made Golem inapplicable in many key application areas, including the learning of chemical properties from atom and bond descriptions. In this paper we show that with Asymmetric Relative Minimal Generalisations (or ARMGs) relative to a bottom clause, clause length is bounded by the length of the initial bottom clause. ARMGs, therefore do not need the determinacy restrictions used in Golem. An algorithm is described for constructing ARMGs and this has been implemented in an ILP system called ProGolem which combines bottom-clause construction in Progol with a Golem control strategy which uses ARMG in place of determinate RLGG. ProGolem has been evaluated on several well-known ILP datasets. It is shown that ProGolem has a similar or better predictive accuracy and learning time compared to Golem on two determinate real-world applications where Golem was originally tested. Moreover, ProGolem was also tested on several non-determinate real-world applications where Golem is inapplicable. In these applications, ProGolem and Aleph have comparable times and accuracies. The experimental results also suggest that ProGolem significantly outperforms Aleph in cases where clauses in the target theory are long and complex.

1 Introduction

There are two key tasks at the heart of ILP systems: 1) enumeration of clauses which explain one or more of the positive examples and 2) evaluation of the numbers of positive and negative examples covered by these clauses. Top-down refinement techniques such as those found in [25,22,23], use a generate-and-test approach to problems 1) and 2). A new clause is first generated by application of a refinement step and then tested for coverage of positive and negative examples.

L. De Raedt (Ed.): ILP 2009, LNAI 5989, pp. 131–148, 2010.
© Springer-Verlag Berlin Heidelberg 2010

It has long been appreciated in AI [20] that generate-and-test procedures are less efficient than ones based on test-incorporation. The use of the bottom clause in Progol [15] represents a limited form of test-incorporation in which, by construction, all clauses in a refinement graph search are guaranteed to cover at least the example associated with the bottom clause. The use of Relative Least General Generalisation (RLGG) in Golem [13] provides an extended form of test-incorporation in which constructed clauses are guaranteed to cover a given set of positive examples. However, in order to guarantee polynomial-time construction the form of RLGG in Golem was constrained to ij-determinate clauses. Without this constraint Plotkin [21] showed that the length of RLGG clauses grows exponentially in the number of positive examples covered.

In this paper we explore variants of Plotkin's RLGG which are based on subsumption order relative to a bottom clause [28]. We give a definition for Asymmetric Relative Minimal Generalisation (ARMGs) and show that the length of ARMGs is bounded by the length of the initial bottom clause. Hence, unlike in Golem, we do not need the determinacy restrictions to guarantee polynomial-time construction. However, we show that the resulting ARMG is not unique and that the operation is asymmetric. ARMGs can easily be extended to the multiple example case by iteration. We describe an ILP system called ProGolem which combines bottom-clause construction in Progol with a Golem control strategy which uses ARMG in place of determinate RLGG. The use of top-down ILP algorithms such as Progol, tends to limit the maximum complexity of learned clauses, due to a search bias which favours simplicity. Long clauses generally require an overwhelming amount of search for systems like Progol and Aleph [27]. In this paper we also explore whether ProGolem will have any advantages in situations when the clauses in the target theory are long and complex.

ProGolem has been evaluated on several well-known ILP datasets. These include two determinate real-world applications where Golem was originally tested and several non-determinate real-world applications where Golem is inapplicable. ProGolem has also been evaluated on a set of artificially generated learning problems with large concept sizes.

The paper is arranged as follows. In Section 2 we review some of basic concepts from the ILP systems Golem and Progol which are used in the definitions and theorems in this paper. In Section 3 we discuss subsumption relative to a bottom clause. ARMG is introduced in Section 4 and some of its properties are demonstrated. An algorithm for ARMG is given in Section 5. This algorithm is implemented in the ILP system ProGolem which is described in Section 6. Empirical evaluation of ProGolem on several datasets is given in Section 7. Related work is discussed in Section 8. Section 9 concludes the paper.

2   Preliminaries

We assume the reader to be familiar with the basic concepts from logic programming and inductive logic programming [19]. This section is intended as a brief reminder of some of the concepts from the ILP systems Golem [13] and Progol [15] which are the basis for the system ProGolem described in this paper.

The general subsumption order on clauses, also known as θ-subsumption, is defined in the following.

Definition 1 (Subsumption). *Let C and D be clauses. We say C subsumes D, denoted by $C \succeq D$, if there exists a substitution θ such that $C\theta$ is a subset of D. C properly subsumes D, denoted by $C \succ D$, if $C \succeq D$ and $D \not\succeq C$. C and D are subsume-equivalent, denoted by $C \sim D$, if $C \succeq D$ and $D \succeq C$.*

Proposition 1 (Subsumption lattice). *Let $\mathcal{C}$ be a clausal language and $\succeq$ be the subsumption order as defined in Definition 1. Then the equivalence classes of clauses in $\mathcal{C}$ and the $\succeq$ order define a lattice. Every pair of clauses C and D in the subsumption lattice have a least upper bound called* least general generalisation (lgg), *denoted by $lgg(C, D)$ and a greatest lower bound called* most general specialisation (mgs), *denoted by $mgs(C, D)$.*

Plotkin [21] investigated the problem of finding the least general generalisation (lgg) for clauses ordered by subsumption. The notion of lgg is important for ILP since it forms the basis of generalisation algorithms which perform a bottom-up search of the subsumption lattice. Plotkin also defined the notion of relative least general generalisation of clauses ($rlgg$) which is the lgg of the clauses relative to clausal background knowledge B. The cardinality of the lgg of two clauses is bounded by the product of the cardinalities of the two clauses. However, the $rlgg$ is potentially infinite for arbitrary B. When B consists of ground unit clauses only the $rlgg$ of two clauses is finite. However the cardinality of the $rlgg$ of m clauses relative to n ground unit clauses has worst-case cardinality of order $O(n^m)$, making the construction of such $rlgg$'s intractable.

The ILP system Golem [13] is based on Plotkin's notion of relative least general generalisation of clauses ($rlgg$). Golem uses extensional background knowledge to avoid the problem of non-finite $rlggs$. Extensional background knowledge B can be generated from intensional background knowledge B' by generating all ground unit clauses derivable from B' in at most h resolution steps. The parameter h is provided by the user. The $rlggs$ constructed by Golem were forced to have only a tractable number of literals by requiring the ij-determinacy.

An ij-determinate clause is defined as follows.

Definition 2 (ij-determinate clause). *Every unit clause is $0j$-determinate. An ordered clause $h \leftarrow b_1, .., b_m, b_{m+1}, .., b_n$ is ij-determinate if and only if a) $h \leftarrow b_1, .., b_m$ is $(i-1)j$-determinate, b) every literal b_k in $b_{m+1}, .., b_n$ contains only determinate terms and has arity at most j.*

The ij-determinacy is equivalent to requiring that predicates in the background knowledge must represent functions. This condition is not met in many real-world applications, including the learning of chemical properties from atom and bond descriptions.

One of the motivations of the ILP system Progol [15] was to overcome the determinacy limitation of Golem. Progol extends the idea of inverting resolution proofs used in the systems Duce [14] and Cigol [16] and uses the general case of Inverse Entailment which is based on the model-theory which underlies proof.

Progol uses Mode-Directed Inverse Entailment (MDIE) to develop a most specific clause $\perp$ for each positive example, within the user-defined mode language, and uses this to guide an A^*-like search through clauses which subsume $\perp$.

The Progol algorithm is based on successive construction of definite clause hypotheses H from a language $\mathcal{L}$. H must explain the examples E in terms of background knowledge B. Each clause in H is found by choosing an uncovered positive example e and searching through the graph defined by the refinement ordering $\succeq$ bounded below by a bottom clause $\perp$ associated with e. In general $\perp$ can have infinite cardinality. Progol uses mode declarations to constrain the search for clauses which subsume $\perp$. Progol's mode declaration (M), definite mode language $(\mathcal{L}(M))$ and depth-bounded mode language $(\mathcal{L}_i(M))$ are defined in Appendix A.

Progol searches a bounded sub-lattice for each example e relative to background knowledge B and mode declarations M. The sub-lattice has a most general element which is the empty clause, $\square$, and a least general element $\perp$ which is the most specific element in $\mathcal{L}_i(M)$ such that $B \wedge \perp \wedge \overline{e} \vdash_h \square$ where $\vdash_h \square$ denotes derivation of the empty clause in at most h resolutions. The following definition describes a bottom clause for a depth-bounded mode language $\mathcal{L}_i(M)$.

Definition 3 (Most-specific clause or bottom clause $\perp_e$). *Let h and i be natural numbers, B be a set of Horn clauses, E be a set of positive and negative examples with the same predicate symbol a, e be a positive example in E, M be a set of mode declarations, as defined in Definitions 13, containing exactly one modeh m such that $a(m) \succeq a$, $\mathcal{L}_i(M)$ be a depth-bounded mode language as defined in Definitions 16 and $\hat{\perp}_e$ be the most-specific (potentially infinite) definite clause such that $B \wedge \hat{\perp}_e \wedge \overline{e} \vdash_h \square$. $\perp_e$ is the most-specific clause in $\mathcal{L}_i(M)$ such that $\perp_e \succeq \hat{\perp}_e$. C is the most-specific clause in $\mathcal{L}$ if for all C' in $\mathcal{L}$ we have $C' \succeq C$. $\overrightarrow{\perp}_e$ is $\perp_e$ with a defined ordering over the literals.*

In this paper, we refer to $\perp_e$ as $\overrightarrow{\perp}$ or $\perp$ depending on whether we use the ordering of the literals or not. Progol's algorithm for constructing the bottom clause is given in [15]. The Proposition below follows from Theorem 26 in [15].

Proposition 2. *Let $\perp_e$ be as defined in Definition 3, M be a set of mode declarations as defined in Definitions 13, $\mathcal{L}_i(M)$ be a depth-bounded mode language as defined in Definitions 16, i the maximum variable depth in $\mathcal{L}_i(M)$ and j be the maximum arity of any predicate in M. Then the length of $\perp_e$ is polynomially bounded in the number of mode declarations in M for fixed values of i and j.*

3   Subsumption Relative to a Bottom Clause

In a previous paper [28] we introduced a subsumption order relative to a bottom clause and demonstrated how clause refinement in a Progol-like ILP system can be characterised with respect to this order. It was shown that, unlike for the general subsumption order, efficient least general generalisation operators can be designed for subsumption order relative to a bottom clause (i.e. $lgg_\perp$). In this section we briefly review the notion of subsumption order relative to bottom clause which is essential for the definition of Asymmetric Relative Minimal Generalisations (ARMGs) in this paper.

Clauses which are considered by Progol, i.e. clauses in $\mathcal{L}(M)$ (Definition 14), as well as determinate clauses considered by Golem (Definition 2), are defined with a total ordering over the literals. Moreover, the subsumption order which characterises clause refinement in a Progol-like ILP system is defined on ordered clauses. In the following we adopt an explicit representation for ordered clauses. We use the same notion used in [19] and an ordered clause is represented as a disjunction of literals (i.e. $L_1 \vee L_2 \vee \ldots \vee L_n$). The set notation (i.e. $\{L_1, L_2, \ldots, L_n\}$) is used to represent conventional clauses.

Definition 4 (Ordered clause). *An ordered clause $\overrightarrow{C}$ is a sequence of literals $L_1, L_2, \ldots, L_n$ and denoted by $\overrightarrow{C} = L_1 \vee L_2 \vee \ldots \vee L_n$. The set of literals in $\overrightarrow{C}$ is denoted by C.*

Unlike conventional clauses, the order and duplication of literals matter for ordered clauses. For example, $\overrightarrow{C} = p(X) \vee \neg q(X)$, $\overrightarrow{D} = \neg q(X) \vee p(X)$ and $\overrightarrow{E} = p(X) \vee \neg q(X) \vee p(X)$ are different ordered clauses while they all correspond to the same conventional clause, i.e. $C = D = E = \{p(X), \neg q(X)\}$.

Selection of two clauses is defined as a pair of compatible literals and this concept was used by Plotkin to define least generalisation for clauses [21]. Here we use selections to define mappings of literals between two ordered clauses.

Definition 5 (Selection and selection function). *Let $\overrightarrow{C} = L_1 \vee L_2 \vee \ldots \vee L_n$ and $\overrightarrow{D} = M_1 \vee M_2 \vee \ldots \vee M_m$ be ordered clauses. A selection of $\overrightarrow{C}$ and $\overrightarrow{D}$ is a pair (i, j) where L_i and M_j are compatible literals, i.e. they have the same sign and predicate symbol. A set s of selections of $\overrightarrow{C}$ and $\overrightarrow{D}$ is called a selection function if it is a total function of $\{1, 2, \ldots, n\}$ into $\{1, 2, \ldots, m\}$.*

Definition 6 (Subsequence). *Let $\overrightarrow{C} = L_1 \vee L_2 \vee \ldots \vee L_l$ and $\overrightarrow{D} = M_1 \vee M_2 \vee \ldots \vee M_m$ be ordered clauses. $\overrightarrow{C}$ is a subsequence of $\overrightarrow{D}$, denoted by $\overrightarrow{C} \sqsubseteq \overrightarrow{D}$, if there exists a strictly increasing selection function $s \subseteq \{1, \ldots, l\} \times \{1, \ldots, m\}$ such that for each $(i, j) \in s$, $L_i = M_j$.*

Example 1. Let $\overrightarrow{B} = p(x, y) \vee q(x, y) \vee r(x, y) \vee r(y, x)$, $\overrightarrow{C} = p(x, y) \vee r(x, y) \vee r(y, x)$ and $\overrightarrow{D} = p(x, y) \vee r(y, x) \vee r(x, y)$ be ordered clauses. $\overrightarrow{C}$ is a subsequence of $\overrightarrow{B}$ because there exists increasing selection function $s_1 = \{(1, 1), (2, 3), (3, 4)\}$ which maps literals from $\overrightarrow{C}$ to equivalent literals from $\overrightarrow{D}$. However, $\overrightarrow{D}$ is not a subsequence of $\overrightarrow{B}$ because an increasing selection function does not exist for $\overrightarrow{D}$ and $\overrightarrow{B}$.   ◇

As shown in [28], clause refinement in Progol-like ILP systems cannot be described by the general subsumption order. However, subsumption order relative to $\perp$ (i.e. $\succeq_\perp$) can capture clause refinement in these systems. In the following we first define $\overrightarrow{\mathcal{L}}_\perp^s$ which can be used to represent the hypotheses language of a Progol-like ILP system.

Definition 7 ($\overrightarrow{\mathcal{L}}^s_\perp$). *Let $\overrightarrow{\perp}$ be the bottom clause as defined in Definition 3 and $\overrightarrow{C}$ a definite ordered clause. $\overrightarrow{\top}$ is $\overrightarrow{\perp}$ with all variables replaced with new and distinct variables. $\theta_\top$ is a variable substitution such that $\overrightarrow{\top}\theta_\top = \overrightarrow{\perp}$. $\overrightarrow{C}$ is in $\overrightarrow{\mathcal{L}}^s_\perp$ if $\overrightarrow{C}\theta_\top$ is a subsequence of $\overrightarrow{\perp}$.*

Example 2. Let $\overrightarrow{\perp} = p(X) \leftarrow q(X), r(X), s(X,Y), s(Y,X)$ and according to Definition 7, we have $\overrightarrow{\top} = p(V_1) \leftarrow q(V_2), r(V_3), s(V_4, V_5), s(V_6, V_7)$ and $\theta_\top = \{V_1/X, V_2/X, V_3/X, V_4/X, V_5/Y, V_6/Y, V_7/X\}$. Then $\overrightarrow{C} = p(V_1) \leftarrow r(V_2), s(V_6, V_7)$, $\overrightarrow{D} = p(V_1) \leftarrow r(V_1), s(V_6, V_1)$ and $\overrightarrow{E} = p(V_1) \leftarrow r(V_1), s(V_4, V_5)$ are in $\overrightarrow{\mathcal{L}}^s_\perp$ as $\overrightarrow{C}\theta_\top$, $\overrightarrow{D}\theta_\top$ and $\overrightarrow{E}\theta_\top$ are subsequences of $\overrightarrow{\perp}$.   ◇

Definition 8 (Subsumption relative to $\perp$). *Let $\overrightarrow{\perp}$, $\theta_\top$ and $\overrightarrow{\mathcal{L}}^s_\perp$ be as defined in Definition 7 and $\overrightarrow{C}$ and $\overrightarrow{D}$ be ordered clauses in $\overrightarrow{\mathcal{L}}^s_\perp$. We say $\overrightarrow{C}$ subsumes $\overrightarrow{D}$ relative to $\perp$, denoted by $\overrightarrow{C} \succeq_\perp \overrightarrow{D}$, if $\overrightarrow{C}\theta_\top$ is a subsequence of $\overrightarrow{D}\theta_\top$. $\overrightarrow{C}$ is a proper generalisation of $\overrightarrow{D}$ relative to $\perp$, denoted by $\overrightarrow{C} \succ_\perp \overrightarrow{D}$, if $\overrightarrow{C} \succeq_\perp \overrightarrow{D}$ and $\overrightarrow{D} \not\succeq_\perp \overrightarrow{C}$. $\overrightarrow{C}$ and $\overrightarrow{D}$ are equivalent with respect to subsumption relative to $\perp$, denoted by $\overrightarrow{C} \sim_\perp \overrightarrow{D}$, if $\overrightarrow{C} \succeq_\perp \overrightarrow{D}$ and $\overrightarrow{D} \succeq_\perp \overrightarrow{C}$.*

Example 3. Let $\overrightarrow{\perp}$, $\theta_\top$, $\overrightarrow{\mathcal{L}}^s_\perp$, $\overrightarrow{C}$, $\overrightarrow{D}$ and $\overrightarrow{E}$ be as in Example 2. Then, $\overrightarrow{C}$ subsumes $\overrightarrow{D}$ relative to $\perp$ since $\overrightarrow{C}\theta_\top$ is a subsequence of $\overrightarrow{D}\theta_\top$. However, $\overrightarrow{C}$ does not subsume $\overrightarrow{E}$ relative to $\perp$ since $\overrightarrow{C}\theta_\top$ is not a subsequence of $\overrightarrow{E}\theta_\top$. Note that $\overrightarrow{C}$ subsumes $\overrightarrow{E}$ with respect to normal subsumption.   ◇

The following Proposition is a special case of Lemma 5 in [28] and follows directly from Definition 8.

Proposition 3. *Let $\overrightarrow{\perp}$ be as defined in Definition 3 and $\overrightarrow{C}$ be an ordered clause obtained from $\overrightarrow{\perp}$ by removing some literals without changing the order of the remaining literals. Then, $\overrightarrow{C} \succ_\perp \overrightarrow{\perp}$.*

The subsumption order relative to $\perp$ was studied in [28]. It was shown that the refinement space of a Progol-like ILP system can be characterised using $\langle \overrightarrow{\mathcal{L}}_\perp, \succeq_\perp \rangle$. It was also shown that $\langle \overrightarrow{\mathcal{L}}_\perp, \succeq_\perp \rangle$ is a lattice which is isomorphic to an atomic lattice and that the most general specialisation relative to $\perp$ ($mgs_\perp$) and the least general generalisation relative to $\perp$ ($lgg_\perp$) can be defined based on the most general specialisation and the least general generalisation for atoms.

4   Asymmetric Relative Minimal Generalisations

The construction of the least general generalisation (lgg) of clauses in the general subsumption order is inefficient as the cardinality of the lgg of two clauses can grow very rapidly (see Section 2). For example, with Plotkin's Relative Least General Generalisation (RLGG), clause length grows exponentially in the number of examples [21]. Hence, an ILP system like Golem [13] which uses RLGG is constrained to ij-determinacy to guarantee polynomial-time construction. However, the determinacy restrictions make an ILP system inapplicable in many key application areas, including the learning of chemical properties from atom and bond descriptions. On the other hand, as shown in [28], efficient operators can be implemented for least generalisation and greatest specialisation in the subsumption order relative to a bottom clause. In this section we define a variant of Plotkin's RLGG which is based on subsumption order relative to a bottom clause and does not need the determinacy restrictions. The relative least general generalisation ($lgg_\perp$) in [28] is defined for a lattice bounded by a bottom clause $\perp_e$. This bottom clause is constructed with respect to a single positive example e and as in Progol we need a search guided by coverage testing to explore the hypotheses space. However, the asymmetric relative minimal generalisation (ARMG) described in this paper is based on pairs of positive examples and as in Golem, by construction it is guaranteed to cover all positive examples which are used to construct it. Hence, ARMGs have the same advantage as RLGGs in Golem but unlike RLGGs the length of ARMGs is bounded by the length of $\perp_e$. The asymmetric relative minimal generalisation of examples e' and e relative to $\perp_e$ is denoted by $armg_\perp(e'|e)$ and in general $armg_\perp(e'|e) \neq armg_\perp(e|e')$. In the following we define asymmetric relative minimal generalisation and study some of their properties. It is normal in ILP to restrict attention to clausal hypotheses which are "head-connected" in the following sense.

Definition 9 (Head-connectness). *A definite ordered clause $h \leftarrow b_1, .., b_n$ is said to be head-connected if and only if each body atom b_i contains at least one variable found either in h or in a body atom b_j, where $1 \leq j < i$.*

Definition 10 (Asymmetric relative common generalisation). *Let E, B and $\perp_e$ be as defined in Definition 3, e and e' be positive examples in E and $\overrightarrow{C}$ is a head-connected definite ordered clause in $\overrightarrow{\mathcal{L}}^s_\perp$. $\overrightarrow{C}$ is an asymmetric common generalisation of e' and e relative to $\perp_e$, denoted by $\overrightarrow{C} \in arcg_\perp(e'|e)$, if $\overrightarrow{C} \succeq_\perp \perp_e$ and $B \wedge C \vdash e'$.*

Example 4. Let $M = \{p(+), q(+, -), r(+, -)\}$ be mode definition, $B = \{q(a, a), r(a, a), q(b, b), q(b, c), r(c, d)\}$ be background knowledge and $e = p(a)$ and $e' = p(b)$ be positive examples. Then we have $\perp_e = p(X) \leftarrow q(X, X), r(X, X)$ and clauses $\overrightarrow{C} = p(V_1) \leftarrow q(V_1, V_1)$, $\overrightarrow{D} = p(V_1) \leftarrow q(V_1, V_3), r(V_3, V_5)$ and $\overrightarrow{E} = p(V_1) \leftarrow q(V_1, V_3)$ are all in $arcg_\perp(e'|e)$.   ◇

Definition 11 (Asymmetric relative minimal generalisation). *Let E and $\perp_e$ be as defined in Definition 3, e and e' be positive examples in E and $arcg_\perp(e'|e)$ be as defined in Definition 10. $\overrightarrow{C}$ is an asymmetric minimal generalisation of e' and e relative to $\perp_e$, denoted by $\overrightarrow{C} \in armg_\perp(e'|e)$, if $\overrightarrow{C} \in arcg_\perp(e'|e)$ and $\overrightarrow{C} \succeq_\perp \overrightarrow{C'} \in arcg_\perp(e'|e)$ implies $\overrightarrow{C}$ is subsumption-equivalent to $\overrightarrow{C'}$ relative to $\perp_e$.*

Example 5. Let B, $\perp_e$, e and e' be as in Example 4. Then clauses $\overrightarrow{C} = p(V_1) \leftarrow q(V_1, V_1)$ and $\overrightarrow{D} = p(V_1) \leftarrow q(V_1, V_3), r(V_3, V_5)$ are both in $armg_\perp(e'|e)$.   ◇

This example shows that ARMGs are not unique.

Theorem 1. *The set $armg_\perp(e'|e)$ can contain more than one clause which are not subsumption-equivalent relative to $\perp_e$.*

Proof. In Example 4, clauses $\overrightarrow{C} = p(V_1) \leftarrow q(V_1, V_1)$ and $\overrightarrow{D} = p(V_1) \leftarrow q(V_1, V_3)$, $r(V_3, V_5)$ are both in $armg_\perp(e'|e)$ but not subsumption-equivalent relative to $\perp_e$. $\square$

The following theorem shows that the length of ARMG is bounded by the length of $\perp_e$.

Theorem 2. *For each $\overrightarrow{C} \in armg_\perp(e'|e)$ the length of $\overrightarrow{C}$ is bounded by the length of $\perp_e$.*

Proof. Let $\overrightarrow{C} \in armg_\perp(e'|e)$. Then by definition $\overrightarrow{C} \succeq_\perp \perp_e$ and according to Definition 8, $\overrightarrow{C}$ is a subsequence of $\perp_e$. Hence, the length of $\overrightarrow{C}$ is bounded by the length of $\perp_e$. $\square$

It follows from Theorem 2 that the number of literals in an ARMG is bounded by the length of $\perp_e$, which according to Proposition 2 is polynomially bounded in the number of mode declarations for fixed values of i and j, where i is the maximum variable depth and j is the maximum arity of any predicate in M. Hence, unlike the RLGGs used in Golem, ARMGs do not need the determinacy restrictions and can be used in a wider range of problems including those which are non-determinate. In Section 7 we apply ARMGs to a range of determinate and non-determinate problems and compare the results with Golem and Aleph. But first we give an algorithm for constructing ARMGs in Section 5 and describe an implementation of ARMGs in Section 6.

5   Algorithm for ARMGs

It was shown in the previous section that ARMGs do not have the limitations of RLGGs and that the length of ARMGs is bounded by the length of $\perp_e$. In this section we show that there is also an efficient algorithm for constructing ARMGs. The following definitions are used to describe the ARMG algorithm.

Definition 12 (Blocking atom). *Let B be background knowledge, E^+ the set of positive examples, $e \in E^+$ and $\overrightarrow{C} = h \leftarrow b_1, \ldots, b_n$ be a definite ordered clause. b_i is a blocking atom if and only if i is the least value such that for all θ, $e = h\theta, B \nvdash (b_1, \ldots, b_i)\theta$.*

An algorithm for constructing ARMGs is given in Figure 1. Given the bottom clause $\perp_e$ associated with a particular positive example e, this algorithm works by dropping a minimal set of atoms from the body to allow coverage of a second example. Below we prove the correctness of the ARMG algorithm.

Theorem 3 (Correctness of ARMG algorithm). *Let E and $\perp_e$ be as defined in Definition 3, e and e' be positive examples in E, $armg_\perp(e'|e)$ be as defined in Definition 11 and $ARMG(\perp_e, e')$ as given in Figure 1. Then $\overrightarrow{C} = ARMG(\perp_e, e')$ is in $armg_\perp(e'|e)$.*

Asymmetric Relative Minimal Generalization (ARMG) Algorithm
Input: Bottom clause $\bot_e$, Positive example e'
$\overrightarrow{C}$ is $\bot_e = h \leftarrow b_1, .., b_n$
While there is a blocking atom b_i wrt e' in the body of $\overrightarrow{C}$
Remove b_i from $\overrightarrow{C}$
Remove atoms from $\overrightarrow{C}$ which are not head-connected
Repeat
Output: $\overrightarrow{C}$

Fig. 1. ARMG algorithm

Proof. Assume $\overrightarrow{C} \notin armg_{\bot}(e'|e)$. In this case, either $\overrightarrow{C}$ is not an asymmetric common generalisation of e and e' or it is not minimal. However, by construction $\overrightarrow{C}$ is a subsequence of $\bot_e$ in which all blocking literals with respect to e' are removed. Then according to Proposition 3, $\overrightarrow{C} \succeq_{\bot} \bot_e$ and by construction $B \wedge C \vdash e'$. Hence, $\overrightarrow{C}$ is an asymmetric common generalisation of e and e'. So, $\overrightarrow{C}$ must be non-minimal. If $\overrightarrow{C}$ is non-minimal then $\overrightarrow{C} \succ_{\bot} \overrightarrow{C'}$ for $\overrightarrow{C'} \in armg_{\bot}(e'|e)$ which must either have literals not found in $\overrightarrow{C}$ or there is a substitution θ such that $\overrightarrow{C}\theta = \overrightarrow{C'}$. But we have deleted the minimal set of literals. This is a minimal set since leaving a blocking atom would mean $B \wedge C \nvdash e'$ and leaving a non-head-connected literal means $\overrightarrow{C} \notin armg_{\bot}(e'|e)$. So it must be the second case. However, in the second case θ must be a renaming since the literals in $\overrightarrow{C}$ are all from $\bot_e$. Hence, $\overrightarrow{C}$ and $\overrightarrow{C'}$ are variants which contradicts the assumption and completes the proof. □

The following example shows that the ARMGs algorithm is not complete.

Example 6. Let B, $\bot_e$, e and e' be as in Example 4. Then clauses $\overrightarrow{C} = p(V_1) \leftarrow q(V_1, V_1)$ and $\overrightarrow{D} = p(V_1) \leftarrow q(V_1, V_3), r(V_3, V_5)$ are both in $armg_{\bot}(e'|e)$. However, the ARMGs algorithm given in Figure 1 cannot generate clause $\overrightarrow{D}$. ◊

Example 6 shows that the ARMGs algorithm does not consider hypotheses which require 'variable splitting'. As shown in [28] (Example 2), there are some group of problems which cannot be learned by a Progol-like ILP system without variable splitting. The concept of variable splitting and the ways it has been done in Progol and Aleph were discussed in [28]. Similar approaches could be adopted for ProGolem, however, the current implementation does not support variable splitting.

Figure 2 gives a comparison between Golem's determinate RLGG and the ARMGs generated by the ARMG algorithm on Michalski's trains dataset from [12]. Note that Golem's RLGG cannot handle the predicate *has_car* because it is non-determinate. The first ARMG (2) subsumes the target concept which is eastbound(A) ← has_car(A,B), closed(B), short(B). Note that in this example RLGG (1) is shorter than ARMGs (2,3) since it only contains determinant literals.

1. $RLGG(e_1, e_2)$ = $RLGG(e_2, e_1)$ = eastbound(A) ← infront(A,B), short(B), open(B), shape(B,C), load(B,triangle,1), wheels(B,2), infront(B,D), shape(D, rectangle), load(D,E,1), wheels(D,F), infront(D,G), closed(G), short(G), shape(G,H), load(G,I,1), wheels(G,2).
2. $ARMG(\perp_{e_1}, e_2)$ = eastbound(A) ← has_car(A,B), has_car(A,C), has_car(A,D), has_car(A,E), infront(A,E), closed(C), short(B), short(C), short(D), short(E), open(D), open(E), shape(B,F), shape(C,G), shape(D,F), shape(E,H), load(D,I,J),2), wheels(E,2)
3. $ARMG(\perp_{e_2}, e_1)$ = eastbound(A) ← has_car(A,B), has_car(A,C), has_car(A,D), infront(A,D), closed(C), short(B), short(D), open(D), shape(B,E), shape(D,E), load(B,F,G), load(D,H,G), wheels(B,2), wheels(D,2)

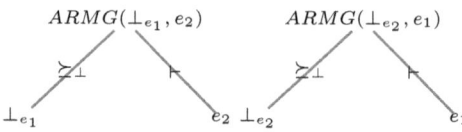

Fig. 2. A comparison between Golem's determinate RLGG (1) and the non-determinate ARMGs (2,3). Note that Golem's RLGG cannot handle the predicate *has_car* because it is non-determinate. The first ARMG (2) subsumes the target concept which is eastbound(A) ← has_car(A,B), closed(B), short(B).

6   Implementation

This section describes ProGolem's implementation. As in Golem and Progol, ProGolem uses the cover set approach to construct a theory consisting of more than one clause. ProGolem's cover set algorithm is shown in Fig. 3. This algorithm repeatedly constructs a clause from a set of best ARMGs, uses negative examples to reduce the clause (see below), adds this clause to the theory and removes the positive examples which are covered.

As in Golem, ProGolem uses negative-based reduction to generalise clauses with respect to negative examples. This algorithm is described in section 6.1. ProGolem uses a greedy beam search to select the best ARMG with respect to $\perp_e$. This algorithm is shown in Fig. 4. The basic idea is to repeatedly extend ARMGs using positive examples and keep the best ARMGs at each iteration to

ProGolem's Cover Set Algorithm
Input: Examples E, mode declarations M, background knowledge B

Let $T = \{\}$
Let S = all positive examples in E
While $S \neq \{\}$
    Let e be the first example in S
    Construct the bottom clause $\perp_e$ from e, M and B ([15])
    Let $\vec{C}$ = Best_ARMG($\perp_e$, E) (see Fig. 4)
    Let $\vec{C'}$ = Negative_based_reduction($\vec{C}$, E) (see section 6.1)
    $T = T \cup \vec{C'}$
    Let S' = all examples from S which are covered by $\vec{C'}$
    $S = S - S'$
Repeat
Output: T

Fig. 3. ProGolem's cover set algorithm

Best ARMG Algorithm
    Input: $\perp_e$, Examples E
    sample size K, beam width N

    Let $best\_armgs = \{\perp_e\}$
    Repeat
      Let $best\_score =$ highest score from $best\_armgs$
      Let $Ex = K$ random positive examples from E
      Let $new\_armgs = \{\}$
      for each $\overrightarrow{C} \in best\_armgs$ do
        for each $e' \in Ex$ do
          Let $\overrightarrow{C'} = \mathrm{ARMG}(\overrightarrow{C}, e')$ (see Fig. 1)
          if $score(\overrightarrow{C'}) > best\_score$ then
            $new\_armgs = new\_armgs \cup \overrightarrow{C'}$
          end if
        end for
      end for
      if $(new\_armgs \neq \{\})$ then
        $best\_armgs =$ highest scoring N clauses from new_armgs
    Until $new\_armgs = \{\}$
    Output: highest scoring clause from $best\_armgs$

Fig. 4. Best ARMG algorithm

be extended in the next iteration until the ARMGs' score no longer increases. The score of an ARMG is computed in the same way a normal clause is evaluated in ProGolem. The evaluation function can be selected by the user (e.g. compression, accuracy, precision, coverage). By default it is compression, that is, the positives covered minus negatives covered minus length of the clause (i.e. its number of literals). At each iteration and for each ARMG in the set of ARMGs under consideration, K examples which are not covered by the current ARMG are selected and used to extend it. The best N (beam width) ARMGs of each iteration are selected to be used as the initial set for the next iteration. The initial set of ARMGs at iteration 0 is the bottom clause $\perp_e$. K and N are user defined parameters with default values of 10 and 2 respectively.

ProGolem is a bottom-up ILP system and unlike in a top-down system such as Progol, the intermediate clauses considered may be very long. The coverage testing of long non-determinate clauses in ProGolem could be very inefficient as it involves a large number of backtracking. Note that clauses considered by Golem are also relatively long but these clauses are determinate which makes the subsumption testing less expensive. In order to address this problem, efficient subsumption testing algorithms are implemented in ProGolem. The following sections describe the negative-based clause reduction and the efficient coverage testing algorithms.

6.1   Negative-Based Clause Reduction

ProGolem implements a negative-based clause reduction algorithm which is similar to the reduction algorithms used in QG/GA [18] and Golem [13]. The aim of negative-based reduction is to generalise a clause by keeping only literals which block negative examples from being proved. The negative-based reduction algorithm works as follows. Given a clause $h \leftarrow b_1, \ldots, b_n$, find the first literal, b_i such that the clause $h \leftarrow b_1, \ldots, b_i$ covers no negative examples. Prune all literals after b_i and move b_i and all its supporting literals to the front, yielding a clause $h \leftarrow S_i, b_i, T_i$, where S_i is a set of supporting literals needed to introduce the input variables of b_i and T_i is $b_1, \ldots, b_{i-1}$ with S_i removed. Then reduce this new clause in the same manner and iterate until the clause length remains the same within a cycle.

6.2   Efficient Coverage Testing

The intermediate clauses considered by ProGolem can be non-determinate and very long. Prolog's standard left-to-right depth-first literal evaluation is extremely inefficient for testing the coverage of such clauses. An efficient algorithm for testing the coverage of long non-determinant clauses is implemented in Pro-Golem. This algorithm works by selecting at each moment the literal which has fewest solutions, from the ones which had their input variables instantiated. This algorithm was further improved by an approach inspired by constraint satisfaction algorithms of [11] and [9]. The core idea is to enumerate variables (rather than literals as before) based on the ones which have the smallest domains. The domain of a variable is the intersection of all the values a variable can assume in the literals it appears. This works well because normally clauses have much fewer distinct variables than literals.

7   Empirical Evaluation

In this section we evaluate ProGolem on several well-known determinate and non-determinate ILP datasets and compare the results with Golem and Aleph. Aleph [27] is a well-known ILP system which works in different modes and can emulate the functionality of several other ILP systems including Progol. Pro-Golem and Aleph are both implemented in YAP Prolog which makes the time comparison between them more accurate.

The materials to reproduce the experiments in this section, including datasets and programs are available from http://ilp.doc.ic.ac.uk/ProGolem/.

7.1   Experiment 1 – Determinate and Non-determinate Applications

Materials and Methods. Several well-known ILP datasets have been used: Proteins [17], Pyrimidines [6], DSSTox [24], Carcinogenesis [26], Metabolism [5] and Alzheimers-Amine [7]. The two determinate datasets, Proteins and Pyrimidines, were used with a hold out test strategy. The data split between training and test sets was done by considering 4/5 for Proteins and 2/3 for Pyrimidines as training data and the remaining for test. For the Carcinogenesis, Metabolism and Alzheimers-Amine datasets a 10-fold cross-validation was performed and for DSSTox it was a 5-fold cross validation. Whenever cross-validation was used the accuracy's standard deviation over all the folds is also reported. Both Aleph and ProGolem were executed in YAP Prolog version 6 with $i = 2$,

Table 1. Predictive accuracies and learning times for *Golem*, *ProGolem* and *Aleph* on different datasets. Golem can only be applied on determinate datasets, i.e. Proteins and Pyrimidines.

dataset	Golem		ProGolem		Aleph	
	$A(\%)$	$T(s)$	$A(\%)$	$T(s)$	$A(\%)$	$T(s)$
Alz-Amine	N/A	N/A	76.1±4.4	36	76.2±3.8	162
Carcino	N/A	N/A	63.0±7.2	649	59.7±6.3	58
DSSTox	N/A	N/A	68.6±4.5	993	72.6±6.9	239
Metabolism	N/A	N/A	63.9±11.6	691	62.1±6.2	32
Proteins	62.3	3568	62.3	2349	50.5	4502
Pyrimidines	72.1	68	75.3	19	73.7	23

$maxneg = 10$ (except for Carcinogenesis and Proteins where $maxneg = 30$) and $evalfn = compression$ (except for DSSTox where $evalfn = coverage$). Aleph was executed with $nodes = 1000$ and $clauselength = 5$ (except for Proteins where $nodes = 10000$ and $clauselength = 40$). ProGolem was executed with $N = 2$ (beam-width) and $K = 5$ (sample size at each iteration). ProGolem's coverage testing was Prolog's standard left-to-right strategy on all these datasets (the same as Aleph). All experiments were performed on a 2.2 Ghz dual core AMD Opteron processor (275) with 8gb RAM.

Results and discussion. Table 1 compares predictive accuracies and average learning times for Golem, ProGolem and Aleph. ProGolem is competitive with Golem on the two determinate datasets. On the Proteins dataset which requires learning long target concepts, Aleph cannot generate any compressive hypothesis and is slower. This is the type of problems where a bottom-up ILP system has an advantage over a top-down one. Golem is inapplicable on the remaining non-determinate problems and ProGolem and Aleph have comparable times and accuracies.

Fig. 5 compares the length and positive coverage of ARMGs in ProGolem. In Fig. 5.(a) the ARMG length (as a fraction of the bottom clause size) is plotted against the number of examples used to construct the ARMG. In Fig. 5.(b) the ARMG positive coverage is plotted against the same X axis. For number of examples equal to 1, the ARMG (i.e. bottom clause) coverage is almost invariably the example which has been used to construct the bottom clause and has the maximum length. The coverage increases with the number of examples used to construct the ARMGs. The ARMGs clause lengths follow an exponential decay and, symmetrically, the positive coverage has an exponential growth since shorter clauses are more general.

7.2   Experiment 2 – Complex Artificial Target Concepts

The results of Experiment 1 suggested that for the Proteins dataset which requires learning long clauses, the performance of Aleph is significantly worse than Golem and ProGolem. In this experiment we further examine whether ProGolem will have any advantages in situations when the clauses in the target theory are long and complex.

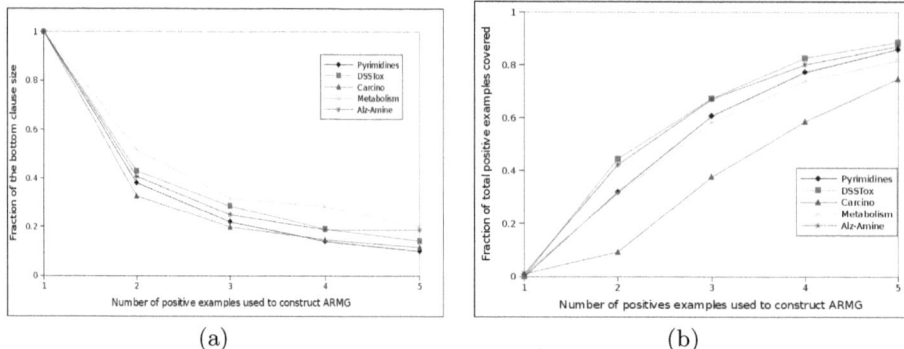

(a)   (b)

Fig. 5. (a) ARMGs length and (b) ARMGs positive coverage as number of examples used to construct the ARMGs increases

Materials and Methods. In this experiment we use a set of eight artificially generated learning problems with varying concept sizes from 6 to 17. These problems are selected from the phase transition study [4] and correspond to problems $m6.l12$ to $m17.l12$. There are two parameters that characterise each problem in the dataset: m the target concept size, and L the number of distinct constants occurring in an example. These problems are selected from the first row of the (m, L) plane, i.e. $L = 12$ so that they only approach the phase transition region. Each problem has 200 training and 200 test examples and the positive and negative examples are balanced in both partitions (100 for each). We use a hold-out test strategy and compare the performance of ProGolem and Aleph. This dataset is noise free and highly non-determinate having about 100 solutions per predicate. In order to curb the combinatorial explosion the recall for constructing the bottom clauses was set to 2 for both Aleph and ProGolem.

Aleph and ProGolem were executed with $evalfn = compression$, $i = 2$, $noise = 0$. Other Aleph parameters are $clauselength = 20$, $search = heuristic$ and $nodes = 100,000$. ProGolem was used with $N = 2$, $K = 10$ and the efficient coverage testing algorithm was used in place of Prolog's left-to-right evaluation. All the experiments were performed, as before, on a 2.2 Ghz dual core AMD Opteron processor (275) with 8gb RAM.

Results and discussion. Table 2 shows predictive accuracies and average learning times for ProGolem and Aleph. Aleph fails to find any solution in four out of eight problems whereas ProGolem can find good approximations of the target concepts. Moreover, ProGolem is significantly faster than Aleph. This is partly because long clauses generally require an overwhelming amount of search in top-down systems like Progol and Aleph, due to a search bias which favours simplicity. This tends to limit the maximum complexity of learned clauses. Note that Golem is inapplicable on this phase transition dataset as it is non-determinate.

Table 2. Predictive accuracies and learning times for *ProGolem* and *Aleph* on a set of learning problems with varying concept sizes from 6 to 17

m	*ProGolem*		*Aleph*	
	$A(\%)$	$T(s)$	$A(\%)$	$T(s)$
6	98.0	7	99.5	1
7	99.5	15	99.5	254
8	97.5	40	100	23
10	97.0	45	50.0	3596
11	99.0	36	50.0	3708
14	93.5	47	96.0	365
16	76.0	501	50.0	4615
17	71.0	485	50.0	4668

8   Related Work and Discussion

It was shown in a previous paper [28] that, unlike for the general subsumption order, efficient least generalisation operators can be designed for the subsumption order relative to a bottom clause. This idea is the basis for ProGolem which implements efficient asymmetric relative minimal generalisations for the subsumption order relative to a bottom clause. The lattice structure and refinement operators for the subsumption order relative to a bottom clause were studied in [28]. The relationship between this subsumption order and some of related subsumption orders including weak subsumption [2], ordered subsumption [8] and sequential subsumption in SeqLog [10] were also discussed.

The least and minimal generalisations relative to a bottom clause can be compared with other approaches which use lgg-like operators but instead of considering all pairs of compatible literals they only consider one pair. For example, LOGAN-H [1] is a bottom-up system which is based on inner products of examples which are closely related to lgg operator. This system constructs lgg-like clauses by considering only those pairs of literals which guarantee an injective mapping between variables. In other words, it assumes one-one object mappings. Other similar approaches use the same idea of simplifying the lgg-like operations by considering only one pair of compatible literals but they select this pair arbitrarily (e.g. [3]).

As already mentioned in the previous sections, ProGolem is closely related to Golem which is based on generalisation relative to background knowledge B. ProGolem is based on generalisation relative to a bottom clause $\perp_e$ which is the result of compiling background knowledge B. Hence, subsumption relative to a bottom clause can be viewed as subsumption relative to a compilation of B which makes it more efficient than subsumption relative to B. Moreover, as already shown in this paper, generalisation relative to a bottom clause allows ProGolem to be used for non-determinate problems where Golem is inapplicable.

9   Conclusions

In this paper we have proposed an asymmetric variant of Plotkin's RLGG, called ARMG. In comparison to the determinate RLGGs used in Golem, ARMGs are capable of representing non-determinate clauses. Although this is also possible using Plotkin's RLGG, the cardinality of the Plotkin RLGG grows exponentially in the number of examples. By contrast, an ARMG is built by constructing a bottom clause for one example and then dropping a minimal set of literals to allow coverage of a second example. By construction the clause length is bounded by the length of the initially constructed bottom clause.

An algorithm is described for constructing ARMGs and this has been implemented in an ILP system called ProGolem which combines bottom-clause construction in Progol with a Golem control strategy which uses ARMG in place of determinate RLGG. It is shown that ProGolem has a similar or better predictive accuracy and learning time compared to Golem on two determinate real-world applications where Golem was originally tested. Moreover, ProGolem was also tested on several non-determinate real-world applications where Golem is inapplicable. In these applications, ProGolem and Aleph have comparable times and accuracies. ProGolem has also been evaluated on a set of artificially generated learning problems with large concept sizes. The experimental results suggest that ProGolem significantly outperforms Aleph in cases where clauses in the target theory are long and complex. These results suggest that while ProGolem has the advantages of Golem for learning large target concepts, it does not suffer from the determinacy limitation and can be used in problems where Golem is inapplicable.

The use of top-down ILP algorithms such as Progol, tends to limit the maximum complexity of learned clauses, due to a search bias which favours simplicity. Long target clauses generally require an overwhelming amount of search for systems like Progol and Aleph. We believe that such targets should be more effectively learned by a bottom-up systems such as ProGolem since long clauses are easier to construct using bottom-up search.

Acknowledgments

We would like to thank Gerson Zaverucha, Erick Alphonse and Filip Zelezny for their helpful discussions and advice on different subjects related to this paper. The first author thanks the Royal Academy of Engineering and Microsoft for funding his present 5 year Research Chair. The second author was supported by a Wellcome Trust Ph.D. scholarship. The third author was supported by the BBSRC grant BB/C519670/1.

References

1. Arias, M., Khardon, R.: Bottom-up ILP using large refinement steps. In: Camacho, R., King, R., Srinivasan, A. (eds.) ILP 2004. LNCS (LNAI), vol. 3194, pp. 26–43. Springer, Heidelberg (2004)
2. Badea, L., Stanciu, M.: Refinement operators can be (weakly) perfect. In: Džeroski, S., Flach, P.A. (eds.) ILP 1999. LNCS (LNAI), vol. 1634, pp. 21–32. Springer, Heidelberg (1999)

3. Basilio, R., Zaverucha, G., Barbosa, V.C.: Learning logic programs with neural networks. In: Rouveirol, C., Sebag, M. (eds.) ILP 2001. LNCS (LNAI), vol. 2157, pp. 15–26. Springer, Heidelberg (2001)
4. Botta, M., Giordana, A., Saitta, L., Sebag, M.: Relational learning as search in a critical region. J. Mach. Learn. Res. 4, 431–463 (2003)
5. Cheng, J., Hatzis, C., Hayashi, H., Krogel, M., Morishita, S., Page, D., Sese, J.: Kdd cup 2001 report. SIGKDD Explorations 3(2), 47–64 (2002)
6. King, R.D., Muggleton, S.H., Lewis, R., Sternberg, M.: Drug design by machine learning. Proceedings of the National Academy of Sciences 89(23), 11322–11326 (1992)
7. King, R.D., Srinivasan, A., Sternberg, M.J.E.: Relating chemical activity to structure: an examination of ILP successes. New Generation Computing 13, 411–433 (1995)
8. Kuwabara, M., Ogawa, T., Hirata, K., Harao, M.: On generalization and subsumption for ordered clauses. In: Washio, T., Sakurai, A., Nakajima, K., Takeda, H., Tojo, S., Yokoo, M. (eds.) JSAI Workshop 2006. LNCS (LNAI), vol. 4012, pp. 212–223. Springer, Heidelberg (2006)
9. Kuzelka, O., Zelezný, F.: Fast estimation of first-order clause coverage through randomization and maximum likelihood. In: Proceedings of the 25th International Conference (ICML 2008), pp. 504–511 (2008)
10. Lee, S.D., De Raedt, L.: Constraint Based Mining of First Order Sequences in SeqLog. In: Database Support for Data Mining Applications, pp. 155–176 (2003)
11. Maloberti, J., Sebag, M.: Fast theta-subsumption with constraint satisfaction algorithms. Machine Learning 55(2), 137–174 (2004)
12. Muggleton, S.: Progol datasets (1996), http://www.doc.ic.ac.uk/~shm/software/progol4.2/
13. Muggleton, S., Feng, C.: Efficient induction of logic programs. In: Muggleton, S. (ed.) Inductive Logic Programming, pp. 281–298. Academic Press, London (1992)
14. Muggleton, S.H.: Duce, an oracle based approach to constructive induction. In: IJCAI 1987, pp. 287–292. Kaufmann, San Francisco (1987)
15. Muggleton, S.H.: Inverse entailment and Progol. New Generation Computing 13, 245–286 (1995)
16. Muggleton, S.H., Buntine, W.: Machine invention of first-order predicates by inverting resolution. In: Proceedings of the 5th International Conference on Machine Learning, pp. 339–352. Kaufmann, San Francisco (1988)
17. Muggleton, S.H., King, R., Sternberg, M.: Protein secondary structure prediction using logic-based machine learning. Protein Engineering 5(7), 647–657 (1992)
18. Muggleton, S.H., Tamaddoni-Nezhad, A.: QG/GA: A stochastic search for Progol. Machine Learning 70(2-3), 123–133 (2007), doi:10.1007/s10994-007-5029-3
19. Nienhuys-Cheng, S.-H., de Wolf, R.: Foundations of Inductive Logic Programming. LNCS (LNAI), vol. 1228, pp. 168–169. Springer, Heidelberg (1997)
20. Nilsson, N.J.: Principles of Artificial Intelligence. Tioga, Palo Alto (1980)
21. Plotkin, G.D.: Automatic Methods of Inductive Inference. PhD thesis, Edinburgh University (August 1971)
22. Quinlan, J.R.: Learning logical definitions from relations. Machine Learning 5, 239–266 (1990)
23. De Raedt, L., Bruynooghe, M.: A theory of clausal discovery. In: Proceedings of the 13th International Joint Conference on Artificial Intelligence. Morgan Kaufmann, San Francisco (1993)
24. Richard, A.M., Williams, C.R.: Distributed structure-searchable toxicity (DSSTox) public database network: A proposal. Mutation Research 499, 27–52 (2000)

25. Shapiro, E.Y.: Algorithmic program debugging. MIT Press, Cambridge (1983)
26. Srinivasan, A., King, R.D., Muggleton, S.H., Sternberg, M.: Carcinogenesis predictions using ILP. In: Džeroski, S., Lavrač, N. (eds.) ILP 1997. LNCS, vol. 1297, pp. 273–287. Springer, Heidelberg (1997)
27. Srinivasan, A.: The Aleph Manual. University of Oxford (2007)
28. Tamaddoni-Nezhad, A., Muggleton, S.H.: The lattice structure and refinement operators for the hypothesis space bounded by a bottom clause. Machine Learning 76(1), 37–72 (2009)

Appendix A   Progol's Definite Mode Language

The following definitions describe Progol's mode declaration (M), definite mode language $(\mathcal{L}(M))$ and depth-bounded mode language $(\mathcal{L}_i(M))$.

Definition 13. Mode declaration M. *A mode declaration has either the form modeh(n,atom) or modeb(n,atom) where n, the recall, is either an integer, $n > 1$, or '*' and atom is a ground atom. Terms in the atom are either normal or place-marker. A normal term is either a constant or a function symbol followed by a bracketed tuple of terms. A place-marker is either +type, -type or #type, where type is a constant. If m is a mode declaration then $a(m)$ denotes the atom of m with place-markers replaced by distinct variables. The sign of m is positive if m is a modeh and negative if m is a modeb.*

Definition 14. Definite mode language $\mathcal{L}(M)$. *Let C be a definite clause with a defined total ordering over the literals and M be a set of mode declarations. $C = h \leftarrow b_1, .., b_n$ is in the definite mode language $\mathcal{L}(M)$ if and only if 1) h is the atom of a modeh declaration in M with every place-marker +type and -type replaced by variables and every place-marker #type replaced by a ground term and 2) every atom b_i in the body of C is the atom of a modeb declaration in M with every place-marker +type and -type replaced by variables and every place-marker #type replaced by a ground term and 3) every variable of +type in any atom b_i is either of +type in h or of -type in some atom b_j, $1 \leq j < i$.*

Definition 15. Depth of variables. *Let C be a definite clause and v be a variable in C. Depth of v is defined as follows:*

$$d(v) = \begin{cases} 0 & \text{if } v \text{ is in the head of } C \\ (max_{u \in U_v} d(u)) + 1 & \text{otherwise} \end{cases}$$

where U_v are the variables in atoms in the body of C containing v.

Definition 16. Depth-bounded mode language $\mathcal{L}_i(M)$. *Let C be a definite clause with a defined total ordering over the literals and M be a set of mode declarations. C is in $\mathcal{L}_i(M)$ if and only if C is in $\mathcal{L}(M)$ and all variables in C have depth at most i according to Definition 15.*

An Inductive Logic Programming Approach to Validate Hexose Binding Biochemical Knowledge

Houssam Nassif[1,2], Hassan Al-Ali[3], Sawsan Khuri[4,5],
Walid Keirouz[6], and David Page[1,2]

[1] Department of Computer Sciences,
[2] Department of Biostatistics and Medical Informatics,
University of Wisconsin-Madison, USA
[3] Department of Biochemistry and Molecular Biology,
[4] Center for Computational Science, University of Miami,
[5] The Dr. John T. Macdonald Foundation Department of Human Genetics,
University of Miami Miller School of Medicine, Florida, USA
[6] Department of Computer Science, American University of Beirut, Lebanon

Abstract. Hexoses are simple sugars that play a key role in many cellular pathways, and in the regulation of development and disease mechanisms. Current protein-sugar computational models are based, at least partially, on prior biochemical findings and knowledge. They incorporate different parts of these findings in predictive black-box models. We investigate the empirical support for biochemical findings by comparing Inductive Logic Programming (ILP) induced rules to actual biochemical results. We mine the Protein Data Bank for a representative data set of hexose binding sites, non-hexose binding sites and surface grooves. We build an ILP model of hexose-binding sites and evaluate our results against several baseline machine learning classifiers. Our method achieves an accuracy similar to that of other black-box classifiers while providing insight into the discriminating process. In addition, it confirms wet-lab findings and reveals a previously unreported TRP-GLU amino acids dependency.

Keywords: ILP, Aleph, rule generation, hexose, protein-carbohydrate interaction, binding site, substrate recognition.

1 Introduction

Inductive Logic Programming (ILP) has been shown to perform well in predicting various substrate-protein bindings (e.g., [9,26]). In this paper we apply ILP to a different and well studied binding task.

Hexoses are 6-carbon simple sugar molecules that play a key role in different biochemical pathways, including cellular energy release, signaling, carbohydrate synthesis, and the regulation of gene expression [24]. Hexose binding proteins belong to diverse functional families that lack significant sequence or, often, structural similarity [16]. Despite this fact, these proteins show high specificity to their hexose ligands. The few amino acids (also called residues) present at the binding

L. De Raedt (Ed.): ILP 2009, LNAI 5989, pp. 149–165, 2010.
© Springer-Verlag Berlin Heidelberg 2010

site play a large role in determining the binding site's distinctive topology and biochemical properties and hence the ligand type and the protein's functionality.

Wet-lab experiments discover hexose-protein properties. Computational hexose classifiers incorporate different parts of these findings in black-box models as the base of prediction. No work to date has taken the opposite approach: given hexose binding sites data, what biochemical rules can we extract with no prior biochemical knowledge, and what is the performance of the resulting classifier based solely on the extracted rules?

This work presents an ILP classifier that extracts rules from the data without prior biochemical knowledge. It classifies binding sites based on the extracted biochemical rules, clearly specifying the rules used to discriminate each instance. Rule learning is especially appealing because of its easy-to-understand format. A set of if-then rules describing a certain concept is highly expressive and readable [18]. We evaluate our results against several baseline machine learning classifiers. This inductive data-driven approach validates the biochemical findings and allows a better understanding of the black-box classifiers' output.

2 Previous Work

Although no previous work tackled data-driven rule generation or validation, many researchers studied hexose binding.

2.1 Biochemical Findings

From the biochemical perspective, Rao et al. [21] fully characterized the architecture of sugar binding in the Lectin protein family and identified conserved loop structures as essential for sugar recognition. Later, Quiocho and Vyas [20] presented a review of the biochemical characteristics of carbohydrate binding sites and identified the planar polar residues (ASN, ASP, GLN, GLU, ARG) as the most frequently involved residues in hydrogen bonding. They also found that the aromatic residues TRP, TYR, and PHE, as well as HIS, stack against the apolar surface of the sugar pyranose ring. Quiocho and Vyas also pinpointed the role of metal ions in determining substrate specificity and affinity. Ordered water molecules bound to protein surfaces are also involved in protein-ligand interaction [15].

Taroni et al. [29] analyzed the characteristic properties of sugar binding sites and described a residue propensity parameter that best discriminates sugar binding sites from other protein-surface patches. They also note that simple sugars typically have a hydrophilic side group which establishes hydrogen bonds and a hydrophobic core that is able to stack against aromatic residues. Sugar binding sites are thus neither strictly hydrophobic nor strictly hydrophilic, due to the dual nature of sugar docking. In fact, as García-Hernández et al. [11] showed, some polar groups in the protein-carbohydrate complex behave hydrophobically.

2.2 Computational Models

Some of this biochemical information has been used in computational work with the objective of accurately predicting sugar binding sites in proteins.

Taroni et al. [29] devised a probability formula by combining individual attribute scores. Shionyu-Mitsuyama et al. [23] used atom type densities within binding sites to develop an algorithm for predicting carbohydrate binding. Chakrabarti et al. [5] modeled one glucose binding site and one galactose binding site by optimizing their binding affinity under geometric and folding free energy constraints. Other researchers formulated a signature for characterizing galactose binding sites based on geometric constraints, pyranose ring proximity and hydrogen bonding atoms [27,28]. They implemented a 3D structure searching algorithm, COTRAN, to identify galactose binding sites.

More recently, researchers used machine learning algorithms to model hexose binding sites. Malik and Ahmad [17] used a Neural Network to predict general carbohydrate as well as specific galactose binding sites. Nassif et al. [19] used Support Vector Machines to model and predict glucose binding sites in a wide range of proteins.

3   Data Set

The Protein Data Bank (PDB) [2] is the largest repository of experimentally determined and hypothetical three-dimensional structures of biological macromolecules. We mine it for proteins crystallized with the most common hexoses: galactose, glucose and mannose [10]. We ignore theoretical structures and files older than PDB format 2.1. We eliminate redundant structures using PISCES [30] with a 30% overall sequence identity cut-off. We use Swiss-PDBViewer [14] to detect and discard sites that are glycosylated or within close proximity to other ligands. We check the literature to ensure that no hexose-binding site also binds non-hexoses. The final outcome is a non-redundant positive data set of 80 protein-hexose binding sites (Table 1).

We also extract an equal number of negative examples. The negative set is composed of non-hexose binding sites and of non-binding surface grooves. We choose 22 binding-sites that bind hexose-like ligands: hexose or fructose derivatives, 6-carbon molecules, and molecules similar in shape to hexoses (Table 2). We also select 27 other-ligand binding sites, ligands who are bigger or smaller than hexoses (Table 2). Finally, we specify 31 non-binding sites: protein surface grooves that look like binding-sites but are not known to bind any ligand (Table 3).

We use 10-folds cross-validation to train, test and validate our approach. We divide the data set in 10 stratified folds, thus preserving the proportions of the original set labels and sub-groups.

4   Problem Representation

In this work, we first extract multiple chemical and spatial features from the binding site. We then apply ILP to generate rules and classify our data set.

Table 1. Inventory of the hexose-binding positive data set

Hexose	PDB ID	Ligand	PDB ID	Ligand	PDB ID	Ligand
Glucose	1BDG	GLC-501	1ISY	GLC-1471	1SZ2	BGC-1001
	1EX1	GLC-617	1J0Y	GLC-1601	1SZ2	BGC-2001
	1GJW	GLC-701	1JG9	GLC-2000	1U2S	GLC-1
	1GWW	GLC-1371	1K1W	GLC-653	1UA4	GLC-1457
	1H5U	GLC-998	1KME	GLC-501	1V2B	AGC-1203
	1HIZ	GLC-1381	1MMU	GLC-1	1WOQ	GLC-290
	1HIZ	GLC-1382	1NF5	GLC-125	1Z8D	GLC-901
	1HKC	GLC-915	1NSZ	GLC-1400	2BQP	GLC-337
	1HSJ	GLC-671	1PWB	GLC-405	2BVW	GLC-602
	1HSJ	GLC-672	1Q33	GLC-400	2BVW	GLC-603
	1I8A	GLC-189	1RYD	GLC-601	2F2E	AGC-401
	1ISY	GLC-1461	1S5M	AGC-1001		
Galactose	1AXZ	GLA-401	1MUQ	GAL-301	1R47	GAL-1101
	1DIW	GAL-1400	1NS0	GAL-1400	1S5D	GAL-704
	1DJR	GAL-1104	1NS2	GAL-1400	1S5E	GAL-751
	1DZQ	GAL-502	1NS8	GAL-1400	1S5F	GAL-104
	1EUU	GAL-2	1NSM	GAL-1400	1SO0	GAL-500
	1ISZ	GAL-461	1NSU	GAL-1400	1TLG	GAL-1
	1ISZ	GAL-471	1NSX	GAL-1400	1UAS	GAL-1501
	1JZ7	GAL-2001	1OKO	GLB-901	1UGW	GAL-200
	1KWK	GAL-701	1OQL	GAL-265	1XC6	GAL-9011
	1L7K	GAL-500	1OQL	GAL-267	1ZHJ	GAL-1
	1LTI	GAL-104	1PIE	GAL-1	2GAL	GAL-998
Mannose	1BQP	MAN-402	1KZB	MAN-1501	1OUR	MAN-301
	1KLF	MAN-1500	1KZC	MAN-1001	1QMO	MAN-302
	1KX1	MAN-20	1KZE	MAN-1001	1U4J	MAN-1008
	1KZA	MAN-1001	1OP3	MAN-503	1U4J	MAN-1009

4.1 Binding Site Representation

We view the binding site as a sphere centered at the ligand. We compute the center of the hexose-binding site as the centroid of the coordinates of the hexose pyranose ring's six atoms. For negative sites, we use the center of the cavity or the ligand's central point. The farthest pyranose-ring atom from the ring's centroid is located 2.9 Å away. Bobadilla et al. [4] consider atomic interactions to be significant within a 7 Å range. We thereby fix the binding site sphere radius to 10 Å. Given the molecule and the binding site centroid, we extract all atoms within the sphere. We include water molecules and ions present in the binding groove [15,19,20]. We discard hydrogen atoms since most PDB entries lack them. We do not extract residues.

For every extracted atom we record its PDB-coordinates, its charge, hydrogen bonding, and hydrophobicity properties, and its atomic element and name. Every PDB file has orthogonal coordinates and all atom positions are recorded

Table 2. Inventory of the non-hexose-binding negative data set

PDB ID	Cavity Center	Ligand	PDB ID	Cavity Center	Ligand
Hexose-like ligands					
1A8U	4320, 4323	BEZ-1	1AI7	6074, 6077	IPH-1
1AWB	4175, 4178	IPD-2	1DBN	pyranose ring	GAL-102
1EOB	3532, 3536	DHB-999	1F9G	5792, 5785, 5786	ASC-950
1G0H	4045, 4048	IPD-292	1JU4	4356, 4359	BEZ-1
1LBX	3941, 3944	IPD-295	1LBY	3944, 3939, 3941	F6P-295
1LIU	15441, 15436, 15438	FBP-580	1MOR	pyranose ring	G6P-609
1NCW	3406, 3409	BEZ-601	1P5D	pyranose ring	G1P-658
1T10	4366, 4361, 4363	F6P-1001	1U0F	pyranose ring	G6P-900
1UKB	2144, 2147	BEZ-1300	1X9I	pyranose ring	G6Q-600
1Y9G	4124, 4116, 4117	FRU-801	2B0C	pyranose ring	G1P-496
2B32	3941, 3944	IPH-401	1PBG	pyranose ring	BGP-469
Other ligands					
11AS	5132	ASN-1	11GS	1672, 1675	MES-3
1A0J	6985	BEN-246	1A42	2054, 2055	BZO-555
1A50	4939, 4940	FIP-270	1A53	2016, 2017	IGP-300
1AA1	4472, 4474	3PG-477	1AJN	6074, 6079	AAN-1
1AJS	3276, 3281	PLA-415	1AL8	2652	FMN-360
1B8A	7224	ATP-500	1BO5	7811	GOL-601
1BOB	2566	ACO-400	1D09	7246	PAL-1311
1EQY	3831	ATP-380	1IOL	2674, 2675	EST-400
1JTV	2136, 2137	TES-500	1KF6	16674, 16675	OAA-702
1RTK	3787, 3784	GBS-300	1TJ4	1947	SUC-1
1TVO	2857	FRZ-1001	1UK6	2142	PPI-1300
1W8N	4573, 4585	DAN-1649	1ZYU	1284, 1286	SKM-401
2D7S	3787	GLU-1008	2GAM	11955	NGA-502
3PCB	3421, 3424	3HB-550			

Table 3. Inventory of the non-binding surface groove negative data set

PDB ID	Cavity Center	PDB ID	Cavity Center	PDB ID	Cavity Center
1A04	1424, 2671	1A0I	1689, 799	1A22	2927
1AA7	579	1AF7	631, 1492	1AM2	1277
1ARO	154, 1663	1ATG	1751	1C3G	630, 888
1C3P	1089, 1576	1DXJ	867, 1498	1EVT	2149, 2229
1FI2	1493	1KLM	4373, 4113	1KWP	1212
1QZ7	3592, 2509	1YQZ	4458, 4269	1YVB	1546, 1814
1ZT9	1056, 1188	2A1K	2758, 3345	2AUP	2246
2BG9	14076, 8076	2C9Q	777	2CL3	123, 948
2DN2	749, 1006	2F1K	316, 642	2G50	26265, 31672
2G69	248, 378	2GRK	369, 380	2GSE	337, 10618
2GSH	6260				

accordingly. We compute atomic properties as done by Nassif et al. [19]. The partial charge measure per atom is positive, neutral, or negative; atoms can form hydrogen bonds or not; hydrophobicity measures are considered as hydrophobic, hydroneutral, or hydrophilic. Finally, every PDB-atom has an atomic element and a specific name. For example, the residue histidine (His) has a particular Nitrogen atom named ND1. This atom's element is Nitrogen, and name is ND1. Since ND1 atoms only occur in His residues, recording atomic names leaks information about their residues.

4.2   Aleph Settings

We use the ILP engine Aleph [25] to learn first-order rules. We run Aleph within Yap Prolog [22]. To speed the search, we use Aleph's heuristic search. We estimate the classifier's performance using 10-fold cross-validation.

We limit Aleph's running time by restricting the clause length to a maximum of 8 literals, with only one in the head. We set the Aleph parameter *explore* to true, so that it will return all optimal-scoring clauses, rather than a single one, in a case of a tie. The consequent of any rule is $bind(+site)$, where $site$ is predicted to be a hexose binding site. No literal can contain terms pertaining to different binding sites. As a result, $site$ is the same in all literals in a clause.

The literal describing the binding site center is:

$$point(+site, -id, -X, -Y, -Z) \tag{1}$$

where $site$ is the binding site and id is the binding center's unique identifier. X, Y, and Z specify the PDB-Cartesian coordinates of the binding site's centroid.

Literals describing individual PDB-atoms are of the form:

$$point(+site, -id, -X, -Y, -Z, -charge, -hbond, -hydro, -elem, -name) \tag{2}$$

where $site$ is the binding site and id is the individual atom's unique identifier. X, Y, and Z specify the PDB-Cartesian coordinates of the atom. $charge$ is the partial charge, $hbond$ the hydrogen-bonding, and $hydro$ the hydrophobicity. Lastly, $elem$ and $name$ refer to the atomic element and its name (see last paragraph of previous section).

Clause bodies can also use distance literals:

$$dist(+site, +id, +id, \#distance, \#error) . \tag{3}$$

The $dist$ predicate, depending on usage, either computes or checks the $distance$ between two points. $site$ is the binding site and the ids are two unique point identifiers (two PDB-atoms or one PDB-atom and one center). $distance$ is their Euclidean distance apart and $error$ the tolerated distance error, resulting in a matching interval of $distance \pm error$. We set $error$ to 0.5 Å.

We want our rules to refer to properties of PDB-atoms, such as "an atom's name is ND1", or "an atom's charge is not positive". Syntactically we do this by relating PDB-atoms' variables to constants using "equal" and "not equal" literals:

$$equal(+setting, \#setting),\qquad(4)$$

$$not\_equal(+feature, \#feature).\qquad(5)$$

feature is the atomic features *charge*, *hbond* and *hydro*. In addition to these atomic features, *setting* includes *elem* and *name*.

Aleph keeps learning rules until it has covered all the training positive set, and then it labels a test example as positive if *any* of the rules cover that example. This has been noted in previous publications to produce a tendency toward giving more false positives [6,7]. To limit our false positives count, we restrict coverage to a maximum of 5 training-set negatives. Since our approach seeks to validate biological knowledge, we aim for high precision rules. Restricting negative rule coverage also biases generated rules towards high precision.

5 Results

The Aleph testing set error averaged to 32.5% with a standard deviation of 10.54%. The confidence interval is [24.97%, 40.03%] at the 95% confidence level. Refer to Table 5 for the 10-folds cross-validation accuracies.

To generate the final set of rules, we run Aleph over the whole data set. We discard rules with $pos\_cover - neg\_cover \leq 2$. Even though Aleph was only looking at atoms, valuable information regarding amino acids can be inferred. For example ND1 atoms are only present within the amino acid His, and a rule requiring the presence of ND1 is actually requiring His. We present the rules' biochemical translation while replacing specific atoms by the amino acids they imply. The queried site is considered hexose binding if any of these rules apply:

1. It contains a TRP residue and a GLU with an OE1 Oxygen atom that is 8.53 Å away from an Oxygen atom with a negative partial charge (GLU, ASP amino acids, Sulfate, Phosphate, residue C-terminus Oxygen).
   [Pos cover = 22, Neg cover = 4]
2. It contains a TRP, PHE or TYR residue, an ASP and an ASN. ASP and an ASN's OD1 Oxygen atoms are 5.24 Å apart.
   [Pos cover = 21, Neg cover = 3]
3. It contains a VAL or ILE residue, an ASP and an ASN. ASP and ASN's OD1 Oxygen atoms are 3.41 Å apart.
   [Pos cover = 15, Neg cover = 0]
4. It contains a hydrophilic non-hydrogen bonding Nitrogen atom (PRO, ARG) with a distance of 7.95 Å away from a HIS's ND1 Nitrogen atom, and 9.60 Å away from a VAL or ILE's CG1 Carbon atom.
   [Pos cover = 10, Neg cover = 0]

5. It has a hydrophobic CD2 Carbon atom (LEU, PHE, TYR, TRP, HIS), a PRO, and two hydrophilic OE1 Oxygen atoms (GLU, GLN) 11.89 Å apart.
   [Pos cover = 11, Neg cover = 2]
6. It contains an ASP residue B, two identical atoms Q and X, and a hydrophilic hydrogen-bonding atom K. Atoms K, Q and X have the same charge. B's OD1 Oxygen atom share the same Y-coordinate with K and the same Z-coordinate with Q. Atoms X and K are 8.29 Å apart.
   [Pos cover = 8, Neg cover = 0]
7. It contains a SER residue, and two NE2 Nitrogen atoms (GLN, HIS) 3.88 Å apart.
   [Pos cover = 8, Neg cover = 2]
8. It contains an ASN residue and a PHE, TYR or HIS residue, whose CE1 Carbon atom is 7.07 Å away from a Calcium ion.
   [Pos cover = 5, Neg cover = 0]
9. It contains a LYS or ARG, a PHE, TYR or ARG, a TRP, and a Sulfate or a Phosphate ion.
   [Pos cover = 3, Neg cover = 0]

Most of these rules closely reproduce current biochemical knowledge. One in particular is novel. We will discuss rule relevance in Section 7.2.

6   Experimental Evaluation

We evaluate our performance by comparing Aleph to several baseline machine learning classifiers.

6.1   Feature Vector Representation

Unlike Aleph, the implemented baseline algorithms require a constant-length feature vector input. We change our binding-site representation accordingly. We subdivide the binding-site sphere into concentric shells as suggested by Bagley and Altman [1]. Nassif et al. [19] subdivided the sphere into 8 layers centered at the binding-site centroid. The first layer had a width of 3 Å and the subsequent 7 layers where 1 Å each. Their results show that the layers covering the first 5 Å, the subsequent 3 Å and the last 2 Å share several attributes. We thereby subdivide our binding-site sphere into 3 concentric layers, with layer width of 5 Å, 3 Å and 2 Å respectively. For each layer, our algorithm reports the total number of atoms in that layer and the fraction of each atomic property (charge, hydrogen-bonding, hydrophobicity). For example, feature "layer 1 hydrophobic atoms" represents the fraction of the first layer atoms that are hydrophobic.

The ILP predicate representation allows it to implicitly infer residues from atomic names and properties. We use a weakly expressive form to explicitly include amino acids in the feature vector representation. Amino acids are categorized into subgroups, based on their structural and chemical properties [3]. We base our scheme on the representation adopted by Nassif et al. [19], grouping histidine, previously a subclass on its own, with the rest of the aromatic residues.

Histidine can have roles which are unique among the aromatic amino acids. We group it with other aromatics because it was not selected as a relevant feature in our previous work. Gilis et al. [12] report the mean frequencies of the individual amino acids in the proteomes of 35 living organisms. Adding up the respective frequencies, we get the expected percentage p of each residue category. We categorize the residue features into "low", "normal" and "high". A residue category feature is mapped to "normal" if its percentage is within $2 \times \sqrt{p}$ of the expected value p. It is mapped to "low" if it falls below, and to "high" if it exceeds the cut-off. Table 4 accounts for the different residue categories, their expected percentages, and their cut-off values mapping boundaries. Given a binding site, our algorithm computes the percentage of amino acids of each group present in the sphere, and records its nominal value. We ignore the concentric layers, since a single residue can span several layers.

Table 4. Residue grouping scheme, expected percentage, and mapping boundaries

Residue Category	Amino Acids	Expected Percentage	Lower Bound	Upper Bound
Aromatic	HIS, PHE, TRP, TYR	10.81%	4.23%	17.39%
Aliphatic	ALA, ILE, LEU, MET, VAL	34.19%	22.50%	45.88%
Neutral	ASN, CYS, GLN, GLY, PRO, SER, THR	31.53%	20.30%	42.76%
Acidic	ASP, GLU	11.91%	5.01%	18.81%
Basic	ARG, LYS	11.55%	4.75%	18.35%

The final feature vector is a concatenation of the atomic and residue features. It contains the total number of atoms and the atomic property fractions for each layer, in addition to the residue features. It totals 27 continuous and 5 nominal features.

6.2   Baseline Classifiers

This section details our implementation and parametrization of the baseline algorithms. Refer to Mitchell [18] and Duda et al. [8] for a complete description of the algorithms.

k-Nearest Neighbor. The scale of the data has a direct impact on k-Nearest Neighbor's (kNN) classification accuracy. A feature with a high data mean and small variance will a priori influence classification more than one with a small mean and high variance, regardless of their discrimination power [8]. In order to put equal initial weight on the different features, the data is standardized by scaling and centering.

Our implementation handles nominal values by mapping them to ordinal numbers. It uses the Euclidean distance as a distance function. It chooses the best k via a leave-one-out tuning method. Whenever two or more k's yield the same performance, it adopts the larger one. If two or more examples are equally distant from the query, and all may be the kth nearest neighbor, our implementation randomly chooses. On the other hand, if a decision output tie arises, the query is randomly classified.

We also implement feature backward-selection (BSkNN) using the steepest-ascent hill-climbing method. For a given feature set, it removes one feature at a time and performs kNN. It adopts the trial leading to the smaller error. It repeats this cycle until removing any additional feature increases the error rate. This implementation is biased towards shorter feature sets, going by Occam's razor principle.

Naive Bayes. Our Naive Bayes (NB) implementation uses a Laplacian smoothing function. It assumes that continuous features, for each output class, follow the Gaussian distribution. Let X be a continuous feature to classify and Y the class. To compute $P(X|Y)$, it first calculates the normal z-score of X given Y using the Y-training set's mean μ_Y and standard deviation s_Y: $z_Y = (x - \mu_Y)/s_Y$. It then converts the z-score into a $[0, 1]$ number by integrating the portions of the normal curve that lie outside $\pm z$. We use this number to approximate $P(X|Y)$. This method returns 1 if $X = \mu$, and decreases as X steps away from μ:

$$P(X|Y) = \int_{|z_Y|}^{\infty} normalCurve + \int_{-|z_Y|}^{-\infty} normalCurve. \qquad (6)$$

Decision Trees. Our Decision Tree implementation uses information gain as a measure for the effectiveness of a feature in classifying the training data. We incorporate continuous features by dynamically defining new discrete-valued attributes that partition the continuous attribute value into a discrete set of intervals. We prune the resulting tree using a tuning set. We report the results of both pruned (Pr DT) and unpruned decision trees (DT).

Perceptron. Our perceptron (Per) implementation uses linear units and performs a stochastic gradient descent. It is therefore similar to a logistic regression. It automatically adjusts the learning rate, treats the threshold as another weight, and uses a tuning set for early stopping to prevent overfitting. We limit our runs to a maximum of 1000 epochs.

Sequential Covering. Sequential Covering (SC) is a propositional rules learner that returns a set of disjunctive rules covering a subset of the positive examples. Our implementation uses a greedy approach. It starts from the empty set and greedily adds the best attribute that improves rule performance. It discretizes continuous attributes using the same method as Decision Trees. It sets the rule coverage threshold to 4 positive examples and no negative examples. The best attribute to add is the one maximizing:

$$|entropy(parent) - entropy(child)| * numberOfPositives(child). \qquad (7)$$

6.3 Baseline Classifiers Results

We apply the same 10-folds cross-validation to Aleph and all the baseline classifiers. Table 5 tabulates the error percentage per testing fold, the mean, standard deviation and the 95% level confidence interval for each classifier.

Table 5. 10-folds cross-validation test error percentage, mean, standard deviation and the 95% level confidence interval for the baseline algorithms and Aleph

Fold	kNN	BSkNN	NB	DT	Pr DT	Per	SC	Aleph
0	25.0	25.0	43.75	31.25	37.5	43.75	31.25	25.0
1	25.0	25.0	25.0	31.25	25.0	43.75	31.25	37.5
2	18.75	18.75	25.0	12.5	25.0	25.0	25.0	25.0
3	18.75	18.75	37.5	6.25	12.5	31.25	12.5	50.0
4	25.0	37.5	37.5	25.0	37.5	25.0	12.5	31.25
5	31.25	31.25	37.5	31.25	18.75	37.5	31.25	18.75
6	31.25	18.75	25.0	37.5	31.25	37.5	25.0	25.0
7	31.25	25.0	37.5	25.0	31.25	31.25	37.5	43.75
8	18.75	18.75	31.25	25.0	12.5	31.25	31.25	25.0
9	31.25	31.25	50.0	50.0	31.25	43.75	25.0	43.75
mean	25.63	25.0	35.0	27.5	26.25	35.0	26.25	32.5
standard deviation	5.47	6.59	8.44	12.22	9.22	7.34	8.23	10.54
lower bound	21.71	20.29	28.97	18.77	19.66	29.76	20.37	24.97
upper bound	29.54	29.71	41.03	36.23	32.84	40.24	32.13	40.03

Our SC implementation learns a propositional rule that covers at least 4 positive and no negative examples. It then removes all positive examples covered by the learned rule. It repeats the process using the remaining positive examples. Running SC over the whole data set generates the following rules, sorted by coverage. Together they cover 63 positives out of 80. A site is hexose-binding if any of these rules apply:

1. **If** layer 1 negatively charged atoms density > 0.0755
   and layer 2 positively charged atoms density < 0.0155
   and layer 3 negatively charged atoms density > 0.0125
   [Pos cover $= 32$]
2. **If** layer 1 non hydrogen-bonding atoms density < 0.559
   and layer 1 hydrophobic atoms density > 0.218
   and layer 3 hydrophilic atoms density > 0.3945
   [Pos cover $= 14$]
3. **If** layer 1 negatively charged atoms density > 0.0665
   and layer 1 hydroneutral atoms density < 0.2615
   and layer 1 non hydrogen-bonding atoms density > 0.3375
   and layer 3 atoms number < 108.5
   [Pos cover $= 12$]

4. **If** layer 1 negatively charged atoms density > 0.0665
   and layer 2 atoms number > 85.5
   and layer 1 negatively charged atoms density < 0.3485
   [Pos cover $= 5$]

7   Discussion

Despite its average performance, the main advantage of ILP is the insight it provides to the underlying discrimination process.

7.1   Aleph's Performance

Aleph's error rate of 32.5% has a p-value < 0.0002, according to a two-sided binomial test. Random guessing would return 50%, since the number of positives and negatives are equal. According to a paired t-test at the 95% confidence level, the difference between Aleph and each of the baseline algorithms is not statistically significant. Aleph's mean error rate (32.5%) and standard deviation (10.54%) are within the ranges observed for the baseline classifiers, $[25\%, 35\%]$ and $[5.47\%, 12.22\%]$ respectively (see Table 5).

Aleph's error rate is also comparable to other general sugar binding site classifiers, ranging from 31% to 39%, although each was run on a different data set (Table 6). On the other hand, specific sugar binding sites classifiers have a much better performance (Table 6). COTRAN [27] galactose-binding site classifier achieves a 5.09% error while Nassif et al. [19] glucose-binding site classifier reports an error of 8.11%. This may suggest that the problem of recognizing specific sugars is easier than classifying a family of sugars.

Table 6. Error rates achieved by general and specific sugar binding site classifiers. Not meant as a direct comparison since the data sets are different.

Program	Error (%)	Method and Data set
General sugar binding sites classifiers		
ILP hexose predictor	32.50	10-folds cross-validation, 80 hexose and 80 non-hexose or non-binding sites
Shionyu-Mitsuyama et al. [23]	31.00	Test set, 61 polysaccharide binding sites
Taroni et al. [29]	35.00	Test set, 40 carbohydrate binding sites
Malik and Ahmad [17]	39.00	Leave-one-out, 40 carbohydrate and 116 non-carbohydrate binding sites
Specific sugar binding sites classifiers		
COTRAN [27]	5.09	Overall performance over 6-folds, totaling 106 galactose and 660 non-galactose binding sites
Nassif et al. [19]	8.11	Leave-one-out, 29 glucose and 35 non-glucose or non-binding sites

7.2   Aleph Rules Interpretation

Contrary to black-box classifiers, ILP provides a number of interesting insights. It infers most of the established biochemical information about residues and relations just from the PDB-atom names and properties. We hereby interpret Aleph's rules detailed in Section 5.

Rules 1, 2, 5, 8 and 9, rely on the aromatic residues TRP, TYR and PHE. This highlights the docking interaction between the hexose and the aromatic residues [17,27,28]. The aromatic residues stack against the apolar sugar pyranose ring which stabilizes the bound hexose. HIS is mentioned in many of the rules, along-side other aromatics (5, 8) or on its own (4, 7). Histidine provides a similar docking mechanism to TRP, TYR and PHE [20].

All nine rules require the presence of a planar polar residue (ASN, ASP, GLN, GLU, ARG). These residues have been identified as the most frequently involved in the hydrogen-bonding of hexoses [20]. The hydrogen bond is probably the most relevant interaction in protein binding in general.

Rules 1, 2, 3, 5 and 6 call for acidic residues with a negative partial charge (ASP, GLU), or for atoms with a negative partial charge. The relative high negative density observed may be explained by the dense hydrogen-bond network formed by the hexose hydroxyl groups.

Some rules require hydrophobic atoms and residues, while others require hydrophilic ones. Rule 5 requires both and reflects the dual nature of sugar docking, composed of a polar-hydrophilic aspect establishing hydrogen bonds and a hydrophobic aspect responsible for the pyranose ring stacking [29].

A high residue-sugar propensity value reflects a high tendency of that residue to be involved in sugar binding. The residues having high propensity values are the aromatic residues, including histidine, and the planar polar residues [29]. This fact is reflected by the recurrence of high propensity residues in all rules.

Rules 8 and 9 require, and rule 1 is satisfied by, the presence of different ions (Calcium, Sulfate, Phosphate), confirming the relevance of ions in hexose binding [20].

Rule 6 specifies a triangular conformation of three atoms within the binding-site. This highlights the relevance of the binding-site's spatial features. On the other hand, we note the absence of the site's centroid literal from the resulting Aleph rules. The center is merely a geometric parameter and does not have any functional role. In fact, the binding site center feature was not used in most computational classifying approaches. Taroni et al. [29] and Malik and Ahmad [17] ignore it, Shionyu-Mitsuyama et al. [23] use the pyranose ring $C3$ atom instead, and Nassif et al. [19] indirectly use it to normalize distances from atoms within the binding pocket to atoms in the ligand. Only Sujatha and Balaji [27] explicitly refer to the center. These results confirm that the biochemical composition of the binding-site and the relative 3-dimensional positioning of its atoms play a much more important role in substrate specificity than the exact location of the ligand's center.

Rao et al. [21] report a dependency between PHE/TYR and ASN/ASP in the Lectin protein family. This dependency is reflected in rules 2 and 8. Similarly, rule 1 suggests a dependency between TRP and GLU, a link not previously identified in literature. This novel relationship merits further investigation and highlights the rule-discovery potential of ILP.

7.3 Baseline Algorithms Insight

In addition to providing a basis for comparison with Aleph, the baseline algorithms shed additional light on our data set and hexose binding site properties.

Naive Bayes and Perceptron return the highest mean error rates, 35.0%. Naive Bayes is based on the simplifying assumption that the attribute values are conditionally independent given the target value. This assumption does not hold for our data. In fact, charge, hydrogen-bonding and hydrophobicity values are correlated. Like Naive Bayes, Perceptron correctly classifies linearly separable data. Its high error stresses the fact that our data is not linearly separable.

On the other hand, backward-selection kNN algorithm, with the lowest mean error rate (25.0%), outperforms all other classifiers. This provides further evidence that similar sites, in terms of biochemical and spatial properties, are good predictors of binding [13]. kNN's good performance further highlights both the correlation between our features, and the data's non-linearity.

Like Aleph, Sequential Covering's rules provide insight into the discriminating process. Unlike Aleph's first-order logic rules, propositional rules are less expressive and reflect a smaller number of biochemical properties. We hereby interpret SC's rules detailed in Section 6.3.

Although SC uses an explicit representation of residues, it completely ignores them in the retained rules. Only atomic biochemical features influence the prediction. This may be due to the fact that it is the binding-site's atoms, rather than overall residues, that bind to and stabilize the docked hexose. These atoms may not be mapped to conserved specific residues.

Another general finding is that most rule antecedents are layer 1 features. This reflects the importance of the atoms lining the binding-site, which establish direct contact with the docking hexose. Layers 2 and 3 are farther away and hence have weaker interaction forces.

Only four amino acid atoms have a partial negative charge in our representation, in addition to the infrequent Sulfate and Phosphate Oxygens [19]. The first rule, covering most of the positive examples, clearly suggests a binding-site with a high density of negatively charged atoms. The first and third antecedents explicitly specify layers with a negatively charged atomic density above some thresholds. The second one implicitly states so by opting for a non-positively charged layer. The relative high negative density observed may be explained by the dense hydrogen-bond network formed by the hexose hydroxyl groups [20]. The fourth rule is similar. It imposes a bond on the first layer's negative charge, between 0.0665 and 0.3485. Although it is well established in the literature that hexose forms hydrogen bonds through its hydrogen's partial positive charge [31], the binding site itself is not known to be negatively charged. Rule 4 captures this distinction, requiring the binding site to have an above-average density of negatively charged atoms, while still setting upper-bounds.

The second rule requires a high density of hydrophobic atoms in layer 1, and a high density of hydrophilic atoms in layer 3. This reflects the dual hydrophobic-hydrophilic nature of hexose binding [29]. It also indirectly specifies a high hydrogen-bonding density by imposing an upper limit for the non-hydrogen bonding atoms.

The third rule is a combination of the other ones. First it demands a slightly negative first layer. Second, it requires hydroneutral atoms. Third it implicitly asks for hydrogen bonding atoms by setting a relatively low threshold for the much more abundant non-hydrogen bonding atoms. It is worth to note that the number of atoms in the third layer is capped by 108. This may reflect a particular spatial-arrangement of atoms in hexose binding sites.

8   Conclusion

In this work, we present the first attempt to model and predict hexose binding sites using ILP. We investigate the empirical support for biochemical findings by comparing Aleph induced rules to actual biochemical results. Our ILP system achieves a similar accuracy as other general protein-sugar binding sites black-box classifiers, while offering insight into the discriminating process. With no prior biochemical knowledge, Aleph was able to induce most of the known hexose-protein interaction biochemical rules, with a performance that is not significantly different than several baseline algorithms. In addition, ILP finds a previously unreported dependency between TRP and GLU, a novel relationship that merits further investigation.

Acknowledgments. This work was partially supported by US National Institute of Health (NIH) grant R01CA127379-01. We would like to thank Jose Santos for his inquiries that led us to improve our Aleph representation.

References

1. Bagley, S.C., Altman, R.B.: Characterizing the microenvironment surrounding protein sites. Protein Science 4(4), 622–635 (1995)
2. Berman, H.M., Westbrook, J., Feng, Z., Gilliland, G., Bhat, T.N., Weissig, H., Shindyalov, I.N., Bourne, P.E.: The protein data bank. Nucleic Acids Research 28(1), 235–242 (2000)
3. Betts, M.J., Russell, R.B.: Amino acid properties and consequences of substitutions. In: Barnes, M.R., Gray, I.C. (eds.) Bioinformatics for Geneticists, pp. 289–316. John Wiley & Sons, West Sussex (2003)
4. Bobadilla, L., Nino, F., Narasimhan, G.: Predicting and characterizing metal-binding sites using Support Vector Machines. In: Proceedings of the International Conference on Bioinformatics and Applications, Fort Lauderdale, FL, pp. 307–318 (2004)

5. Chakrabarti, R., Klibanov, A.M., Friesner, R.A.: Computational prediction of native protein ligand-binding and enzyme active site sequences. Proceedings of the National Academy of Sciences of the United States of America 102(29), 10153–10158 (2005)
6. Davis, J., Burnside, E.S., de Castro Dutra, I., Page, D., Ramakrishnan, R., Santos Costa, V., Shavlik, J.: View Learning for Statistical Relational Learning: With an application to mammography. In: Proceedings of the 19th International Joint Conference on Artificial Intelligence, Edinburgh, Scotland, pp. 677–683 (2005)
7. Davis, J., Burnside, E.S., de Castro Dutra, I., Page, D., Santos Costa, V.: An integrated approach to learning Bayesian Networks of rules. In: Gama, J., Camacho, R., Brazdil, P.B., Jorge, A.M., Torgo, L. (eds.) ECML 2005. LNCS (LNAI), vol. 3720, pp. 84–95. Springer, Heidelberg (2005)
8. Duda, R.O., Hart, P.E., Stork, D.G.: Pattern Classification, 2nd edn. Wiley-Interscience, New York (2001)
9. Finn, P., Muggleton, S., Page, D., Srinivasan, A.: Pharmacophore discovery using the Inductive Logic Programming system PROGOL. Machine Learning 30(2-3), 241–270 (1998)
10. Fox, M.A., Whitesell, J.K.: Organic Chemistry, 3rd edn. Jones & Bartlett Publishers, Boston (2004)
11. García-Hernández, E., Zubillaga, R.A., Chavelas-Adame, E.A., Vázquez-Contreras, E., Rojo-Domínguez, A., Costas, M.: Structural energetics of protein-carbohydrate interactions: Insights derived from the study of lysozyme binding to its natural saccharide inhibitors. Protein Science 12(1), 135–142 (2003)
12. Gilis, D., Massar, S., Cerf, N.J., Rooman, M.: Optimality of the genetic code with respect to protein stability and amino-acid frequencies. Genome Biology 2(11), research0049 (2001)
13. Gold, N.D., Jackson, R.M.: Fold independent structural comparisons of protein-ligand binding sites for exploring functional relationships. Journal of Molecular Biology 355(5), 1112–1124 (2006)
14. Guex, N., Peitsch, M.C.: SWISS-MODEL and the Swiss-PdbViewer: An environment for comparative protein modeling. Electrophoresis 18(15), 2714–2723 (1997)
15. Kadirvelraj, R., Foley, B.L., Dyekjær, J.D., Woods, R.J.: Involvement of water in carbohydrate-protein binding: Concanavalin A revisited. Journal of the American Chemical Society 130(50), 16933–16942 (2008)
16. Khuri, S., Bakker, F.T., Dunwell, J.M.: Phylogeny, function and evolution of the cupins, a structurally conserved, functionally diverse superfamily of proteins. Molecular Biology and Evolution 18(4), 593–605 (2001)
17. Malik, A., Ahmad, S.: Sequence and structural features of carbohydrate binding in proteins and assessment of predictability using a Neural Network. BMC Structural Biology 7, 1 (2007)
18. Mitchell, T.M.: Machine Learning. McGraw-Hill International Editions, Singapore (1997)
19. Nassif, H., Al-Ali, H., Khuri, S., Keirouz, W.: Prediction of protein-glucose binding sites using Support Vector Machines. Proteins: Structure, Function, and Bioinformatics 77(1), 121–132 (2009)
20. Quiocho, F.A., Vyas, N.K.: Atomic interactions between proteins/enzymes and carbohydrates. In: Hecht, S.M. (ed.) Bioorganic Chemistry: Carbohydrates, ch. 11, pp. 441–457. Oxford University Press, New York (1999)
21. Rao, V.S.R., Lam, K., Qasba, P.K.: Architecture of the sugar binding sites in carbohydrate binding proteins—a computer modeling study. International Journal of Biological Macromolecules 23(4), 295–307 (1998)

22. Santos Costa, V.: The life of a logic programming system. In: de la Banda, M.G., Pontelli, E. (eds.) ICLP 2008. LNCS, vol. 5366, pp. 1–6. Springer, Heidelberg (2008)
23. Shionyu-Mitsuyama, C., Shirai, T., Ishida, H., Yamane, T.: An empirical approach for structure-based prediction of carbohydrate-binding sites on proteins. Protein Engineering 16(7), 467–478 (2003)
24. Solomon, E., Berg, L., Martin, D.W.: Biology, 8th edn. Brooks Cole, Belmont (2007)
25. Srinivasan, A.: The Aleph Manual, 4th edn. (2007), http://www.comlab.ox.ac.uk/activities/machinelearning/Aleph/aleph.html
26. Srinivasan, A., King, R.D., Muggleton, S.H., Sternberg, M.J.E.: Carcinogenesis predictions using ILP. In: Džeroski, S., Lavrač, N. (eds.) ILP 1997. LNCS, vol. 1297, pp. 273–287. Springer, Heidelberg (1997)
27. Sujatha, M.S., Balaji, P.V.: Identification of common structural features of binding sites in galactose-specific proteins. Proteins: Structure, Function, and Bioinformatics 55(1), 44–65 (2004)
28. Sujatha, M.S., Sasidhar, Y.U., Balaji, P.V.: Energetics of galactose and glucose-aromatic amino acid interactions: Implications for binding in galactose-specific proteins. Protein Science 13(9), 2502–2514 (2004)
29. Taroni, C., Jones, S., Thornton, J.M.: Analysis and prediction of carbohydrate binding sites. Protein Engineering 13(2), 89–98 (2000)
30. Wang, G., Dunbrack, R.L.: PISCES: A Protein Sequence Culling Server. Bioinformatics 19(12), 1589–1591 (2003)
31. Zhang, Y., Swaminathan, G.J., Deshpande, A., Boix, E., Natesh, R., Xie, Z., Acharya, K.R., Brew, K.: Roles of individual enzyme-substrate interactions by alpha-1,3-galactosyltransferase in catalysis and specificity. Biochemistry 42(46), 13512–13521 (2003)

Boosting First-Order Clauses
for Large, Skewed Data Sets*

Louis Oliphant[1,3], Elizabeth Burnside[2,3], and Jude Shavlik[1,3]

[1] Computer Sciences Department
[2] Radiology Department
[3] Biostatistics and Medical Informatics Department
University of Wisconsin-Madison

Abstract. Creating an effective ensemble of clauses for large, skewed
data sets requires finding a diverse, high-scoring set of clauses and then
combining them in such a way as to maximize predictive performance.
We have adapted the RankBoost algorithm in order to maximize area
under the recall-precision curve, a much better metric when working with
highly skewed data sets than ROC curves. We have also explored a range
of possibilities for the weak hypotheses used by our modified RankBoost
algorithm beyond using individual clauses. We provide results on four
large, skewed data sets showing that our modified RankBoost algorithm
outperforms the original on area under the recall-precision curves.

Keywords: Learning to Rank, Ensembles, Boosting.

1 Introduction

Research over the past 15 years has shown an improvement in predictive accuracy
by using an ensemble of classifiers [4] over individual classifiers. In the Inductive
Logic Programming [6] domain ensembles have been successfully used to increase
performance [5,9,10]. Successful ensemble approaches must both learn individual classifiers that work well with a set of other classifiers as well as combine
those classifiers in a way that maximizes performance. AdaBoost [8] is a well
known ensemble method that does both of these things. AdaBoost learns weak
hypotheses iteratively, increasing the weight on previously misclassified examples so successive learners focus on misclassified examples. AdaBoost combines
weak hypotheses into a single classifier by using a weighted sum, where each
weak hypothesis is weighted according to its accuracy.

While AdaBoost focuses on improving accuracy of the final classifier, other
boosting algorithms have been created that maximize other metrics. The objective of Freund et al.'s RankBoost algorithm [7] is to maximize the correct
ordering of all possible pairs of examples in a list of examples. RankBoost maintains a probability distribution over all pairs of examples. The weak learner uses
this distribution and finds a hypothesis that minimizes the weighted misorder-
ings from the correct ordering of the examples.

---

* Appears In the ILP-2009 Springer LNCS Post-conference Proceedings.

L. De Raedt (Ed.): ILP 2009, LNAI 5989, pp. 166–177, 2010.

One version of RankBoost, named RankBoost.B, is designed to work with binary classification problems. Weights are only assigned to pairs of examples if the examples are from different classes. This focuses learning on ordering examples so that all positive examples will be ranked before the negative examples and ignoring the ordering of examples if they are of the same class. Cortes and Mohri [1] showed RankBoost.B maximizes the area under the receiver operator characteristic (AUROC) curve.

AUROC is a common metric used to discriminate between classifiers. Davis and Goadrich [3] however demonstrated that AUROC is not a good metric for discriminating between classifiers when working with highly skewed data where the negatives outnumber the positives. They recommend using area under the recall-precision curve (AURPC) when working with skewed data.

We present a modified version of the RankBoost.B algorithm that works well with skewed data which we name PRankBoost for *precision-recall RankBoost*. Its objective function seeks to maximize AURPC. We implement a top-down, heuristic-guided search to find high-scoring rules for the weak hypotheses and then use this modified RankBoost algorithm to combine them into a single classifier. We also evaluate several other possibilities for weak hypotheses that use sets of the best-scoring rules found during search.

2   PRankBoost – A Modified RankBoost Algorithm

PRankBoost, a modified version of Freund et al.'s RankBoost.B algorithm, appears in Table 1. We have modified the sum of the weights on the negative set to the skew between the size of the negative set and the size of the positive set. We make this change to expose enough information to the weak learner so that it can optimize the AURPC.

PRankBoost initializes weights on the positive examples uniformly to $\frac{1}{|X_1|}$ where X_1 is the set of positive examples. Negative examples are also uniformly initialized so that the sum of their weights is equal to the skew between positives and negatives. These initial weights preserve the same distribution between positive and negative examples as what exists in the unweighted data set. Calculating recall and precision for a model on the initial-weighted data set will be identical to calculating recall and precision on the unweighted version of the data set.

After PRankBoost initializes example weights, the algorithm enters a loop to learn a set of T weak learners. A weak learner is trained using the weighted examples. We have explored using several different weak learners which we will discuss shortly. The objective function used during training is the weighted AURPC. After training, PRankBoost assigns a weight to the weak learner. The weight is calculated analogous to the *third method* discussed by Freund et al. In this method α is an upper bound on the normalization factor, Z. Cortes and Mohri show that the r parameter used to calculate α is equivalent to a weighted version of the area under the ROC curve. We modify this approach for PRankBoost so that the r is a weighted version of AURPC.

PrankBoost updates weights using the parameter α, the weak learner $h(x)$, and a factor Z, which maintains the same weight distribution between the positive and negative examples as exists with the initial weights. An example's weight is decreased relative to how well the weak learner scores the example. The higher a positive example is scored by the weak learner the smaller the weight while be, while the opposite is true for negative examples. The effect is to place more weight on examples which the weak learner has difficulty classifying.

The final classifier, $H(x)$, assigns a score to a new example, x, as a weighted sum of the individual weak learners. We designed PRankBoost to be analogous to RankBoost. While RankBoost's final classifier maximizes AUROC, our modified version attempts to maximize AURPC. We hypothesize that this modified version will outperform RankBoost when comparing AURPC.

Table 1. PRankBoost–A modified RankBoost algorithm for optimizing area under the recall-precision curve

---

Modified RankBoost Algorithm

Given: disjoint subsets of negative, X_0, and positive, X_1, examples

Initialize:
$$skew = \frac{|X_0|}{|X_1|}, \qquad w_1(x) = \begin{cases} \frac{skew}{|X_0|} & if\ x \in X_0 \\ \frac{1}{|X_1|} & if\ x \in X_1 \end{cases}$$

for $t = 1, ..., T$:

Train weak learner, h_t, using w_t and $skew$.

Get weak ranking $h_t : X \longrightarrow \mathbb{R}$.

Choose $\alpha_t = 0.5 \ln\left(\frac{1+r}{1-r}\right)$ where $r = AURPC$(see text).

Update
$$w_{t+1}(x) = \begin{cases} \frac{w_t(x)\exp(-\alpha_t h_t(x))}{Z_t^1} & if\ x \in X_1 \\ \frac{w_t(x)\exp(\alpha_t h_t(x))}{Z_t^0} & if\ x \in X_0 \end{cases}$$

where
$$Z_t^1 = \sum_{x \in X_1} w_t(x)\exp(-\alpha_t h_t(x))$$
$$Z_t^0 = \frac{1}{skew} \times \sum_{x \in X_0} w_t(x)\exp(\alpha_t h_t(x))$$

Output the final ranking: $H(x) = \sum_{t=1}^{T} \alpha_t h_t(x)$.

---

3   Weak Learners

As shown in Table 1, a weak learner, $h_t(x)$ is a function that maps an example to a real value. A perfect weak learner maps all positive examples to higher values than negative examples. Often it is not possible to find a perfect weak learner and some objective function is used to decide among possible weak learners. In Adaboost the object function guides learning towards models that minimize a weighted version of misclassification error. In RankBoost the objective function maximizes a weighted area under the ROC curve. Our PRankBoost algorithm for finding weak learners uses area under the recall-precision curve as the object function.

When deciding what search algorithm to use for finding a weak learner we had several goals in mind. First, we wanted the search algorithm to find a clause that worked well with highly skewed data. This is the reason we use AURPC as the objective function. Second, we wanted to apply this algorithm to large data sets. Evaluation of clauses in large data sets is a costly time step and limits the number of weak learners that can be considered in a reasonable amount of time. Because of this we use a greedy hill-climbing algorithm to find weak learners.

We consider several possibilities for weak learners. The simplest weak learner we use consists of a single first-order rule. To find this rule we select a random positive example as a seed and saturate it to build the bottom clause. We begin with the most general rule from this bottom clause. All legal literals are considered to extend the rule. The extension that improves the AURPC the most is selected and added to the rule. The process repeats until no improvement can be found or some time limit or rule-length limit is reached. Each weak hypothesis, $h_t(x)$, is the best scoring individual rule found during this search.

This weak learner maps an example, x, to the range $\{0, 1\}$ where the mapping is 1 if the example is predicted as true, 0 otherwise. We call this learner PRankBoost.Clause.

We have also explored other possibilities for the weak learner and how the AURPC is calculated for the objective function. Our goal in developing other weak learners was to create more accurate models without increasing the number of rules evaluated on training data. One method of developing more complex first-order models is to retain more than just the best clause found during search. Taking an idea from the Gleaner algorithm [9] which retains an entire set of rules found during search that span the range of recall values, we have developed a second weak learner that retains a set of the best rules found during search. This weak learner, PRankBoost.Path, contains all rules along the path from the most general rule to the highest-scoring rule found during search. This set of rules will contain short, general rules that cover many examples and longer, more specific rules that have higher accuracy but lower coverage on the positive examples.

For example consider the rules that appear in Figure 1. A set of rules would contain the highest-scoring rule, h(X):-p(X),q(X,Y),r(Y), along with the subsets of the rule from the most general rule to this rule, h(X):-p(X,Y),p(Y,Z) and h(X):-p(X,Y). This weak hypothesis, $h_t(x)$, maps an example, x, to the range $[0, 1]$ by finding the most specific of these rules that covers the example. If the highest-scoring rule did not cover some new example then the next most specific rule would be considered until a rule is found that covers the example. $h_t(x)$ is the fraction of the total AURPC covered by this rule as illustrated in Figure 1. The total AURPC, r, is the area under the entire path from the most specific rule to the most general rule (the total grayed area in Figure 1).

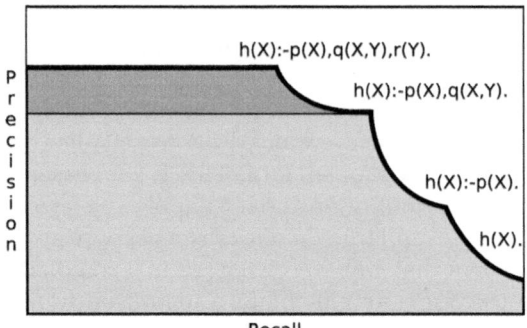

Fig. 1. Area under the recall-precision curve for a path of clauses learned during hill climbing. The total grayed area is the total AURPC, r. If h(X) :- p(X), q(X,Y) is the most specific clause in the path to cover an example then $h_t(x)$ maps the example to the value (light gray area / total grayed area).

4   Calculating AURPC

We use a weighted version of AURPC as both the objective function used to find weak learners as well as to weight weak learners when combining them into an ensemble. In general we follow the algorithm outlined by Goadrich et al. [9] to calculate AURPC, however We made two modifications to work in this ensemble setting and to improve accuracy and increase speed. First, we use a weighted version of recall and precision. Second, when calculating the area between two points in recall-precision space, A and B, Davis and Goadrich use a discretized version that estimates the area under the curve. We calculate the area under the curve exactly using a closed form solution to the integral for the curve between the two points,

$$\int_{TP_A}^{TP_B} \frac{x}{x + FP_A + s(x - TP_A)}\, dx$$

where TP is the true positive weight and FP is the false positive weight. Parameter s is the local skew of false positives to true positives between the two points A and B, $s = \frac{FP_B - FP_A}{TP_B - TP_A}$. The total AURPC is a piece-wise integral between each of the points in recall-precision space that correspond to the rules of a weak learner. For PRankBoost.Clause, which consists of a single clause, this would be a single point in recall-precision space. We use Goadrich et al.'s method for extending this point down to zero recall and up to 100% recall by using the most general clause. For PrankBoost.Path we perform the same extension down to zero recall and up to 100% recall but we use all point that correspond to the clauses in the set retained by the weak learner. This curve is shown in Figure 1.

5   Experimental Methodology and Results

We modified Aleph [12] to incorporate RankBoost and my modified versions, PRankBoost.Clause and PRankBoost.Path. RankBoost uses the same hill-climbing algorithm for finding weak learners as my two variants use. We used individual clauses for the weak learners in RankBoost. This makes the RankBoost algorithm directly comparable to PRankBoost.Clause. We compared these algorithms using AUROC and AURPC on four large, skewed data sets, two from the information-extraction domain and two from the mammography domain.

Protein Localization data set consists of text from 871 abstracts taken from the Medline database. The task is to find all phrase pairs that specify a protein and where it localizes in a cell. The data set comes from Ray and Craven [11]. Additional hand annotation was done by Goadrich et al. [9] The data set contains 281,071 examples with a positive/negative skew of 1:149.

Gene Disease data set also comes from Ray and Craven [11]. We utilized the ILP implementation by Goadrich et al. [9] The task is to find all phrase pairs showing genes and their associated disorder. the data set contains 104,192 examples with a positive/negative skew of 1:446.

Mammography1 data set is described by Davis et al. [2] It contains 62,219 findings. The objective with this data set is to determine if a finding is benign or malignant given descriptors of the finding, patient risk factors, and radiologist's prediction. The positive/negative skew is 1:121.

Mammography2 is a new data set that has the same task as Mammography1, however the data was collected from mammograms from a second institution, the University of Wisconsin Hospital and Clinics. The data set consists of 30,405 findings from 18,375 patients collected from mammograms at the radiology department. The positive/negative skew is 1:86.

We ran 10-fold, cross-validation for the mammography data sets and 5-fold for the IE data sets. We ran each fold 10 times using a different random seed to average out differences due to random effects such as seed selection. We calculated average AURPC, average AUROC, and standard deviations across the different runs and folds. Also, to compare how quickly the ensembles converged, we created learning curves with the x-axis showing the number of rules evaluated and the y-axis showing the average AURPC.

Table 2 shows average AURPC and AUROC results with standard deviations for ensembles containing 100 weak learners for RankBoost and PRankBoost.Clause. RankBoost outperforms PRankBoost.Clause when comparing AUROC on three of the four data sets. The AUROC scores are high and close together. This makes it more difficult to visually distinguish ROC curves from each other. However when comparing AURPC the difference between the two algorithms is large. PRankBoost.Clause outperforms RankBoost on three of the four data sets. The variance is much larger for AURPC scores than for AUROC scores because when recall is close to zero variance in precision values is high.

Table 2. Average AUROC and AURPC percentages with standard deviations for several large, skewed data sets using the RankBoost and PRankBoost.Clause algorithms. Bold indicates statistically significant improvement at 5% confidence level.

Data set	AUROC		AURPC	
	RankBoost	PRankBoost.Clause	RankBoost	PRankBoost.Clause
Mammography 1	**89.9 ± 4.2**	88.1 ± 5.8	18.5 ± 5.7	**32.9 ± 7.6**
Mammography 2	92.5 ± 2.0	**96.7 ± 1.1**	16.2 ± 7.4	**41.3 ± 10.6**
Protein Localization	**98.9 ± 0.1**	97.9 ± 0.7	40.4 ± 7.9	40.5 ± 8.6
Gene Disease	**98.2 ± 0.9**	95.4 ± 2.4	32.9 ± 10.7	**46.6 ± 11.9**

Learning curves on the four data sets appear in Figure 2. Each graph shows the AURPC on the y-axis by the number of rules considered during training on the x-axis. Each curve extends until 100 weak hypotheses have been found. We do this as a way of showing that the various algorithms do different amounts of work to produce 100 hypotheses, a fact that would be lost if we simply extended all three to the full width of the x-axis.

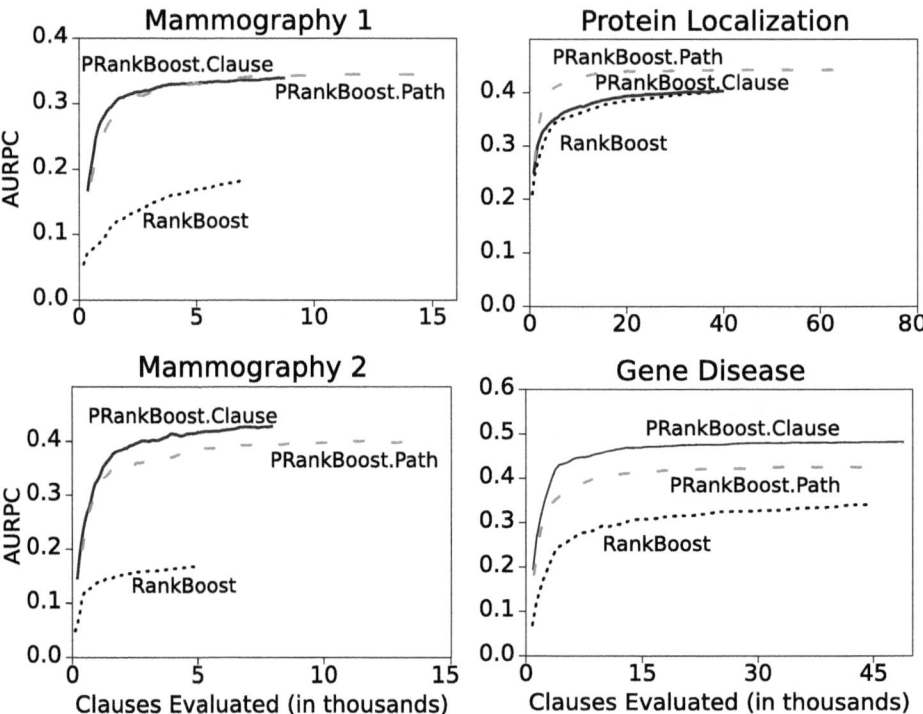

Fig. 2. Learning curves for Freund et al.'s RankBoost algorithm, our PRankBoost.Clause and PRankBoost.Path algorithms on four large, skewed data sets. Learning curves extend until 100 weak hypotheses are learned. This makes some curves extend farther than others.

My PRankBoost.Path algorithm reaches an AURPC of 0.44 on the Protein Localization data set after less than 20,000 clauses searched. The Gleaner algorithm takes over 100,000 clauses to surpass this level of performance [9]. On the Gene Disease data set my PRankBoost.Clause algorithm reaches 0.48 AURPC after 45,000 clauses searched, while the Gleaner algorithm does not reach this level of performance even after 10 million clauses searched.

The more complex weak learner, PRankBoost.Path does not appear to dominate the simple learner, PRankBoost.Clause, on all of the data sets. We believe this is because PRankBoost.Clause learns a very specific clause and reduces the weights on just a few positive examples. This forces search to focus on other positive examples and find other specific clauses that perform well on those positive examples. PRankBoost.Path on the other hand learns both specific and general clauses in the set of clauses used as a model for the weak learner. This means many positive examples will be down-weighted rather quickly. The remaining positive examples may consist of very difficult examples where it is not easy to find a good clause that covers those positive examples without also covering many negatives. After observing these characteristics We designed other weak learners that try to find a mix of models somewhere between PRankBoost.Clause and PRankBoost.Path.

6   Additional Experiments with Variations on Weak Learners

We have additional results using other weak learners that combine variations of PRankBoost.Clause and PRankBoost.Path. Remember that PRankBoost.Clause retains the single rule that is the best seen during search. Its score, α, is a weighted version of the area under the recall-precision curve of that single rule. PRankBoost.Path retains a set of rules along the trajectory from the most general rule to the best rule found during hill climbing. The weak learner's score is based upon the area under the entire path of rules. Figure 3 shows the two methods of scoring a weak learner based upon the single rule or the entire trajectory. The solid curve uses the entire trajectory from the most general rule to the rule itself while the dashed curve uses only the rule itself.

These two scoring methods create a very different search pattern. Consider scoring rules based upon the entire path from the most general rule. A portion of the score is fixed based upon the portion of the rule that has already been chosen. Any extension to the rule will only decrease recall or at best leave recall unchanged. The score will change only the left-most portion of the recall-precision curve. Any extension that increases precision will also increase the rule's overall score. This is not true when scoring a rule based upon only the rule itself. Adding a literal to a rule, even though it may increase the precision of the rule, may still decrease the overall rule's score because the curve to the single rule will also change. No portion of the curve is fixed. The difference between these two scoring methods means that using the entire path to score a rule will

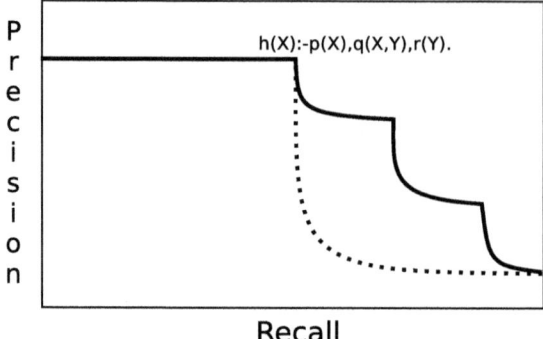

Fig. 3. Two scoring methods for a weak learner. One scoring method (solid curve) used by the PRankBoost.Path and Mix1 weak learners is based upon the entire trajectory of rules from the most general rule to the best rule. The second scoring method (dashed curve) used by the PRankBoost.Clause and Mix2 weak learners is based upon the single best rule alone. Mix3 alternates between using these two scoring methods.

search more deeply in the search space and discover longer rules with higher precision but lower recall.

As variations on PRankBoost.Path and PRankBoost.Clause we have created three other weak learners. The first retains the entire set of rules like PRank-Boost.Path, but the scoring function of the learner is based upon the single best rule like PRankBoost.Clause. The second does just the reverse by retaining only the single best rule, but scoring it based upon the entire trajectory. As a final variation we have also alternated between PRankBoost.Clause and PRankBoost.Path for each weak learner created.

We ran experiments using the same experimental setup as my previous experiments. Results for these three new weak learners appear in Figure 4. It appears that these variations do not find models whose precision-recall performance is consistently higher than PRankBoost.Clause and PRankBoost.Path models when measuring AURPC. However the first mixed model (dashed line) does show some interesting properties. Its initial performance is very low compared to the other models. It has a more shallow learning curve and it does not appear to have reached its asymptotic performance after 100 weak learners have been included in the model. All of these observations make sense when considering the type of weak learner. Each weak learner is an individual clause that will have high precision but low recall due to the scoring function being the area under the entire path. After each weak hypothesis is learned the few positive examples that are covered will be down-weighted and a new weak hypothesis will be learned that covers new examples. Because of the small coverage of each individual clause, learning will be slow and consistent, showing improvement even after many clauses have been learned.

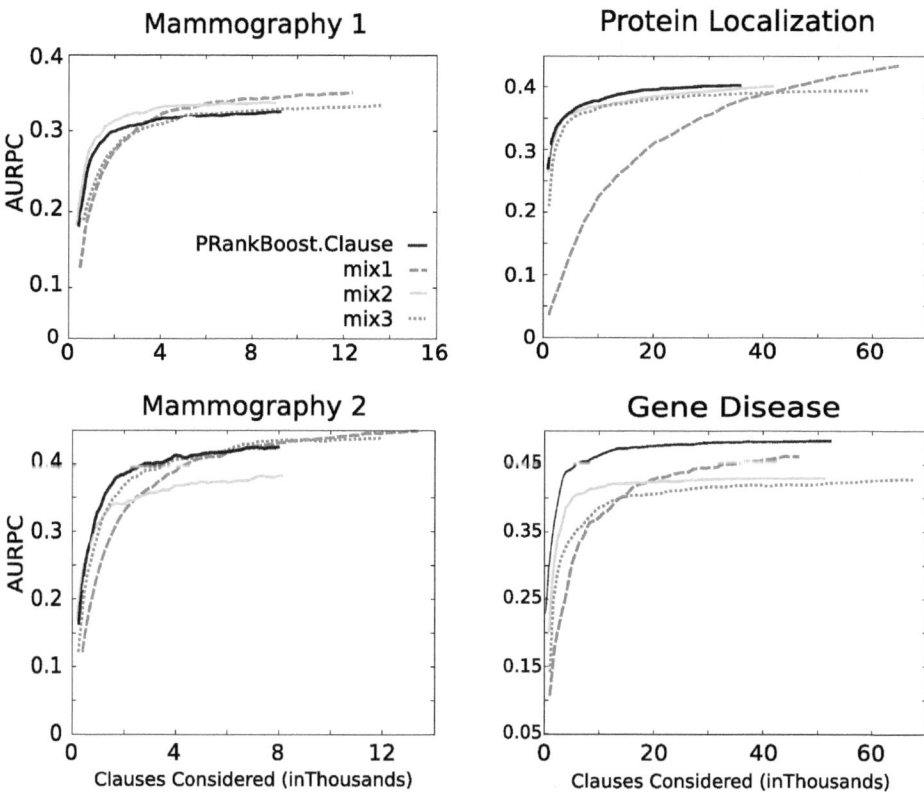

Fig. 4. Learning curves for three models that mix components of PRankBoost.Path and PRankBoost.Clause on four large data sets. *Mix1* includes clauses as weak learners like PRankBoost.Clause but scores them like PRankBoost.Path. *Mix2* includes entire paths of clauses as PRankBoost.Path but scores the path like PRankBoost.Clause. *Mix3* alternates between the method used in PRankBoost.Path and the one used in PRankBoost.Clause. PRankBoost.Clause is graphed for comparison purposes only.

7   Conclusion and Future Work

When working with skewed data sets metrics such as area under the recall-precision curve have been shown to discriminate well between competing models. We designed a modified RankBoost algorithm to maximize area under the recall-precision curve. We compared the original RankBoost algorithm, which is designed to maximize area under the ROC curve, with our modified version. When comparing AUROC on four large, skewed data sets the original Rank-Boost algorithm outperforms our modified PRankBoost version. However when comparing AURPC PRankBoost outperforms the original algorithm.

We created several first-order logic weak learners. The simplest weak learner, PRankBoost.Clause, consists of an individual rule. A second, more complex weak learner, PRankBoost.Path, consists of all rules along the path to the best rule. This more complex learner does not require any additional rules be evaluated on training data. This is especially important when working with large data sets because evaluation is a costly time step. Both weak learners have different strengths with neither learner dominating in performance across all data sets. In addition to these two weak learners We created several other weak learners that are a combination of these two. The most promising, Mix1, consists of the highest-scoring clause found during search, but its score is calculated using the entire trajectory of rules from the most general rule to this best rule.

For future work we would like to create additional mixed models that begin by learning more general clauses as seen in PRankBoost.Clause and then switching to learning more specific clauses as seen in the first mixed model. We believe this type of model will show good initial performance and will continue to show predictive improvement reaching a higher asymptote. We would also like to perform theoretical analysis to support our empirical work showing that PRankBoost maximizes AURPC following Freund et al.'s proof that RankBoost maximizes AUROC [7].

Acknowledgements

We gratefully acknowledge the funding from USA grants R01 LM07050 and 1 R01 CA127379, Houssam Nassif, David Page, and Ryan Woods and our anonymous reviewers for their comments and suggestions.

References

1. Cortes, C., Mohri, M.: AUC optimization vs. error rate minimization. In: Neural Information Processing Systems (NIPS), MIT Press, Cambridge (2003)
2. Davis, J., Burnside, E., Dutra, I., Page, D., Costa, V.: An integrated approach to learning Bayesian networks of rules. In: Gama, J., Camacho, R., Brazdil, P.B., Jorge, A.M., Torgo, L. (eds.) ECML 2005. LNCS (LNAI), vol. 3720, pp. 84–95. Springer, Heidelberg (2005)
3. Davis, J., Goadrich, M.: The relationship between precision-recall and ROC curves. In: Proceedings of the 23rd International Conference on Machine Learning, pp. 233–240 (2006)
4. Dietterich, T.: Ensemble methods in machine learning. In: Kittler, J., Roli, F. (eds.) MCS 2000. LNCS, vol. 1857, pp. 1–15. Springer, Heidelberg (2000)
5. Dutra, I., Page, D., Costa, V., Shavlik, J.: An empirical evaluation of bagging in inductive logic programming. In: Matwin, S., Sammut, C. (eds.) ILP 2002. LNCS (LNAI), vol. 2583, pp. 48–65. Springer, Heidelberg (2002)
6. Džeroski, S., Lavrac, N.: An introduction to inductive logic programming. In: Proceedings of Relational Data Mining, pp. 48–66 (2001)
7. Freund, Y., Iyer, R., Schapire, R., Singer, Y.: An efficient boosting algorithm for combining preferences. In: Proceedings of 15th International Conference on Machine Learning, pp. 170–178 (1998)

8. Freund, Y., Schapire, R.: Experiments with a new boosting algorithm. In: Proceedings of the 13th International Conference on Machine Learning, pp. 148–156 (1996)
9. Goadrich, M., Oliphant, L., Shavlik, J.: Gleaner: Creating ensembles of first-order clauses to improve recall-precision curves. Machine Learning 64(1-3), 231–261 (2006)
10. Quinlan, J.R.: Relational learning and boosting. In: Relational Data Mining, pp. 292–306 (2001)
11. Ray, S., Craven, M.: Representing sentence structure in hidden Markov models for information extraction. In: Proceedings of the 17th International Joint Conference on Artificial Intelligence (2001)
12. Srinivasan, A.: The Aleph manual version 4 (2003),
    http://web.comlab.ox.ac.uk/oucl/research/areas/machlearn/Aleph/

Incorporating Linguistic Expertise Using ILP for Named Entity Recognition in Data Hungry Indian Languages

Anup Patel, Ganesh Ramakrishnan, and Pushpak Bhattacharya

Department of Computer Science and Engineering, IIT Bombay, Mumbai – 400076, India
{anuppatel,ganesh,pb}@cse.iitb.ac.in

Abstract. Developing linguistically sound and data-compliant rules for named entity annotation is usually an intensive and time consuming process for any developer or linguist. In this work, we present the use of two Inductive Logic Programming (ILP) techniques to construct rules for extracting instances of various named entity classes thereby reducing the efforts of a linguist/developer. Using ILP for rule development not only reduces the amount of effort required but also provides an interactive framework wherein a linguist can incorporate his intuition about named entities such as in form of mode declarations for refinements (suitably exposed for ease of use by the linguist) and the background knowledge (in the form of linguistic resources). We have a small amount of tagged data - approximately 3884 sentences for Marathi and 22748 sentences in Hindi. The paucity of tagged data for Indian languages makes manual development of rules more challenging, However, the ability to fold in background knowledge and domain expertise in ILP techniques comes to our rescue and we have been able to develop rules that are mostly linguistically sound that yield results comparable to rules hand-crafted by linguists. The ILP approach has two advantages over the approach of hand-crafting all rules: (i) the development time reduces by a factor of 240 when ILP is used instead of involving a linguist for the entire rule development and (ii) the ILP technique has the computational edge that it has a complete and consistent view of all significant patterns in the data at the level of abstraction specified through the mode declarations. The point (ii) enables the discovery of rules that could be missed by the linguist and also makes it possible to scale the rule development to a larger training dataset. The rules thus developed could be optionally edited by linguistic experts and consolidated either (a) through default ordering (as in TILDE[1]) or (b) with an ordering induced using [2] or (c) by using the rules as features in a statistical graphical model such a conditional random field (CRF) [3]. We report results using WARMR [4] and TILDE to learn rules for named entities of Indian languages namely Hindi and Marathi.

Keywords: Named Entity Recognition, WARMR, TILDE, ILP.

1 Introduction

Identifying entities from unstructured text forms a very important part of information extraction systems. These entities are typically noun phrases and comprise of one to a

L. De Raedt (Ed.): ILP 2009, LNAI 5989, pp. 178–185, 2010.

few tokens in the unstructured text. Named entities like names of persons, locations, and companies are the most popular form of entities as popularized in the MUC [5][6], ACE [7][8], and CoNLL [9] competitions. Named entity recognition was first introduced in the sixth MUC [6] and consisted of three detection subtasks: proper names and acronyms of persons, locations, and organizations (ENAMEX), absolute temporal terms (TIMEX) and monetary and other numeric expressions (NUMEX). Early named entity recognition systems were rule-based with hand-crafted rules [10][11]. Since hand-crafting rules is tedious, algorithms for automatically learning rules were developed [12][13], but these approaches did not provide adequate mechanism for incorporating linguistic knowledge. This paper is organized as follows: Section 2 describes the complexity of Named Entity Recognition for Indian Languages, the motivation for using an ILP approach for this task and some specifics of the ILP approach. In Section 3, we explain our way of representing named entity corpus in first order logic. Lastly in Section 4, we show our experimental results for the ILP and other approaches on Indian Language NER.

2   NER for Indian Languages Using ILP

There has been a lot of work in NER for English and European Languages with claims for high precision and recall. The reason for success in these languages is a very rich tagged corpus and good linguistic insight about the usage of named entities. For Indian languages we do not have this privilege of huge tagged corpus which makes it difficult to have a good linguistic insight about named entities. The table below shows current status of tagged corpus for NER in Hindi and Marathi:

Table 1. Hindi and Marathi named entity corpus

Language	Words	Person Tags	Organization Tags	Location Tags
Marathi	54340	3025	833	997
Hindi	547138	5253	2473	6041

For further analyzing the efforts required for NER in Hindi and Marathi, we analyzed the tagged corpus and recorded some ambiguous cases which create problems in manually developing rules for named entities. Since both languages show similar ambiguities, we have listed some of ambiguities only for Marathi:

Table 2. Ambiguities in named entities found in Indian languages

Ambiguity	Examples
Variations of Proper Nouns	• डॉ. काशिनाथ घाणेकर, डॉ. घाणेकर, डॉक्टर *(Dr. Kashinath Ghanekar, Dr. Ghanekar, Doctor)* • भारतीय जनता पार्टी, भा. ज. पा. *(Bhartiya Janta Party, B. J. P.)*

Table 2. (*continued*)

Person v/s Adjective v/s Verb	• डॉ. लागू/PER यांनी मनोगत मांडले (*Dr. Lagu expressed his thoughts*) • ही योजना संपूर्ण शहरात लागू/JJ करण्यात येणार आहे. *(This scheme will be applicable in the whole city.)* • ..... पण अजिबात झोप लागू/VM दिली नाही. *(..... but he didn't allow me fall asleep at all.)*
Person v/s Common Noun	• मुंबईला आल्यावर डॉक्टरांना/PER फोन करणे भागच होते. *(After coming to Mumbai it was must to call the Doctor.)* • तू डॉक्टर/NN की मी? *(Are you doctor or me?)*
Person v/s Organization	• नेताजींच्या/PER गूढ मृत्यूचा मागोवा ..... *(Following Netaji's suspicious death .....)* • "मिशन नेताजी/ORG' या स्वयंसेवी संस्थेने ..... *("Mission Netaji" is a voluntary organization that .....)*
Organization v/s Location	• पाक/ORG संघ/ORG शनिवारी लंडनमार्ग पाकला/LOC प्रयाण करणार आहे. *(The Pakistan team will go to Pakistan via London on Saturday)*
Person v/s Facility	• सरस्वती आणि लक्ष्मीची/PER एकत्रित उपासना केल्यास ..... *(If Saraswati and Laxmi are worshiped together .....)* • श्रीकृष्ण,सुंदर, लक्ष्मी/FACअशी नाट्य मंदिरे होती. *(There were Drama Theaters like Shri Krishna, Sundar, Laxmi.)*
Location v/s Person	• निगडी येथील भक्ती शक्ती चौक, टिळक/LOC चौक/LOC, ... *(Bhakti Chauk, Tilak Chauk, ... from Nigdi)* • टिळक/PER व डॉ. बाबासाहेब आंबेडकर ..... *(Tilak and Dr. Ambedkar .....)*

Note: The abbreviations ORG=Organization, PER=Person, FAC=Facility, LOC=Location, NN=Noun, JJ=Adjective, and VM=Verb

The above ambiguous cases motivate us to use ILP for learning named entity rules for Indian languages. Following are the benefits of using ILP for inducing named entity rules in Indian language:

i. **Incorporating linguistic expertise using mode declaration:** Developing hand-crafted rules for named entities in the presence of ambiguities could lead to rules that may produce false positives (in other words imprecise). This makes it difficult for a linguist to get a named entity rule correct in the first shot; (s)he has to undergo a number of iterations of manually refining each rule until the rule is precise enough. On the other hand, if the linguist used an ILP technique then (s)he needs to only give high-level specification for the search space of rules in the form of mode declaration for refinements of rules. The onus is then on the ILP technique to produce rules with good confidence and/or support resulting in good overall precision. Our experience with NER for Hindi and Marathi shows that ILP techniques have a computational advantage in coming up with a good and consistent set of named entity rules in considerably less time compared to process of hand-crafting rules.

ii. **Incorporating linguistic expertise using background knowledge:** Since most of the Indian languages currently have very small tagged corpus, the linguist has to apply apriori knowledge about named entities while hand-crafting rules to cover cases not occurring in the tagged corpus. ILP techniques provide a principled approach of applying such apriori knowledge in the form of the background knowledge.

iii. **Covering all significant rules:** There is always a possibility of human error in covering all hand-crafted rules. Consequently a significant rule may be missed out. However, ILP techniques (such as WARMR) will never miss out any such rule that can be generated by successive application of mode declarations provided by the linguist. If the mode declarations are complete enough the ILP approach can yield all possible significant rules.

The above benefits illustrate that ILP does not substitute a linguist but it is excellently complements the linguist by helping him save efforts and also by improving his ability to come up with a complete set of significant rules. There are a number of ways in which we can use the rules learned by ILP, but for simplicity we suggest three ways of consolidating the learned rules in a named entity annotator:

a) Retain the default ordering of learned rules in the rule firing engine.
b) Induce an ordering on the learned rules using greedy heuristics such as [2].
c) Construct a feature corresponding to each rule, with the feature value 1 if the rule covers an instance and 0 otherwise. The features (which can be functions of both the head as well as the body of the rules) can be used in a statistical graphical model such as CRF [3].

We could use several ILP techniques for learning rules, but we shall experiment with only two techniques:

1. **WARMR:** This is an extension of the apriori algorithm to first-order logic. Typically apriori based techniques are computationally expensive. The resulting rules are not ordered and we need to explicitly induce ordering using some heuristic or greedy approach since ordering a decision list is a NP-hard problem [2]. Consolidation techniques **b)** and **c)** are suitable in this case.

2. **TILDE:** This is an extension of traditional C4.5 decision tree learner to first-order logic. Decision tree induction algorithms are usually greedy and hence computationally faster than WARMR like algorithms. Since a decision tree can be serialized to an equivalent ordered set of rules (decision list). Consolidation technique **a)** is suitable in this case.

3 Representing Named Entity Corpus in First Order Logic

Most ILP systems require input examples, background knowledge and mode declarations in the form of first order predicates. Therefore, in order to apply WARMR or

डॉ. काशिनाथ घाणेकर भारावल्यासारखे माझ्या आईसमोर नमस्कारासाठी वाकले.
Getting carried away, Dr. Kashinath Ghanekar bowed in front of my mother.

(a) Sample Marathi Sentence

b_entity(PER). b_entity(ORG). b_entity(LOC).   ...
b_word(डॉ.). b_word(काशिनाथ). b_word(घाणेकर).   ...
b_postag(NNP). b_postag(NN). b_postag(DT).   ...
b wordcollections(titles). b wordcollections(firstnames).   ...

(b) Sample Background Knowledge

p_entity(X,0,PER) :- p_word(X,-1,डॉ.), p_postag(X,0,NNP).

(c) Example of learned rule based on mode declarations

p_entity(d0s10w1,-1, PER).
p_postag(d0s10w1,-1,NNP).
p_word(d0s10w1,-1, डॉ.).
p_wordcollections(d0s10w1,-1,titles).
p_entity(d0s10w1,0, PER).
p_postag(d0s10w1,0, NNP).
p_word(d0s10w1,0, काशिनाथ).
p_wordcollections(d0s10w1,0, firstnames).
p_entity(d0s10w1,1, PER).
p_postag(d0s10w1,1,NNP)
p_word(d0s10w1,1, घाणेकर).
p_wordcollections(d0s10w1,1, lastnames).

(d) Input Example for word काशिनाथ

Fig. 1. An input example for word in the sample sentence

TILDE systems to learn rules for NER we first need to convert tagged data into appropriate first order logic. We convert Hindi and Marathi named entity tagged data into first order logic in the following manner:

i. **Input Examples:** We will have one input example for each word of each sentence from the corpus. Each input example is a set of predicates describing a set of properties of the word and surrounding words in a window of size one. Each example will have unique identifier and properties of words are represented by 3-ary predicates. The first argument of each predicate is the unique identifier for example, second argument is relative position of word whose property we are describing and third argument is value of the property we are describing. As an illustration, consider the input example shown in Figure 1 (d) for word काशिनाथ in the sample Marathi sentence shown in Figure 1 (a). For simplicity we have shown only four predicates describing properties of words, but in our implementation we have used many more predicates.

ii. **Background Knowledge:** In background knowledge we assert more facts about the constants appearing as third argument of the predicates used in input examples. For simplicity we have used only unary predicates in our representation but in general any horn clause can be used. Figure 1 (b) shows a sample background knowledge created for the sample sentence shown in Figure 1 (a).

iii. **Mode declarations:** In most of the ILP systems mode declarations are represented using built-in predicates, which vary from system to system. These mode declarations restrict hypothesis search space for ILP system and also control the predicates appearing in the learned rules. In our case predicate $p\_entity(X,0,...)$ should appear the head of learned rule and all other predicates in the body of learned rule. As an example of the rules we desire to learn consider rule in Figure 1(c).

4   Experimental Results

We have use a hand-crafted rule based named-entity recognizer for Marathi developed by a linguist using the GATE [14] system. The rules were hand-crafted over a period of 1 month (240 hours for 8 hours per day). We measured the performance of hand-crafted rule based system on a completely tagged corpus (3884 sentences and 54340 words).

We learnt Marathi named entity rules using the WARMR and TILDE systems available as a part of ACE [15] data mining system. For induction of rules using 80% (3195 sentences and 42635 words) of tagged corpus, TILDE took 1 hour and WARMER took 140 hours (5 days and 20 hours). This gives us and reduction in time for rule development by factor of 240 for TILDE and by a factor of 1.7 for WARMR. To compare the quality of the learnt rules we consolidated the rules and applied them over the remaining 20% (689 sentences and 11705 words) of the tagged corpus in following ways:

TILDE Rule Based NER: Rules learned by TILDE are plugged into a rule-based named entity recognizer without altering the order of rules.

WARMR Rule Based NER: Rules learned by WARMR are ordered using simple precision score heuristic and a greedy algorithm mentioned in [2]. These ordered rules are then plugged into a rule-based named entity recognizer.

WARMR CRF Based NER: Rules learned by WARMR plugged into CRF [16] as features ignoring the order of rules.

The performances of the hand-crafted rule based (HR), the TILDE rule based (TR), the WARMR rule based (WR), and the WARMR CRF based (WC) systems are shown below in Table 3 for Marathi.

Table 3. Experimental results for Marathi

Entity	Precision				Recall				F-Measure			
	HR	TR	WR	WC	HR	TR	WR	WC	HR	TR	WR	WC
PER	0.61	0.55	0.60	**0.74**	0.70	**0.99**	0.90	0.91	0.65	0.71	0.72	**0.82**
ORG	0.15	**0.85**	0.19	0.59	0.10	0.37	0.46	**0.52**	0.12	0.51	0.27	**0.55**
LOC	0.51	**0.54**	0.41	0.51	0.24	0.18	0.35	**0.45**	0.33	0.27	0.38	**0.48**

5   Conclusion

We have reported an inductive approach to rule building for two Indian Languages, which generally yields substantially more accurate rules than manual construction, while offering significant reduction in construction time. Further, the ILP approach is easy to maintain, given the clean separation between background knowledge, example set and learning algorithm. Also, probabilistic reasoning on the induced rules is found to be more effective than employing the rules deterministically.

To take the work further, we need to find efficient methods of rule induction on large volumes of annotated data, developing interactive ILP assisted rule development system for linguists, harnessing morphological properties and finally, including the language base to include other languages.

References

[1] Blockeel, H., Raedt, L.D.: Top-down induction of logical decision trees. In: Artificial Intelligence (1998)

[2] Chakravarthy, V., Joshi, S., Ramakrishnan, G., Godbole, S., Balakrishnan, S.: Learning Decision Lists with Known Rules for Text Mining. In: The Third International Joint Conference on Natural Language Processing, IJCNLP 2008 (2008)

[3] Lafferty, J., McCallum, A., Pereira, F.: Conditional Random Fields: Probabilistic Models for Segmenting and Labeling Sequence Data. In: Proceedings of the International Conference on Machine Learning, ICML 2001 (2001)

[4] Dehaspe, L., De Raedt, L.: Mining association rules in multiple relations. In: Džeroski, S., Lavrač, N. (eds.) ILP 1997. LNCS, vol. 1297, pp. 125–132. Springer, Heidelberg (1997)

[5] Chinchor, N.A.: Overview of MUC-7/MET-2 (1998)

[6] Grishman, R., Sundheim, B.: Message understanding. Conference-6: A brief history, pp. 466–471 (1996)

[7] Grishman, R., Sundheim, B.: Automatic content extraction program. In: NIST (1998)

[8] Annotation guidelines for entity detection and tracking (2004),
http://www.ldc.upenn.edu/Projects/ACE/docs/
EnglishEDTV4-2-6.PDF

[9] Tjong Kim Sang, E.F., Meulder, F.D.: Introduction to the conll-2003 shared task: Language. In: Seventh Conference on Natural Language Learning (CoNLL 2003), pp. 142–147 (2003)

[10] Appelt, D.E., Hobbs, J.R., Bear, J., Israel, D.J., Tyson, M.: Fastus: A finite-state processor for information extraction from real-world text. In: IJCAI, pp. 1172–1178 (1993)

[11] Riloff, E.: Automatically constructing a dictionary for information extraction tasks. In: AAAI, pp. 811–816 (1993)
[12] Califf, M.E., Mooney, R.J.: Relational learning of pattern-match rules for information extraction. In: Proceedings of the Sixteenth National Conference on Artificial Intelligence (AAAI 1999), pp. 328–334 (1999)
[13] Soderland, S.: Learning information extraction rules for semi-structured and free text. In: Machine Learning (1999)
[14] Cunningham, H., Maynard, D., Bontcheva, K., Tablan, V.: Gate: An architecture for development of robust HLT applications. In: Recent Advanced in Language Processing, pp. 168–175 (2002)
[15] Blockeel, H., et al.: Machine Learning Group - ACE Dataming System (March 2008), http://www.cs.kuleuven.be/~dtai/ACE/doc/ACEuser-1.2.12-r1.pdf
[16] Sarawagi, S.: CRF Project Page (2004), http://crf.sourceforge.net/

Transfer Learning via Relational Templates

Scott Proper and Prasad Tadepalli

Oregon State University
Corvallis, OR 97331-3202, USA
{proper,tadepall}@eecs.oregonstate.edu

Abstract. Transfer Learning refers to learning of knowledge in one domain that can be applied to a different domain. In this paper, we view transfer learning as generalization of knowledge in a richer representation language that includes multiple subdomains as parts of the same superdomain. We employ relational templates of different specificity to learn pieces of additive value functions. We show significant transfer of learned knowledge across different subdomains of a real-time strategy game by generalizing the value function using relational templates.

Keywords: model-based, afterstates, transfer learning, relational reinforcement learning, Markov decision processes.

1 Introduction

Transfer learning is defined as transferring knowledge learned from one domain to accelerate learning in another domain. In this paper, we argue that transfer learning can be viewed as generalization in a rich representation language over a superdomain that includes multiple subdomains as special cases. In particular, we consider a language that includes relational templates that may each be instantiated to yield pieces of an additive value function. Each piece of the value function may be applicable to a single subdomain, multiple subdomains, or to the entire superdomain based on its structure. The advantage of this approach is that transfer happens naturally through generalization.

In this paper we learn value functions for a model-based multiagent reinforcement learning problem. The value function is a sum of terms, which are entries in multiple tables. Each term is generated by instantiating a relational template with appropriate parameters. By indexing the table entries with types of units, we learn values that correspond to units of only certain types. By not indexing them with any type, one could learn more general terms that apply across all subdomains. Thus controlling the specificity of the relational templates controls the generalization ability and hence the the ability to transfer of the system.

We learn value functions for real-time strategy games that includes multiple instances of multiple types of units. Each subdomain consists of a small number of units of certain type. We train the system on a set of multiple subdomains, and test the system on a test domain. We show that in most cases, there is positive transfer. In other words after training from similar domains, the performance in

L. De Raedt (Ed.): ILP 2009, LNAI 5989, pp. 186–193, 2010.
© Springer-Verlag Berlin Heidelberg 2010

the test domain raises more quickly than it would without such prior training on similar domains. However, we see some instances of negative transfer, and some cases where adding more domains to the training set decreases the rate of learning in the test domain. While the results support the thesis that transfer learning can be successfully viewed as generalization in a suitably rich relational language, it also suggests that if performance in a single test domain is the goal, it is often best to train it on a similar training domain.

The rest of the paper is organized as follows: Section 2 describes our learning approach, namely, Afterstate Temporal Difference Learning. Section 3 describes the value function approximation based on relational templates. Section 4 describes the experimental results, followed by our conclusions in the final section.

2   Afterstate Total Reward Learning

We assume that the learner's environment is modeled by a Markov Decision Process (MDP), defined by a 4-tuple $\langle S, A, p, r \rangle$, where S is a discrete set of states, and A is a discrete set of actions. Action u in a given state $s \in S$ results in state s' with some fixed probability $p(s'|s, u)$ and a finite immediate reward $r(s, u)$. A *policy* μ is a mapping from states to actions. Here, we seek to optimize the total expected reward received until the end of the episode starting from state s. This is denoted by $v(s)$ and satisfies the following Bellman equation for non-absorbing states. The value function is 0 for all absorbing states:

$$v(s) = \max_{u \in A} \left\{ r(s, u) + \sum_{s'=1}^{N} p(s'|s, u)v(s') \right\} \qquad (1)$$

The optimal policy chooses actions maximizing the right hand side of this equation. We can use the above Bellman equation to update $v(s)$. However this is often computationally expensive because of high stochasticity in the domain. Thus, we want to replace this with a sample update as in model-free reinforcement learning. To do this, we base our Bellman equation on the "afterstate," which incorporates the deterministic effects of the action on the state but not the stochastic effects [1,2]. Since conceptually the afterstate can be treated as the state-action pair, it strictly generalizes model-free learning. We can view the progression of states/afterstates as $s \xrightarrow{a} s_a \rightarrow s' \xrightarrow{a'} s'_{a'} \rightarrow s''$. The "a" suffix used here indicates that s_a is the afterstate of state s and action a. The stochastic effects of the environment create state s' from afterstate s_a with probability $P(s'|s_a)$. The agent chooses action a' leading to afterstate $s'_{a'}$ and receiving reward $r(s', a')$. The environment again stochastically selects a state, and so on. We call this variation of afterstate total-reward learning "ATR-learning". ATR-learning is similar to ASH-learning from [2], however we use total reward here instead of average reward. The Bellman equation for ATR-learning is as follows:

$$v(s_a) = \sum_{s'=1}^{N} \left\{ p(s'|s_a) \left[\max_{u \in A} \left\{ r(s', u) + v(s'_u) \right\} \right] \right\} \qquad (2)$$

We use sampling to avoid the expensive calculation of the expectation above. At every step, the ATR-learning algorithm updates the parameters of the value function in the direction of reducing the temporal difference error (TDE), i.e., the difference between the r.h.s. and the l.h.s. of the Bellman equation:

$$TDE(s_a) = \max_{u \in U(s')} \{r(s', u) + v(s'_u)\} - v(s_a) \tag{3}$$

The ATR-learning algorithm is shown in Figure 1.

Initialize afterstate value function $v(.)$
Initialize s to a starting state
for each step **do**
    Find action u that maximizes $r(s, u) + v(s_u)$
    Take an exploratory action or a greedy action in the state s. Let a be the joint action taken, r the reward received, s_a the corresponding afterstate, and s' be the resulting state.
    Update the model parameters $r(s', a)$.
$$v(s_a) \leftarrow v(s_a) + \alpha \left(\max_{u \in A} \{r(s', u) + v(s'_u)\} - v(s_a) \right)$$
    $s \leftarrow s'$
end for

Fig. 1. The ATR-learning algorithm, using the update of Equation 3

3 Function Approximation via Relational Templates

In this paper, we are interested in object-oriented domains where the state consists of multiple objects or units O of different classes, each with multiple attributes. Relational templates generalize the tabular linear value functions (TLFs) of [2] to object-oriented domains [3]. As with TLFs, relational templates also generalize tables, linear value functions, and tile coding. A relational template is defined by a set of relational features over shared variables (see Table 1). Each template is instantiated in a state by binding its variables to units of the correct type. An instantiated template i defines a table θ_i indexed by the values of its features in the current state. In general, each template may give rise to multiple instantiations in the same state. The value $v(s)$ of a state s is the sum of the values represented by all instantiations of all templates.

$$v(s) = \sum_{i=1}^{n} \sum_{\sigma \in \mathcal{I}(i,s)} \theta_i(f_{i,1}(s, \sigma), \ldots, f_{i,m_i}(s, \sigma)) \tag{4}$$

where i is a particular template, $\mathcal{I}(i, s)$ is the set of possible instantiations of i in state s, and σ is a particular instantiation of i that binds the variables of the template to units in the state. The relational features $f_{i,1}(s, \sigma), \ldots, f_{i,m_i}(s, \sigma)$ map state s and instantiation σ to discrete values which index into the table θ_i.

Table 1. Various relational templates used in experiments. See Table 2 for descriptions of relational features.

Template #	Description
#1	$\langle Distance(A, B), UnitHP(B), EnemyHP(A), UnitsInrange(B) \rangle$
#2	$\langle UnitType(B), EnemyType(A), Distance(A, B), UnitHP(B), EnemyHP(A), UnitsInrange(A) \rangle$
#3	$\langle UnitType(B), Distance(A, B), UnitHP(B), EnemyHP(A), UnitsInrange(A) \rangle$
#4	$\langle EnemyType(A), Distance(A, B), UnitHP(B), EnemyHP(A), UnitsInrange(A) \rangle$
#5	$\langle UnitX(A), UnitY(A), UnitX(B), UnitY(B) \rangle$

All instantiations of each template i share the same table θ_i, which is updated for each σ using the following equation:

$$\theta_i(f_{i,1}(s, \sigma), \ldots, f_{i,m_i}(s, \sigma)) \leftarrow \theta_i(f_{i,1}(s, \sigma), \ldots, f_{i,m_i}(s, \sigma)) + \alpha(TDE(s, \sigma)) \quad (5)$$

where α is the learning rate. This update suggests that the value of $v(s)$ would be adjusted to reduce the temporal difference error in state s. In some domains, the number of objects can grow or shrink over time: this merely changes the number of instantiations of a template.

We say a template is more refined than another if it has a superset of features. The refinement relationship defines a hierarchy over the templates with the base template forming the root and the most refined templates at the leaves. The values in the tables of any intermediate template in this hierarchy can be computed from its child template by summing up the entries in its table that refine a given entry in the parent template. Hence, we can avoid maintaining the intermediate template tables explicitly. This adds to the complexity of action selection and updates, so our implementation explicitly maintains all templates.

4   Experimental Results

We performed all experiments on several variations of a real-time strategy game (RTS) simulation. We tested 3 units vs. a single, more powerful enemy unit. As the enemy unit is more powerful than the friendly units, it requires coordination to defeat. Units also vary in type, requiring even more complex policies and coordination. We show how we can easily deal with differences between unit types by taking advantage of transfer learning and relational templates.

We implemented a simple real-time strategy game simulation on a 10x10 gridworld. The grid is presumed to be a coarse discretization of a real battlefield,

Table 2. Meaning of various relational features

Feature	Meaning
$Distance(A, B)$	Manhattan distance between enemy A and unit B
$UnitHP(B)$	Hit points of an friendly unit B
$EnemyHP(A)$	Hit points of an enemy A
$UnitsInrange(A)$	Count of the number of units in range of (able to attack) enemy A
$UnitX(A)$	X-coordinate of units A
$UnitY(A)$	Y-coordinate of units A
$UnitType(B)$	Type (archery or infantry) of unit B
$EnemyType(A)$	Type (tower, ballista, or knight) of enemy A

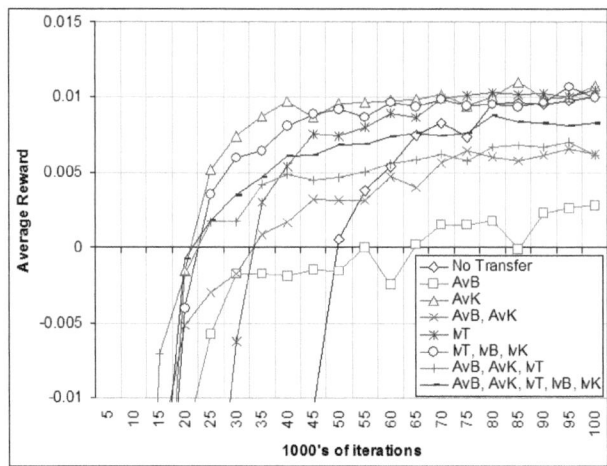

Fig. 2. Comparison of training on various source domains transferred to the Archers vs. Tower domain

and so units are permitted to share spaces. Units, either enemy or friendly, were defined by several features: position (in x and y coordinates, hit points (0-6), and type (see Table 3). We also defined relational features such as distance between agents and the assigned enemy unit, and aggregation features such as a count over the number of opposing units within range. In addition each unit type was defined by how many starting hit points it had, how much damage it did, the range of its attack (in manhattan or "city block" distance), and whether it was mobile or not. Friendly units were always created as one of the weaker unit types (archer or infantry), and enemy units were created as one of the stronger types (tower, ballista, or knight).

Friendly units had five actions available each time step: move in one of four directions, wait, or attack the enemy (if in range). Enemy units had the same options (although a choice of whom to attack) and followed predefined policies, either attacking if in range or approaching (if mobile). An attack at a unit within range always hits, damaging it and killing it (removing it from the game) if it is reduced to 0 hit points. Thus, the number of units is reduced over time. Eventually, one side "wins" by destroying the other, or a time limit of 20 steps

Table 3. Different unit types

Unit	HP	Damage	Range	Mobile
Archer	3	1	3	yes
Infantry	6	1	1	yes
Tower	6	1	3	no
Ballista	2	1	5	yes
Knight	6	2	1	yes

is reached. We gave a reward of $+1$ for a successful kill of an enemy unit, a reward of -1 if an friendly unit is killed, and a reward of $-.1$ each time step to encourage swift completion. Thus, to receive positive reward, it is necessary for units to coordinate with each other to quickly kill enemy units without any losses of their own.

We used relational templates to define the value function (see Table 1). Template #1 is a "base" template which does not distinguish between types of objects and captures a generic function. We adapt this base template to create more refined templates which take into account differences between types of objects (see Tables 2 and 1). Templates #3-4 generalize across enemy unit types and friendly unit types respectively. Template #2 learns specialized domain-specific functions. Template #5 facilitates coordination between friendly units. Together, these templates create a "superdomain" capable of specializing to specific domains and generalizing across domains.

The results of our experiments are shown in Figures 2 and 3. Both figures show the influence that learning on various combinations of source domains has on the final performance of a different target domain (Archers vs. Tower or Infantry vs. Knight). We trained a value function for 10^6 steps on the various combinations of source domains indicated in each figure. Abbreviations such as "AvK" indicates a single kind of domain – Archers vs. Knights for example. Likewise I,T,B indicate Infantry, Towers, and Ballista respectively. When training multiple domains at once, each episode was randomly initialized to one of the allowable combinations of domains. We transferred the parameters of the relational templates learned in these source domains to the target "AvT" or "IvK" domains, and tested each target domain for 30 runs of 10^5 steps each, averaging the results of each run together. With all experiments conducted here, we used $\epsilon = .1$, $\alpha = .1$ and the ATR-learning algorithm.

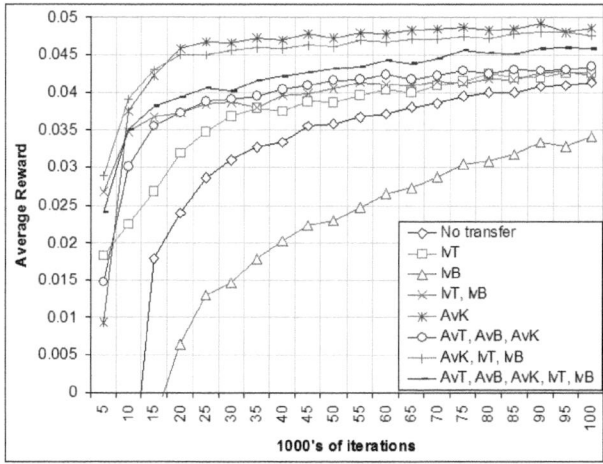

Fig. 3. Comparison of training on various source domains transferred to the Infantry vs. Knight domain

Our results show that additional relevant knowledge (in the form of training on source domains that share a unit type with the target domain) is usually helpful, though not always. For example, in the IvK target domain, training on the IvB domain alone performs worse than not using transfer learning at all. However, training IvB and IvT together is better than training on IvT alone, and training on IvT is much better than no transfer at all. These results also show that irrelevant information – training on the AvT and AvB domains, which do not share a unit type with the IvK domain – always harms transfer.

For the AvT target domain, transfer from any domain initially performs better than no transfer at all, but only a few source domains continue to perform better than no transfer by the end of each run. In both target domains the "AvK" source domain provides the best possible training for both target domains. The IvT, and IvT, IvB, and IvK source domains also perform well here.

5   Discussion

We have shown how relational templates may be refined from a "base" template that is applicable to all subdomains to more detailed templates that can specialize to particular subdomains based on the type features. By using several templates with different combinations of type features, we create a function approximator that generalizes between similar subdomains and also specializes to particular subdomains. This process allows for easy transfer of knowledge between subdomains.

Transfer learning using relational templates has limitations: for example, it can be difficult to know how particular combinations of source domains will influence the performance on the target domain. Additionally, it is not clear what the best way to train on source domain(s) might be - is it best to select randomly between available source domains each episode as I did in this paper, or should the value function be trained on these domains in some particular order? A random selection was used to ensure that general, transferable knowledge was not "forgotten" as the value function was trained on new source domains, but there may be a more efficient way to accomplish this.

Relational templates resemble the "class-based local value subfunctions" of [3], however in this paper we do not assume that relationships between objects, or even the exact numbers of objects in a domain, remain fixed throughout time. We also allow templates to reference multiple objects in the world. This flexibility makes relational templates a powerful and general function approximator.

Relational templates may also be used to transfer knowledge from small domains (such as that shown here) to larger domains with many units. We can do this using a technique called assignment-based decomposition [4], which decomposes the action-selection step of any reinforcement learning algorithm into an assignment level that assigns units to particular tasks, and a task execution level that chooses actions for units given their assigned task. This is similar to hierarchical reinforcement learning [5]. However we replace a joint value function at the root level with an approximate search technique over possible assignments.

Also introduced in this paper is ATR-learning. ATR-learning is a new model-based total-reward reinforcement learning algorithm that uses afterstates to eliminate the explosion in stochastic branching factor, or number of possible next states. This is important in domains with many agents controlled by the environment (or opposing team) such as most RTS games. The domain shown in this paper has only one enemy unit and so does not suffer from a high stochastic branching factor, but this would not be the case were the number of enemy units to be increased.

Acknowledgments

We gratefully acknowledge the support of the Defense Advanced Research Projects Agency under grant number FA8750-05-2-0249 and the National Science Foundation for grant number IIS-0329278.

References

1. Sutton, R.S., Barto, A.G.: Reinforcement learning: an introduction. MIT Press, Cambridge (1998)
2. Proper, S., Tadepalli, P.: Scaling model-based average-reward reinforcement learning for product delivery. In: Fürnkranz, J., Scheffer, T., Spiliopoulou, M. (eds.) ECML 2006. LNCS (LNAI), vol. 4212, pp. 735–742. Springer, Heidelberg (2006)
3. Guestrin, C., Koller, D., Gearhart, C., Kanodia, N.: Generalizing plans to new environments in relational mdps. In: IJCAI 2003: International Joint Conference on Artificial Intelligence, pp. 1003–1010 (2003)
4. Proper, S., Tadepalli, P.: Solving multiagent assignment markov decision processes. In: AAMAS 2009: Proceedings of the 8th International Joint Conference on Autonomous Agents and Multiagent Systems, pp. 681–688 (2009)
5. Makar, R., Mahadevan, S., Ghavamzadeh, M.: Hierarchical multi-agent reinforcement learning. In: AGENTS 2001: Proceedings of the 5th International Conference on Autonomous Agents, Montreal, Canada, pp. 246–253. ACM Press, New York (2001)

Automatic Revision of Metabolic Networks through Logical Analysis of Experimental Data

Oliver Ray[1], Ken Whelan[2], and Ross King[2]

[1] University of Bristol, Bristol, BS8 1UB, UK
oray@cs.bris.ac.uk
[2] University of Aberystwyth, Ceredigion, SY23 3DB, UK
{knw,rdk}@aber.ac.uk

Abstract. This paper presents a nonmonotonic ILP approach for the automatic revision of metabolic networks through the logical analysis of experimental data. The method extends previous work in two respects: by suggesting revisions that involve both the addition and removal of information; and by suggesting revisions that involve combinations of gene functions, enzyme inhibitions, and metabolic reactions. Our proposal is based on a new declarative model of metabolism expressed in a nonmonotonic logic programming formalism. With respect to this model, a mixture of abductive and inductive inference is used to compute a set of minimal revisions needed to make a given network consistent with some observed data. In this way, we describe how a reasoning system called XHAIL was able to correctly revise a state-of-the-art metabolic pathway in the light of real-world experimental data acquired by an autonomous laboratory platform called the Robot Scientist.

1 Introduction

Metabolic networks are formal descriptions of the enzyme-catalysed biochemical transformations that mediate the breakdown and synthesis of molecules within a living cell. Logic programs are useful for representing and reasoning about such networks as they provide an expressive relational language with efficient computational support for deduction, abduction and induction. Moreover, the recent development of nonmonotonic learning systems such as XHAIL (eXtended Hybrid Abductive Inductive Learning) [9] means that the full potential of logic programs with both classical and default negation can be now exploited for expressing or inferring defaults and exceptions under uncertainty.

This paper presents a nonmonotonic model of metabolism which can be used to compute a set of minimal revisions that make a given network consistent with some observed data. We explain how this model was used by XHAIL to revise a state-of-the-art metabolic pathway in the light of real-world data acquired by an autonomous laboratory platform called the Robot Scientist [2]. Our method goes beyond earlier work by suggesting revisions that involve the addition and removal of hypotheses about gene functions, enzyme inhibitions and metabolic reactions. The resulting hypotheses have been tested and thereby led to theoretical and experimental benefits in the Robot Scientist project.

L. De Raedt (Ed.): ILP 2009, LNAI 5989, pp. 194–201, 2010.

2 Background

Metabolic networks [4] are collections of interconnected biochemical reactions that mediate the synthesis and breakdown of essential compounds within a cell. These reactions are catalysed by specific enzymes whose amino acid sequences are specified in regions of the host genome called Open Reading Frames (ORFs). The activity of particular pathways within such networks are controlled by regulating the expression of those genes on which the associated ORFs are located. One such pathway is exemplified in Figure 1, below, which shows the Aromatic Amino Acid (AAA) biosynthesis pathway of the yeast *S. cerevisiae*.

Nodes in the graph are the chemical names of metabolites involved in the transformation of the start compound Glycerate-2-phosphate into the amino acids Tyrosine, Phenylalanine, and Tryptophan. Arrows are chemical reactions from substrates to products. Each node is labelled with a KEGG identifier (e.g., C00082); and each arrow is annotated with an EC number (e.g., 2.5.1.54) and a set of ORF identifiers (e.g., YBR249C). The single dashed line shows the inhibition of an ORF by a metabolite, while the double dashed line is the cellular membrane, which separates the cell cytosol from the growth medium;

ORFs appearing on top of each other are iso-enzymes, which independently catalyse a reaction; while ORFs appearing next to each other are enzyme-complexes, which jointly catalyse a reaction. All reactions take place in the cell cytosol using nutrients imported from the growth medium; and they proceed at a standard rate (less than 1 day), except for the importation of two italicised compounds *C01179* and *C00166*, which take longer (more than 1 day).

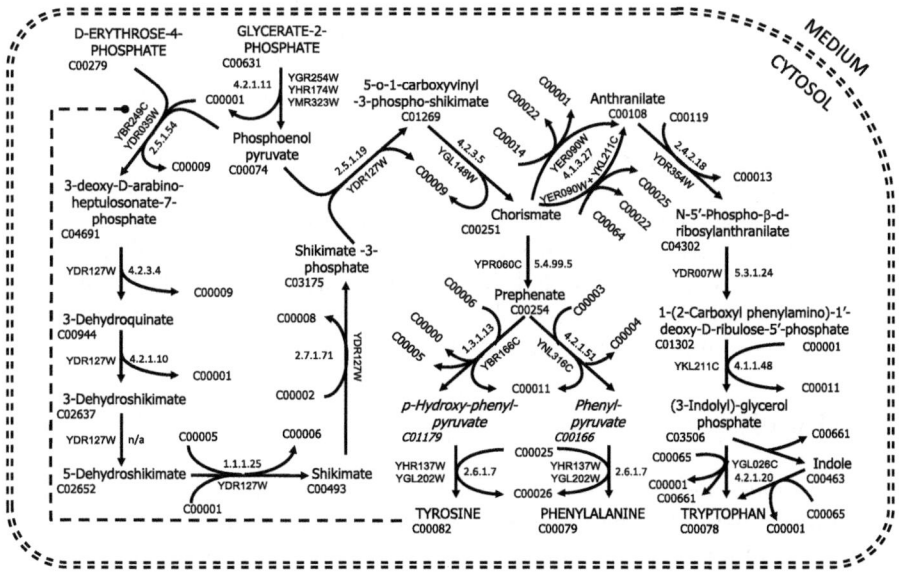

Fig. 1. Aromatic Amino Acid (AAA) biosynthesis pathway of the yeast *S. cerevisiae*

Logic programs [5] are well suited to representing and reasoning about biological networks as they provide an expressive relational language supported by efficient tools for deductive (consequence-based) [5], abductive (assumption-based) [1], and inductive (generalisation-based) [8] inference. XHAIL (eXtended Hybrid Abductive Inductive Learning) [9] is a nonmonotonic ILP system that combines all three inference types within a common logical framework. The system takes as input a background theory B and a set of examples E, to return as output a set of hypotheses H that entail E with respect to B. The hypothesis space is constrained by a set of user-defined mode declarations [6] and is filtered by a compression heuristic [6] that prefers hypotheses with fewer literals.

Each hypothesis H is obtained by constructing and generalising a preliminary ground theory K, called a Kernel Set, which bounds the search space according to the prescribed bias. First, the head atoms of K are found by an abductive procedure which returns a set of ground atoms Δ that entail E with respect to B. Then, the body atoms of K are found by a deductive procedure which returns a set of ground atoms Γ that are entailed by B with respect to Δ. Finally, K is generalised by an inductive procedure which returns a compressive hypothesis H that subsumes K. A key feature of XHAIL is its ability to exploit traditional forms of machine learning bias in a nonmonotonic reasoning setting; and thereby increase the class of learning problems that are soluble in practice [9].

3   Objective

Our work is part of a project called the Robot Scientist [3], of which a key aim is to mechanise the revision of metabolic networks through the integration of laboratory automation and artificial intelligence. This project has developed hardware to autonomously conduct a class of microbial growth experiments whereby strains of yeast (from which certain ORFs are knocked out) are cultured in growth media (to which certain nutrients are added) and the resulting growth is regularly monitored for a number of days [2].

Over time, such experiments tend to produce results that contradict the best available models, which must therefore be revised. So far, this has been done by hand — as in the case of the AAA pathway above, which was derived from the KEGG database, but was manually tuned to better explain preliminary results from a semi-automated Robot Scientist prototype [3]. Now, our goal is to mechanise this revision process by using XHAIL to analyse new growth data acquired by an improved fully-automated Robot Scientist platform [2].

4   Model

The metabolic model we have developed to fulfil our objectives has the following basic types: ORFs and metabolites, which are denoted by their KEGG identifiers; enzyme-complexes and reactions, which are given unique integer identifiers; days and experiments, which are also represented by integers; and extra-cellular or intra-cellular compartments, of which we only use the medium and cytosol.

The additional nutrients and knockout strains used in each growth experiment are represented by ground atoms of the form `additional_nutrient(e,m)` and `knockout(e,o)`, for some particular experiment e, ORF o, and metabolite m. In addition, a minimal set of growth nutrients common to all experiments are represented by ground atoms of the form `start_compound(m)`.

By definition, any metabolite `Met` that is a start compound or additional nutrient is in the compartment `medium` on any `Day` in any experiment `Exp`:

```
in_compartment(Exp,Met,medium,Day) :- start_compound(Met).

in_compartment(Exp,Met,medium,Day) :- additional_nutrient(Exp,Met).
```

Each enzyme-complex is given an integer identifier c. Then, for each reaction catalysed by c, one fact is added to the model of the form `catalyst(r,c)`, where r is the corresponding reaction identifier. Also, for each ORF o needed in the complex c, one fact is added to the model of the form `component(o,c)`.

A ground atom of the form `inhibitor(c,m)` indicates that the complex c is inhibited by the metabolite m. Metabolites that are essential to cell growth, like the three amino acids, are specified as such by ground atoms of the form `essential_compound(m)`.

Cell development is arrested if an essential metabolite is not in the cytosol but growth is predicted otherwise. An enzyme-complex is said to be deleted if a component ORF is knocked out; and it is inhibited if some inhibitor is present (in high concentration) as an additional nutrient:

```
arrested(Exp,Day) :-
    essential_compound(Met), not in_compartment(Exp,Met,cytosol,Day).

predicted_growth(Exp,Day) :- not arrested(Exp,Day).

deleted(Exp,Cid) :- component(Orf,Cid), knockout(Exp,Orf).

inhibited(Exp,Cid) :- inhibitor(Cid,Met), additional_nutrient(Exp,Met).
```

To complete our background theory, it remains to give a logical encoding of the metabolic reactions. To facilitate the addition and removal of reactions, they are each given one of three degrees of belief: *certain* (i.e., definitely in the model), *retractable* (i.e., initially in the model, but can later be excluded), or *assertable* (i.e., initially out of the model, but can later be included). Note that this allows us to consider reactions from related pathways or organisms for inclusion in a revised network. This approach is common practice as it ensures all newly introduced reactions are at least biologically feasible.

For every reaction, one rule is added to the theory for each product. Each rule states that the product will be in its compartment if (i) all substrates are in their respective compartments, (ii) there is an enzyme-complex catalysing the reaction whose activity is not inhibited and whose ORFs are not deleted, (iii) sufficient time has passed for the reaction to complete, and (iv) the reaction has not been excluded (if it is retractable) or it has been included (if it is assertable). As an example, the following is one of two rules produced for reaction 2.5.1.19, assuming it has been given the identifier 31 and a retractable status:

```
in_compartment(Exp,"C01269",cytosol,Day) :-
    in_compartment(Exp,"C00074",cytosol,Day),
    in_compartment(Exp,"C03175",cytosol,Day),
    catalyst(31,Cid),
    not inhibited(Exp,Cid),
    not deleted(Exp,Cid),
    Day >= 1,
    not exclude(31).
```

For every start compound and additional nutrient, m, we assume there is an import reaction which takes m from the medium into the cytosol; and to each reaction with no known catalysts, we attribute an unknown catalyst (so all reactions are assumed to proceed in the absence of evidence to the contrary).

Positive and negative examples, which correspond to results about the growth and non-growth of the yeast in an experiment e on a day d, are denoted by ground literals of the form observed_growth(e,d) or ¬observed_growth(e,d) where, for convenience, we use classical negation to denote a negative result.

The learning task itself is triggered by two integrity constraints requiring that the predicted growth should be made to coincide with observed growth:

```
:- observed_growth(Exp,Day), not predicted_growth(Exp,Day).
```

```
:- predicted_growth(Exp,Day), ¬observed_growth(Exp,Day).
```

The hypothesis space is specified by a set of mode declarations, like those shown below, which allow the inclusion/exclusion of reactions to/from the model:

```
modeh(*,include(#assertable_reaction)).
```

```
modeh(*,exclude(#retractable_reaction)).
```

5   Validation

To validate our approach, we used the AAA model in Figure 1 to generate a set of artificial growth results in some hypothetical growth experiments. We then introduced some changes into the model (so it no longer predicted the same growth results) and used XHAIL to try and repair the model (so it once again predicted the original growth results). We carried out several such proof-of-principle tests by varying both the number of experiments generated and also the type of changes introduced. As explained in [10], these tests confirmed that XHAIL was able to consistently add or remove reactions, discover new enzyme-complexes or enzyme-inhibitions, and perform different combinations of these tasks at the same time. In all cases, XHAIL was able to successfully revise the initial model, although, as might be expected, its execution time was negatively correlated with the complexity of the learning task.

6   Results

The main contribution of this paper comes from applying our approach to real data acquired by the Robot Scientist in order to try and improve the state-of-the-art

AAA pathway model in Figure 1. The growth data was obtained from statistical processing of 40 high quality trials conducted by the Robot Scientist and repeated several times to ensure significance [2]. Upon submitting the growth results to XHAIL, they were immediately found to be inconsistent with the AAA model, thereby suggesting that a revision of the model was necessary.

After experimenting with the language bias for about one hour, we obtained some half a dozen hypotheses corresponding to various revisions that ensured logical consistency between the predicted and observed results. A closer examination showed that each such hypothesis was a conjunction of two conjectures, concerning the metabolites Anthranilate and Indole, all of which fell into one of the following four categories:

(a) Enzyme Complex: There were 3 alternative abductive conjectures of the form `component("YER090W",7)`, `component("YER090W",8)` or `component("YER090W",9)` stating that YER090W forms a complex in any one of the 3 reactions, 2.4.2.18, 5.3.1.24 or 4.1.1.48 (catalysed by complexes 7,8 and 9, respectively) just downstream of the Anthranilate Synthase step 4.1.3.27 in the Tryptophan branch of the AAA pathway.

(b) Anthranilate Import: There was 1 abductive conjecture `inhibited(Exp,25,1)` stating that the import of Anthranilate (which is mediated by a hypothetical enzyme with id 25) is blocked on day 1 in all experiments. This can be understood as meaning that the importation of Anthranilate into the cell should be modelled as a slow reaction such that Anthranilate would only be available in the cell cytosol after the first day.

(c) Indole Interference: In addition, there was 1 inductive conjecture of the form `predicted_growth(Exp,Day):-additional_nutrient(Exp,"C00463")` which suggests that growth is always reported whenever Indole is added. This led us to consider the possibility that the addition of Indole might be confusing the optical growth reader, as a result of its relative insolubility, and thereby leading to biased results.

(d) Indole Contamination: Finally, there was 1 inductive conjecture of the form `additional_nutrient(Exp,"C00078"):-additional_nutrient(Exp,"C00463")` which states that Tryptophan (C00078) is always added whenever Indole (C00463) is added. This can be understood as meaning that the Robot Scientist's source of Indole is contaminated with Tryptophan.

It turns out any hypothesis consisting of one conjecture from the first pair of categories (a) or (b), along with one more conjecture from the second pair of categories (c) or (d), is sufficient to make a revised metabolic model that is logically consistent with all 40 experimental results.

7   Evaluation

The only way to determine which, if any, of the proposed hypotheses might be correct, was to biologically investigate each of the basic conjectures in turn:

(a) Enzyme Complex: The presence of a YER090W complex downstream of 4.1.3.27 is not unreasonable given that 4.1.1.48 is catalysed by YKL211C, which is already known to form a complex with YER090W in reaction 4.1.3.27. But, a detailed examination of the extensive literature on this pathway (which includes evolutionary investigations of these ORFs in yeast and related organisms together with more recent protein-protein interaction studies of the associated gene products) has not provided any support for this conjecture.

(b) Anthranilate Import: The slow import of Anthranilate is quite plausible given that two related phenyl-pyruvates are already believed to move slowly across the cellular membrane. Moreover, we believe the raw growth curves produced by the Robot Scientist provide indirect support for this conjecture and we have therefore updated our model accordingly.

(c) Indole Interference: The fact that Indole solution is less soluble than the other growth nutrients could conceivably be confusing the optical density sensor that is used to measure cell growth. But further analysis of the raw growth curves showed this was definitely not the case, conclusively ruling out this hypothesis.

(d) Indole Contamination: Contamination of Indole with Tryptophan could well be explained by the inadequate purification of Indole produced by a Tryptophan degradation process. In any event, mass spectrometry analysis of the Indole source used as an input in the Robot Scientist work has conclusively established that it is indeed contaminated with Tryptophan. Now it has been brought to light, we will obviously take care to avoid this source of error in future work.

8   Related Work

Our approach builds upon prior research [3] which used a learning system called Progol5 [7] to rediscover ORF-enzyme mappings removed from the AAA pathway. But, the scope of these earlier studies was greatly constrained by two key limitations of Progol5: namely its inability to reason hypothetically through negated literals; and its inability to infer more than one clause for each seed example. For these reasons, previous work utilised a rather complex nested list representation for reactions and employed a complicated metabolic theory in which all negations are restricted to built-in predicates and where much of the code is devoted to procedural issues such as pruning of the search tree, avoidance of cyclic computations, and efficient searching and sorting of data structures.

   As a result, the earlier work is restricted to learning hypotheses about missing ORF-enzyme mappings using data from single gene deletion experiments.

   By using XHAIL to overcome these limitations of Progol5 our approach adopts a fully declarative metabolic representation (based on [11] but extended to support enzyme-complexes) which imposes no a-priori restrictions on the learning task. Moreover, by the effective use of classical and default negation, the new method can simultaneously add and/or remove information about ORFs, enzymes, and/or reactions as required. It can also be applied to multiple gene deletion experiments.

We are not currently aware of any other systems able perform the range of metabolic network revisions demonstrated in this work.

9   Conclusions

This paper presented a logical method for the automatic revision of metabolic networks through abductive and inductive analysis of experimental data. First we showed how a nonmonotonic logic programming formalism can be used to declaratively model metabolic reactions with arbitrary reactants catalysed by multiple enzymes subject to inhibitory feedbacks. Then, we described how the XHAIL reasoning system was used to revise a state-of-the-art AAA pathway in the light of real-world data obtained by an autonomous Robot Scientist. The results have been tested biologically and have thereby led to improvements in both our current metabolic model and our future experimental setup.

References

1. Kakas, A., Kowalski, R., Toni, F.: Abductive Logic Programming. Journal of Logic and Computation 2(6), 719–770 (1992)
2. King, R., Rowland, J., Oliver, S., Young, M., Aubrey, W., Byrne, E., Liakata, M., Markham, M., Pir, P., Soldatova, L., Sparkes, A., Whelan, K., Clare, A.: The automation of science. Science 324(5923), 85–89 (2009)
3. King, R., Whelan, K., Jones, F., Reiser, P., Bryant, C., Muggleton, S., Kell, D., Oliver, S.: Functional Genomic Hypothesis Generation and Experimentation by a Robot Scientist. Nature 427, 247–252 (2004)
4. Lehninger, A.: Biochemistry: The Molecular Basis of Cell Structure and Function, 2nd edn. Worth Publishers (1979)
5. Lloyd, J.: Foundations of Logic Programming. Springer, Heidelberg (1987)
6. Muggleton, S.: Inverse Entailment and Progol. New Gen. Comp. 13, 245–286 (1995)
7. Muggleton, S., Bryant, C.: Theory Completion Using Inverse Entailment. In: Cussens, J., Frisch, A.M. (eds.) ILP 2000. LNCS (LNAI), vol. 1866, pp. 130–146. Springer, Heidelberg (2000)
8. Muggleton, S., De Raedt, L.: Inductive Logic Programming: Theory and Methods. Journal of Logic Programming 19(20), 629–679 (1994)
9. Ray, O.: Nonmonotonic Abductive Inductive Learning. Journal of Applied Logic 3(7), 329–340 (2009)
10. Ray, O., Whelan, K., King, R.: A nonmonotonic logical approach for modelling and revising metabolic networks. In: Proc. 3rd Int. Conf. on Complex, Intelligent and Software Intensive Systems, pp. 825–829. IEEE, Los Alamitos (2009)
11. Whelan, K., King, R.: Using a logical model to predict the growth of yeast. BMC Bioinformatics 9(97) (2008)

Finding Relational Associations in HIV Resistance Mutation Data

Lothar Richter, Regina Augustin, and Stefan Kramer

Technische Universität München, Institut für Informatik
Boltzmannstr. 3, 85748 Garching bei München, Germany
{richter,kramer}@in.tum.de

Abstract. HIV therapy optimization is a hard task due to rapidly evolving mutations leading to drug resistance. Over the past five years, several machine learning approaches have been developed for decision support, mostly to predict therapy failure from the genotypic sequence of viral proteins and additional factors. In this paper, we define a relational representation for an important part of the data, namely the sequences of a viral protein (reverse transcriptase), their mutations, and the drug resistance(s) associated with those mutations. The data were retrieved from the Los Alamos National Laboratories' (LANL) HIV databases. In contrast to existing work in this area, we do not aim directly for predictive modeling, but take one step back and apply descriptive mining methods to develop a better understanding of the correlations and associations between mutations and resistances. In our particular application, we use the Warmr algorithm to detect non-trivial patterns connecting mutations and resistances. Our findings suggest that well-known facts can be rediscovered, but also hint at the potential of discovering yet unknown associations.

1 Introduction

The optimization of HIV therapy is a crucial task, as the virus rapidly develops mutations to evade drug pressure [4]. Several machine learning approaches have been developed for decision support in this area. The task is typically predictive modeling [5,4,2], for instance, to predict the resistance against one or several drugs from the genotypic sequence of viral proteins and other data. While this is the ultimate goal in this application, we take one step back here and aim to find all possible correlations and associations between mutations and resistance, which has been done so far only in a limited form [5]. To do so, we define a relational representation for an important part of the data, namely the sequences of a viral protein (reverse transcriptase), their mutations, and the drug resistance(s) associated with those mutations. In the following, we present some background of this work, the raw data, the relational data representation, and some highlight results from applying standard relational association rules to the problem.

L. De Raedt (Ed.): ILP 2009, LNAI 5989, pp. 202–208, 2010.

2   Background

The infection with HI virus sooner or later causes the disease AIDS, which is still beyond remedy. HIV belongs to the group of retroviruses which establishes itself permanently in cells of the host's immune system. This causes a chronic infection which leads to a depletion of essential immune system cells, and the AIDS characteristics become observable. Up to now, no treatment for complete remedy or vaccination against HIV is available. Current medication strategies try to defer the point where AIDS symptoms begin to show and extend the survival time of patients. To achieve this, several drugs suppressing different steps of virus replication and infection were developed. These drugs are predominantly active against one of a few target proteins – e.g., reverse transcriptase, HIV protease or the envelope protein – which are all essential for a successful virus propagation. In this work, we present the analysis of gene sequences for reverse transcriptase genes. This is the enzyme which converts the viral RNA into DNA, which is necessary for the integration into the host cell's genome, and hence essential for replication. The integrated copy of the virus DNA acts later on as a template for the production of new virus RNA offspring. Due to a high mutation rate during replication, alterations of the gene sequences occur frequently and some of them confer resistance against a certain drug. If such a mutation occurs during drug treatment, the resistant virus strain becomes selectively propagated, the drug treatment is not effective any longer, and a new drug has to be administered. A strategy to overcome this problem is the concurrent application of several drugs, each acting against a different protein and target site. The rationale for this is that a single point mutation cannot confer resistance against drugs targeting different sites in one step. However, even this strategy does not work indefinitely, but only impedes the selection of a resistant virus strain for an extended period. We studied the occurrence of mutations conferring resistance against a certain number of drugs, and looked for rules correlating mutations and resistances against drugs.

3   Dataset and Preprocessing

In the following, we describe the data we used and how we defined a suitable representation for mining first-order association rules with Warmr [3].

The gene sequence and mutation data we used in this study were retrieved from the Los Alamos National Laboratories (LANL) on November 28th 2006. The main information was derived from the LANL HIV resistance database (for a recent screen shot see Figure 1). Each entry of the database describes an observed resistance mutation and has eight attributes: affected gene, drug class, string identifier, wild type amino acid, position, altered amino acid, underlying nucleotide sequence alteration and literature reference. Wild type in this context means that the sequence was retrieved from the reference virus strain, which is one of the first isolates and which can therefore be regarded as original or wild. Currently, the database contains information about 370 known resistance mutations.

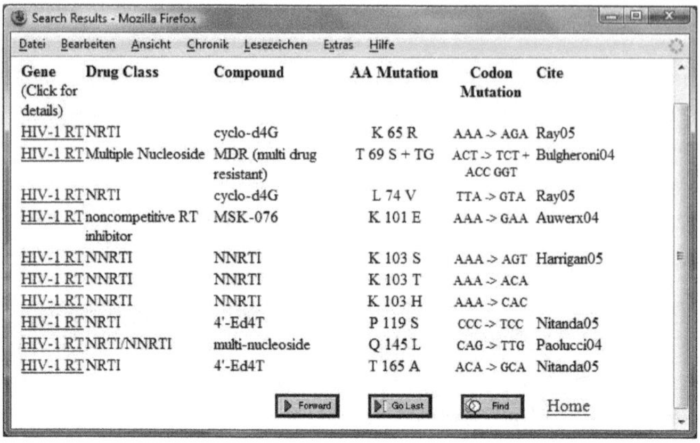

Fig. 1. Recent screenshot of LANL Resistance Database web interface showing the first ten results

In addition, DNA sequences corresponding to the reverse transcriptase gene region were selected via the web interface of LANL, and subsequently re-aligned manually. This part of the data comprises 2,339 DNA sequences of a length from 1,320 to 1,443 nucleotides, which corresponds to protein sequences from 440 to 481 amino acid residues. The DNA sequences were aligned using the program package ARB [6], taking into account the genes' reading frame to avoid unnecessary frame shifts. The alignments are necessary to make sequence positions comparable and to identify mutations, i.e., deviations from the wild type at a given position. A screenshot displaying the DNA alignment and the corresponding amino acid sequence as colored blocks is shown in Figure 2. This alignment was exported to an amino acid alignment consisting of 481 positions and further processed.

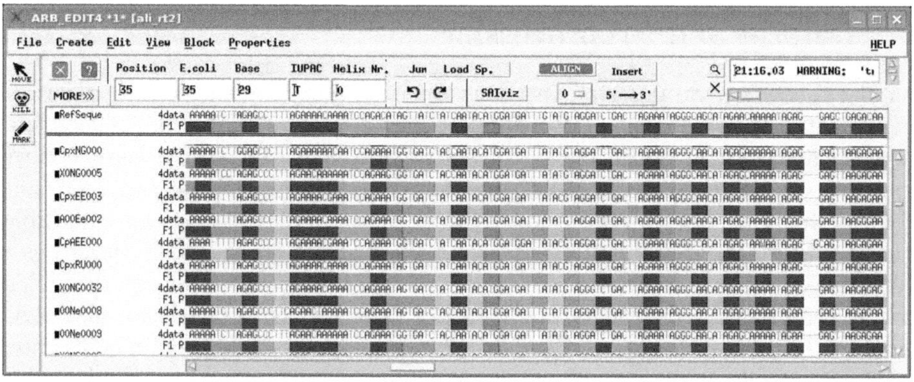

Fig. 2. Screenshot of reverse transcriptase DNA alignment with corresponding amino acid color blocks

Table 1. Tabular representation of amino acid sequence alignment with only a few selected columns and rows

```
0,   WildType, 204,205,206,207,208,209,210,211,212,213,214,215,216,217,218,-,219
1,   RefSeque, E , L , R , Q , H , L , L , R , W , G , L , T , T , P , D ,-, K
796 ,CpxNG000, E , L , R , E , H , L , L , K , W , G , F , A , T , P , D ,-, K
1992,X0NG0000, E , L , R , E , H , L , L , K , W , G , F , T , T , P , D ,-, K
1426,CpxEE003, E , L , R , E , H , L , L , K , W , G , F , T , T , P , D ,-, K
1427,A00Ee002, E , L , R , E , H , L , L , K , W , G , F , T , T , P , D ,-, K
1975,CpAEE000, Q , L , R , E , H , L , L , E , W , G , I , P , X , P , R ,X, K
2175,CpxRU000, E , L , R , E , H , L , L , K , W , G , F , T , T , P , D ,-, K
2023,X0NG0032, E , L , R , E , H , L , L , K , W , G , F , T , T , P , D ,-, K
837, 00Ne0008, E , L , R , E , H , L , L , K , W , G , F , T , T , P , D ,-, K
838, 00Ne0009, E , L , R , E , H , L , L , K , W , G , F , T , T , P , D ,-, K
```

Because positions for mutations are given with respect to the wild type sequence (HXBII), a head line was inserted which holds the corresponding wild type residue number if there is an amino acid in the wild type sequence, or a '−' -sign, if the alignment shows a gap for the wild type sequence at a certain position. The resulting table has the following layout (for an example, see Table 1): Line 1 contains the reference sequence indices in the alignment, followed by the values of the aligned 2,339 gene sequences. For identification, an index starting at 0 is inserted in column 1, and a string identifier in column 2. The remaining columns contain the values of the aligned amino acids of the above mentioned sequences. This table is used to detect mutations (deviations from the wild type), which are stored in the relations has_mutation(Sequence, Mutation) and mutation_property(Mutation, AA1, Position, AA2) (for details, see below).

The resulting relational representation consists of three relations: The base relation has_mutation connects sequences and mutations, the second relation mutation_properties describes the properties of mutations, and the third relation resistance_against connects mutations and their resistance against a certain drug. The relations are summarized in Table 2. Typical instances from these relations could be the following: Sequence number 8 carries mutation number 8 (has_mut(8,8)), where mutation number 8 confers resistance against a substance from the NRTI group (res_against(8, 'Nucleoside RT Inhibitor(NRTI)')). Mutation number 8 is further described as an amino acid change from 'I' to 'R' at position 51 (mut_prop(8,'I',51,'R')).

Table 2. Schemata of Warmr input relations

key(Sequence)	index number to address the sequences
has_mutation(Sequence, Mutation)	connects a sequence identifier with a mutation identifier
resistance_against(Mutation, Drug)	links a mutation with the resistance against a drug
mutation_property(Mutation, AA1, Position, AA2)	defines the properties of a mutation, where an amino acid at a certain position was substituted by another

Table 3. Sample association rules found

Number	Natural language description	*Supp.*	*Conf.*	*Lift*
(1)	If position 177 is changed to E and position 286 changed to A, then a change of position 335 to D conferring resistance against NRTI is found.	0.29	0.83	1.90
(2a)	If 177 E, 292 I and 35 T then a mutation 335 D with NRTI resistance is found.	0.28	0.87	2.00
(2b)	If 177 E, 292 I and 291 D then a mutation 335 D with NRTI resistance is found.	0.29	0.87	2.00
(2c)	If 177 E, 292 I and 293 V, then a mutation 335 D with NRTI resistance is found.	0.30	0.86	1.97
(3)	If 6 D, 11 T, 35 T and 39 K, then 43 E responsible for resistance against NRTI.	0.06	0.96	8.7
(4)	If 41 L, then 215 Y.	0.06	0.82	6.8
(5a)	If 41 L and 215 Y, then 277 K.	0.05	0.79	1.5
(5b)	If 41 L and 215 Y, then 293 V.	0.04	0.75	1.1

The data is available for download from `http://wwwkramer.in.tum.de/research/applications/hiv-resistances`.

4   Results and Discussion

For our analysis, we used Warmr in the implementation of ACE 1.2.11. With a minimum support of 0.1, we obtained 4,931 frequent queries after a running time of two days on a standard Linux machine.[1] The output of Warmr consists of frequent queries and association rules. Because association rules are more interesting than frequent queries in our application, we will focus on the rules here. To estimate the "interestingness" of a rule, we also calculated the lift measure for each of the resulting 5,096 rules. In the rules, each mutation is given with the position relative to the wild type sequence and the altered amino acid.

Remarkably, some of the rules concern resistance mutations that were not known at the time of data retrieval. This is the case for the first rule presented here (see rule (1) in Table 3; resistance mutation 335 D was not part of the analyzed data), which is very similar to three variants of another rule (see rules (2a) to (2c)). These results show the potential of the approach, as patients that are already positive for mutations 177 E and 292 I or 286 A might better not be treated with NRTI, because the development of a resistance caused by mutation 335 D is rather likely. Another interesting rule with a newly discovered resistance is rule (3), which reflects a very tight coupling with mutation 43 E, as the frequency of the body alone is already 0.063.

Additionally to those findings which elucidated correlations between mutations and newly discovered resistances, we also found well-known correlations in

---

[1] For subsequent experiments, the minimum support was lowered.

the data. The mutations 41 L and 215 Y (see rule (4), both linked with resistance against NRTI) have also been described as highly correlated before [7]. In addition to this rule, there is an interesting extension (see rules (5a) and (5b)). Mutation 277 K is an alteration with respect to the wild type sequence and is not described as conferring resistance yet. Nevertheless, 277 K has shown a high correlation to known resistance mutations and may turn out to be a resistance mutation in the future, or may give strong hints for an evolving resistance based on mutations 41 L or 215 Y in the further course of a patient's treatment. For a more detailed analysis, we refer to the diploma thesis of the second author of this paper [1].

5   Conclusion

We presented a relational representation of data derived from the LANL HIV databases, and an analysis of the data using descriptive mining methods, to discover new correlations and associations between mutations and resistances against HIV drugs. Given the relevance of the application and the complex structure of the data, we believe it is a rewarding new field for ILP and relational learning methods. In particular, these methods lend themselves to the discovery of co-occurring mutations, potentially also giving hints for viral evolution paths.

The work presented in this paper could be extended in several ways: First, it would be interesting to consider a richer representation of proteins, for instance, taking into account amino acid properties. However, it has to be noted that only sequence information is known.[2] Second, the chemical structure of the inhibitor could be included. Third, it is straightforward to extend this type of analysis to other viral genes, e.g., protease. Fourth, the EuResist database (see http://www.euresist.org/) contains more detailed information than the LANL HIV databases, including therapy and patient data. To make the proposed approach scalable to such large-scale data, one would have to preprocess them appropriately and use suitable abstractions.

References

1. Augustin, R.: Data Mining in HIV-Resistenzmutationen. Diploma Thesis, Technische Universität München (2008)
2. Bickel, S., Bogojeska, J., Lengauer, T., Scheffer, T.: Multi-task learning for HIV therapy screening. In: Proceedings of the Twenty-Fifth International Conference on Machine Learning (ICML 2008), pp. 56–63 (2008)
3. Dehaspe, L., Toivonen, H.: Discovery of frequent Datalog patterns. Data Min. Knowl. Discov. 3(1), 7–36 (1999)
4. Rosen-Zvi, M., et al.: Selecting anti-HIV therapies based on a variety of genomic and clinical factors. In: Proceedings of the 16th International Conference on Intelligent Systems for Molecular Biology (ISMB 2008), pp. 399–406 (2008)

---

[2] Protein structure is, in general, only known for the wild type.

5. Sing, T., et al.: Characterization of novel HIV drug resistance mutations using clustering, multidimensional scaling and SVM-based feature ranking. In: Jorge, A.M., Torgo, L., Brazdil, P.B., Camacho, R., Gama, J. (eds.) PKDD 2005. LNCS (LNAI), vol. 3721, pp. 285–296. Springer, Heidelberg (2005)
6. Ludwig, W., et al.: ARB: a software environment for sequence data. Nucleic Acids Research 32(4), 1363–1371 (2004)
7. Lengauer, T., Sing, T.: Bioinformatics-assisted anti-HIV therapy. Nature Reviews Microbiology 4, 790–797 (2006)

ILP, the Blind, and the Elephant: Euclidean Embedding of Co-proven Queries

Hannes Schulz[1], Kristian Kersting[2], and Andreas Karwath[1]

[1] Institut für Informatik, Albert-Ludwigs Universität
Georges-Köhler-Allee 79, 79110 Freiburg, Germany
{schulzha,karwath}@informatik.uni-freiburg.de
[2] Dept. of Knowledge Discovery, Fraunhofer IAIS
Schloss Birlinghoven, 53754 St Augustin, Germany
kristian.kersting@iais.fraunhofer.de

Abstract. Relational data is complex. This complexity makes one of the basic steps of ILP difficult: understanding the data and results. If the user cannot easily understand it, he draws incomplete conclusions. The situation is very much as in the parable of the blind men and the elephant that appears in many cultures. In this tale the blind work independently and with quite different pieces of information, thereby drawing very different conclusions about the nature of the beast. In contrast, visual representations make it easy to shift from one perspective to another while exploring and analyzing data. This paper describes a method for embedding interpretations and queries into a single, common Euclidean space based on their co-proven statistics. We demonstrate our method on real-world datasets showing that ILP results can indeed be captured at a glance.

1 Introduction

Once upon a time, there lived six blind men in a village. One day the villagers told them, "Hey, there is an elephant in the village today." They had no idea what an elephant is. They decided, "Even though we would not be able to see it, let us go and feel it anyway." All of them went where the elephant was and touched the elephant. Each man encountered a different aspect of the elephant and drew a different inference as to its essential nature. One walked into its side, concluding that an elephant is like a wall. Another, prodded by the tusk, declared that an elephant is like a spear. The chap hanging onto the tail was convinced that he had found a sort of rope. The essential nature of the elephant remained undiscovered.

The tale is that of "The Blind and the Elephant", which appears in many cultures. It illustrates the problem many ILP users face, to make sense of relational data and models, the elephants, before applying their algorithms or while interpreting the results. Due to the complexity of the data and the models, the user can only touch small parts of them, like specific queries. Hence, he often gets only a narrow and fragmented understanding of their meaning.

L. De Raedt (Ed.): ILP 2009, LNAI 5989, pp. 209–216, 2010.

In contrast, visual representations make it easy to shift from one perspective to another while exploring and analyzing data. How to visually explore relational data and queries jointly, however, has not received a lot of attention within the ILP community. This can be explained by the fact that relational data involves objects of several very different types without a natural measure of similarity. Syntactic similarity measures typically used in ILP (c. f. [8]) cannot be used for two reasons: First, they cannot relate interpretations and queries, and second, syntactically different queries might actually have identical semantics w.r.t. the data, resulting in hard to optimize and hard to interpret embeddings.

Our paper addresses this problem of creating embeddings which visualize interpretations and queries. In our embeddings, query-query and query-interpretation distances are determined by their respective co-occurrence statistics, i. e., queries are placed close to queries which were often co-proven and close to interpretations in which they are true. Properly colored and labeled, our embeddings provide a generic visualization and interactive exploration of relational databases as well as the working of query-generating ILP-algorithms.

2   Euclidean Embedding of Co-proven Queries

Given a set of interpretations $\mathcal{I}$ and queries $\mathcal{Q}$, we assign positions in $\mathbb{R}^d$ to all $i \in \mathcal{I}$ and $q \in \mathcal{Q}$ such that the distances in $\mathbb{R}^d$ reflect the co-occurrence statistics. Computing joint embeddings of $\mathcal{I}$ and $\mathcal{Q}$ essentially requires three steps: (1) collecting embeddable queries, (2) embedding queries and interpretations into a single Euclidean space, and – as an optional postprocessing step – (3) labelling the representation by extracting local representatives. We make the C++ implementation of our method available on our website.[1]

Step 1 – Queries: Given a finite set $\mathcal{I}$ of observed interpretations, any ILP algorithm can be used to preselect embeddable queries $\mathcal{Q}$ for $\mathcal{I}$. Although in this work we concentrate on feature miners, our embeddings can also be used to for example embed features extracted for classification. In this paper, we use *Molfea* [5] and *C-armr*[2] [2] to mine databases of molecules. Both systems are inspired by the Agrawal's *Apriori* algorithm [1]: they construct general queries and specialize them only if they are frequent. Only queries more frequent than some threshold are retained and further expanded, i.e., specialized. While *Molfea* constructs linear fragments only (atom, bond, atom, ...), *C-armr* constructs general queries and can take background knowledge into account as well. In addition to the queries, we also store the interpretations in which they were true. This will prove useful for the next step and can efficiently be represented in a binary matrix $\mathcal{C} \in \{0, 1\}^{|\mathcal{Q}| \times |\mathcal{I}|}$, where $|\mathcal{Q}|$ is the number of queries and $|\mathcal{I}|$ is the number of observed interpretations.

Step 2 – Embedding: We wish to assign positions to elements in $\mathcal{I}$ and $\mathcal{Q}$ through mappings $\phi : \mathcal{I} \mapsto \mathbb{R}^d$ and $\psi : \mathcal{Q} \mapsto \mathbb{R}^d$ for a given dimensionality d.

---

[1] http://www.ais.uni-bonn.de/~schulz/
[2] In the CLASSIC'CL implementation [10].

These mappings should reflect the dependence between $\mathcal{I}$ and $\mathcal{Q}$ such that the co-occurrence $\mathcal{C}_{qi}$ (c. f. Step 1) of some $i \in \mathcal{I}$ and $q \in \mathcal{Q}$ determines the distance between $\phi(i)$ and $\psi(q)$.

For this purpose, we employ Globerson *et al.*'s *CODE* algorithm [4]. *CODE* is a generic scheme for co-occurrence-based embedding algorithms. Its main idea is to represent the empirical joint distribution $\bar{p}(A, B)$ of two random variables A and B in a low dimensional space, such that items with high co-occurrence probability are placed close to each other. In *CODE*, this idea is realized in the assumption that

$$p(i, q) = Z^{-1} \cdot \exp(-\|\phi(i) - \psi(q)\|^2), \qquad (1)$$

where $Z = \sum_{i,q} \exp(-\|\phi(i) - \psi(q)\|^2)$ is a normalization constant. Starting from a random position assignment in ϕ and ψ, *CODE* then minimizes the Kullback-Leibler divergence between $\bar{p}(i, q) \propto \mathcal{C}_{qi}$ and $p(i, q)$ by maximizing the log-likelihood $l(\phi, \psi) = \sum_{i,q} \bar{p}(i, q) \log p(i, q)$. The maximization is performed by gradient ascent on the gradient of the log-likelihood function derived with respect to the axis of the embedding space for each $i \in \mathcal{I}$ and $q \in \mathcal{Q}$. The same problem can also be stated as a convex semi-definite programming problem which is guaranteed to yield the optimal solution, see [4] for details.

Our situation, however, is slightly more complicated. Following suggestions by Globerson *et.al.*, we extend the simple model by marginalization and within-variable co-occurrence measures.

(i) First, we observe that the marginal probability of an interpretation is a quantity artificially created by the number of queries which are true in it. An embedding influenced by this margin is not very intuitive to read. We therefore make the embedding insensitive to $\bar{p}(i)$ by adding it as a factor to (1), yielding

$$p(i, q) = \frac{1}{Z} \cdot \bar{p}(i) \cdot \exp(-\|\phi(i) - \psi(q)\|^2). \qquad (2)$$

(ii) However, the results will still be unsatisfactory: using $p(\mathcal{I}, \mathcal{Q})$ only, *CODE* tends to map the interpretations to a circle. This results from the fact that $\mathcal{C}$ is binary and all distances are therefore enforced to be of the same length. To generate more expressive embeddings, we use the interpretations and queries to generate a non-binary query-query co-occurence matrix $\mathcal{D} = \mathcal{C}\mathcal{C}^T$ and set $\bar{p}(q, q') \propto \mathcal{D}_{qq'}$. This co-proven statistics of queries q and q' should be represented by distances in the embedding (similar to (1)) as

$$p(q, q') = \frac{1}{Z} \cdot \exp\left(-\|\psi(q) - \psi(q')\|^2\right). \qquad (3)$$

As described above, we assign initial positions randomly but now adapt them so that they maximize the "log-likelihood" of the combined and weighted models in (2) and (3):

$$l(\phi, \psi) = \sum_{i,q} \bar{p}(i, q) \log p(i, q) + |\mathcal{I}|/|\mathcal{Q}| \cdot \sum_{q,q'} \bar{p}(q, q') \log p(q, q') \ .$$

Thus, our embeddings reflect the relations between co-proven queries as well as interpretations and queries.

Step 3 – Condensation: Literally thousands of queries and instances can be embedded into a single Euclidean space and can – as a whole – provide useful insights into the structure of the data. However, we would also like to get a grasp on what queries in certain regions focus on. To do so, we propose to single out queries q in an iterated fashion. Specifically, we assign to each query q the weight $\text{w}(q) = F(q)/\text{length}(q)$, where $F(q)$ is q's F_1 or F_2-measure, see e.g. [7], and $\text{length}(q)$ is its description length. We now locally remove queries with a low weight in a two-step process. First, we build the k-nearest neighbour graph of the embedded queries. From the weight, we subtract the weight of its graph neighbours, thereby removing large-scale differences. Second, we increase weights of queries with lower weighted neighbours and decrease weights which have higher weighted neighbours. The last step is repeated until the number of queries q with a positive weight is not changing anymore. In other words, we prefer short queries with high F-measures on a local neighbourhood.

3 Interpretation of Generated Embeddings

When feature miners are used to generate queries, embeddings have a typical shape which looks like a tree as seen from above.

(i) Very common queries are generally placed in the center. For databases where different classes can be distinguished, we note that those central queries typically are not very discriminative. For example, in a molecular database containing organic molecules, we would expect chains of carbon atoms to be in the center of the embedding.

(ii) Queries which hold only on a subset of the interpretations are placed further away from the center. Those queries can therefore be seen as "specializations" of the queries in the center or as a cluster of interpretations. The specializations can describe discriminative or non-discriminative properties. In the database of organic molecules, certain aromatic rings might be common in one group of the molecules, but not in the other.

(iii) The tree-like branching repeats recursively. Neighboring branches most likely represent a meta-cluster of interpretations; far away branches represent interpretations which are very dissimilar w.r.t their queries.

Please note that the embedding reflects two of the most important axis in ILP learning tasks: Learning a theory is always a tradeoff between coverage and simplicity. In our embeddings, it is easy to judge how specific a query is, how many of the total interpretations it covers and how it relates to other, similar queries. Furthermore, the embeddings abstract from the syntax of the queries. They are solely determined by the interaction of the queries with each other and the interpretations, which is essential if the complexity of queries necessary to distinguish interpretations varies over the database.

4   Showcases

We tested our approach on several real-world datasets for the two-dimensional case. To provide a qualitative assessment of our method, we apply it to datasets where some structures or models have already been discovered and point out the properties described in Section 3.

4.1   Mutagenesis

On **Mutagenesis** [9], the problem is to predict the mutagenicity of a set of compounds. In our experiments, we use the atom and bond structure information only (including the predefined predicate like *ball3s*, *ring_size_5s*, and others). The dataset consists of 230 compounds (138 positives, 92 negatives). The 2D Euclidean embedding is shown and in Fig. 1.

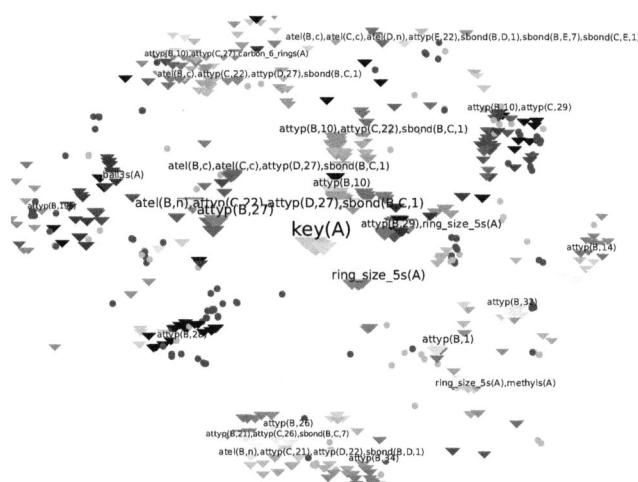

Fig. 1. Mutagenesis embedding. We show all instances (circles) and all frequent queries (triangles) distinct w.r.t. the interpretations. Dark/bright queries have high/low precision, small/large queries have low/high recall. Dark/bright instances are of class "active"/"inactive". The queries with textual descriptions were automatically selected, the trivial **key** attribute was omitted in all but the central queries.

As discussed in Section 1, the most common query, here **key(A)**, is placed in the center. Also notice specializations branching from the center, for example variants of **attyp(B,10)**. The embedding reflects rules we could induce employing Srinivasan's ALEPH on the same dataset such as **active(A) :- attyp(A,B,29)**, **ring_size_5s(A)** or **active(A) :- ball3s(A)**. Queries which primarily hold on positive/negative interpretations are spatially well separated and form clusters which indicate similar coverage of interpretations.

4.2   Estrogen Database

The **Estrogen** database was extracted from the EPA's DSSTox NCTRER Database[3]. The original dataset was published by Fang *et al.* [3], and is specially designed to evaluate QSAR approaches. The NCTRER database provides activity classifications for a total of 232 chemical compounds, which have been tested regarding their binding activities for the estrogen receptor. The database contains a diverse set of natural, synthetic, and environmental estrogens, and is considered to cover most known estrogenic classes spanning a wide range of biological activity [3]. Here, "activity" is an empirically measured value between 0 and 100, which we averaged and used as a query's color. The 2D Euclidean embedding is shown and discussed in Fig. 2.

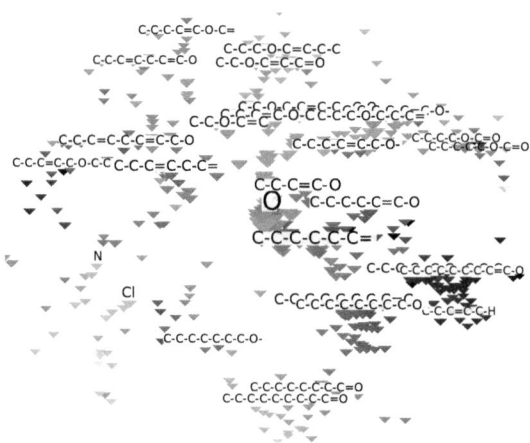

Fig. 2. Estrogen embedding. The coding is as in Fig. 1; only bright/dark queries indicate now low/high mean activity of the respective interpretations.

The embedding shows oxygen (O) as a primary component of organic molecules and a typical branching structure, starting from the center and on a smaller level for example from nitrogen (N) and chlorine (Cl) -based compounds. The activity varies smoothly in the embedding, as extending a chain with certain atoms increases or decreases the chance of higher activity. Even for clearly active molecules, different subtypes can be distinguished.

In their original publication Fang *et al.* have identified that a phenolic ring connected by one to three atoms to another benzene ring is one of the key features that have to be present regarding the likelihood of a compound being an ER ligand. A phenolic ring is a 6-carbon benzene ring with an attached hydroxyl (OH) group. In the embedding, it can be seen that this is reflected in features like C-C-C=C-O, which indicates that there is a path of one carbon atom to (a part of) a ring structure (C-C=C) connected to an oxygen.

---

[3] http://www.epa.gov/ncct/dsstox/sdf_nctrer.html

4.3 AIDS Database

The DTP **AIDS** Antiviral Screening Database originating from the NCI's development therapeutics program NCI/NIH[4] consists of SMILES representations of 41,768 chemical compounds [6]. Each data entry is classified as either active, moderately active, or inactive. A total of 417 compounds are classified as active, 1,069 as moderately active, and 40,282 as inactive. We have converted this dataset into SDF format using the OpenBabel toolkit and randomly sampled 400 active and 400 moderate/inactive compounds. The 2D Euclidean embedding is shown and discussed in Fig. 3.

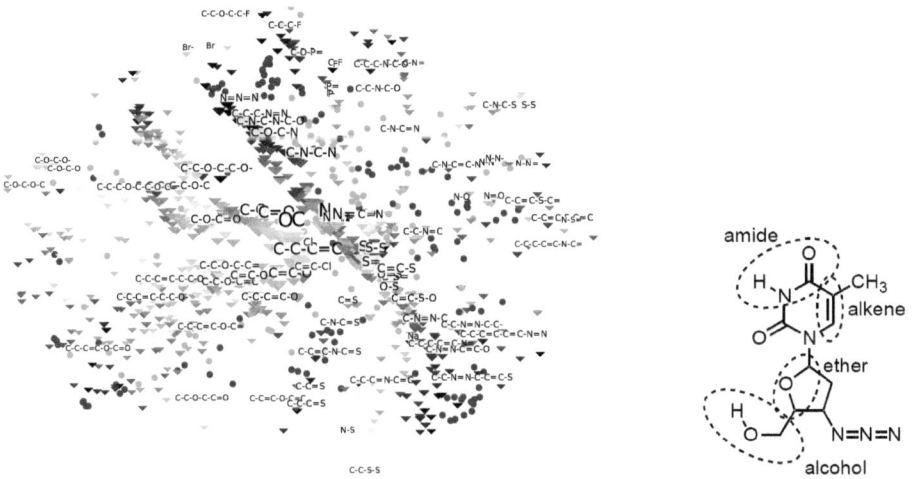

Fig. 3. Left: **AIDS** database embedding. The coding is as in Fig. 1. Queries were subsampled according to their occurrence probabilities. Right: Azidothymidine (AZT), a component known to be very active. AZT and derivatives are represented by the left central branch in the embedding.

In this database we can clearly see a natural distinction of queries: While oxygen (O) and nitrogen (N) are both common, the lower half of the embedding focuses on nitrogen-based molecules. Within this half, further distinctions (left: flourine (F), right: sulfur (S) -based compounds) can be made.

The embedding clearly indicates compounds that are derivatives of Azidothymidine (AZT), a potent inhibitor of HIV-1 replication. In the left central branch, the embedding clearly indicates prominent features of AZT, such as the nitrogen-group N=N=N and the C-N-C-N-C-O chain connecting both rings.

5 Concluding Remarks

In our opinion, to unveil its full power, ILP must incorporate visual analysis methods. With the work presented here, we have made a step in this direction.

---

[4] http://dtp.nci.nih.gov/

We have presented the first method for embedding interpretations and queries into the same Euclidean space based on their co-occurrence statistics. As our experiments demonstrate, the spatial relationships in the resulting embedding are intuitive and can indeed reveal useful and important insights at a glance. Aside from their value for visual analysis, embeddings are also an important tool in unsupervised learning and as a preprocessing step for supervised learning algorithms. In future research, we will explore this direction.

Acknowledgments. The authors would like to thank Amir Globerson for providing the *CODE* code. The research was partially supported by the EU seventh framework program under contract no. Health-F5-2008-200787 (*OpenTox*), and by the Fraunhofer ATTRACT fellowship STREAM.

References

1. Agrawal, R., Srikant, R.: Fast algorithms for mining association rules in large databases. In: Proceedings of the 20th International Conference on Very Large Data Bases, pp. 487–499. Morgan Kaufmann, San Francisco (1994)
2. De Raedt, L., Ramon, J.: Condensed representations for inductive logic programming. In: Proceedings of 9th International Conference on the Principles of Knowledge Representation and Reasoning, pp. 438–446 (2004)
3. Fang, H., Tong, W., Shi, L.M., Blair, R., Perkins, R., Branham, W., Hass, B.S., Xie, Q., Dial, S.L., Moland, C.L., Sheehan, D.M.: Structure-activity relationships for a large diverse set of natural, synthetic, and environmental estrogens. Chem. Res. Tox 14, 280–294 (2001)
4. Globerson, A., Chechik, G., Pereira, F., Tishby, N.: Euclidean Embedding of Co-occurrence Data. The Journal of Machine Learning Research 8, 2265–2295 (2007)
5. Helma, C., Kramer, S., De Raedt, L.: The molecular feature miner MolFea. In: Proceedings of the Beilstein-Institut Workshop (2002)
6. Kramer, S., De Raedt, L., Helma, C.: Molecular feature mining in HIV data. In: Provost, F., Srikant, R. (eds.) Proc. KDD 2001, August 26-29, pp. 136–143. ACM Press, New York (2001)
7. Lewis, D.D.: Evaluating and optimizing autonomous text classification systems. In: Proceedings of the 18th Int. ACM-SIGIR Conference on Research and Development in Information Retrieval, pp. 246–254 (1995)
8. Ramon, J.: Clustering and instance based learning in first order logic. PhD thesis, CS Dept., K.U. Leuven (2002)
9. Srinivasan, A., Muggleton, S.H., King, R.D., Sternberg, M.J.E.: Theories for Mutagenicity: A Study of First-Order and Feature -based Induction. Artificial Intelligence Journal 85, 277–299 (1996)
10. Stolle, C., Karwath, A., De Raedt, L.: CLASSIC'CL: An Integrated ILP System. In: Hoffmann, A.G., Motoda, H., Scheffer, T. (eds.) DS 2005. LNCS (LNAI), vol. 3735, pp. 354–362. Springer, Heidelberg (2005)

Parameter Screening and Optimisation for ILP Using Designed Experiments

Ashwin Srinivasan[1,*] and Ganesh Ramakrishnan[2]

[1] IBM India Research Laboratory, 4-C, Vasant Kunj Institutional Area
New Delhi 110070, India
ashwin.srinivasan@wolfson.oxon.org
[2] Department of Computer Science and Engineering,
Indian Institute of Technology, Bombay, India
ganesh@cse.iitb.ac.in

Abstract. Reports of experiments conducted with an Inductive Logic Programming system rarely describe how specific values of parameters of the system are arrived at when constructing models. Usually, no attempt is made to identify sensitive parameters, and those that are used are often given "factory-supplied" default values, or values obtained from some non-systematic exploratory analysis. The immediate consequence of this is, of course, that it is not clear if better models could have been obtained if some form of parameter selection and optimisation had been performed. Questions follow inevitably on the experiments themselves: specifically, are all algorithms being treated fairly, and is the exploratory phase sufficiently well-defined to allow the experiments to be replicated? In this paper, we investigate the use of parameter selection and optimisation techniques grouped under the study of experimental design. Screening and "response surface" methods determine, in turn, sensitive parameters and good values for these parameters. This combined use of parameter selection and response surface-driven optimisation has a long history of application in industrial engineering, and its role in ILP is investigated using two well-known benchmarks. The results suggest that computational overheads from this preliminary phase are not substantial, and that much can be gained, both on improving system performance and on enabling controlled experimentation, by adopting well-established procedures such as the ones proposed here.

1 Introduction

We are concerned in this paper with Inductive Logic Programming (ILP) primarily as a tool for constructing models. Specifications of the appropriate use of a tool, its testing, and analysis of benefits and drawbacks over others of a similar nature are matters for the engineer concerned with its routine day-to-day use. Much of the literature on the applications of ILP have, to date, been once-off

---

* A.S. is also an Adjunct Professor at the School of CSE, UNSW, Sydney; and Visiting Professor at the Computing Laboratory, Oxford.

L. De Raedt (Ed.): ILP 2009, LNAI 5989, pp. 217–225, 2010.

demonstrations of either the model construction abilities of a specific system, or of the ability of ILP systems to represent and use complex domain-specific relationships [4,6]. It is not surprising, therefore, that there has been little reported on practical issues that arise with the actual use of an ILP system.

Assuming some reasonable solution has been found to difficult practical problems like the appropriateness of the representation, choice of relevant "background knowledge", poor user-interfaces, and efficiency[1], we are concerned here with a substantially simpler issue. Like all model-building methods, an ILP system's performance is affected by values assigned to input parameters (not to be confused with the notion of a parameter, as used by a statistician). For example, the model constructed by an ILP system may be affected by the maximal length of clauses, the minimum precision allowed for any clause in the theory, the maximum number of new variables that could appear in any clause, and so. The ILP practitioner is immediately confronted with two questions: (a) Which of these parameters are relevant for the particular application at hand?; and (b) What should their values be in order to get a good model? As it stands, experimental applications of ILP usually have not used any systematic approach to answer these questions. Typically, parameters are given "factory-supplied" default values, or values obtained from a limited investigation of performance across a few pre-specified values. The immediate consequence of this is that it is not clear if better models could have been obtained if some form of parameter selection and optimisation had been performed.

The work of Bengio [1] is related to that here, in that it presents a methodology to optimize several parameters (the authors call them hyperparameters, to avoid confusion with their statistical counterparts), based on the computation of the gradient of a model selection criterion with respect to the parameters. The main restriction is that the training criterion must be a continuous and differentiable function of the parameters almost everywhere. In almost all ILP settings, the training criterion cannot be even expressed in closed form, let alone being a differentiable and continuous function of the parameters. That is, what can be done at best is to treat the ILP system is a black box and its variation as a function of the parameters can be measured only empirically in terms of the response of the system to changes in the values of the parameters. In this work, we directly approximate the evaluation function as a function of parameters using response surface methodology[3].

Here we take up both selection of parameters and assignment of their values directly with the only restrictions being that parameter and goodness values are quantitative in nature. The methods we use have origins in optimising industrial processes [3] and been developed under the broad area concerned with the design and analysis of experiments. This area is concerned principally concerned with discovering something about a system by designing deliberate changes to the system's input variables, and analysing changes in its output response. The process

---

[1] In [10], experience gained from applications of ILP to problems in biochemistry were used to extract some guiding principles of relevance to these problems for any ILP application.

being modelled transforms some input into an output that is characterised a measurable response y. The system has some controllable factors, and some uncontrollable ones and the goals of an experiment could be to answer questions like: which of the controllable factors are most influential on y; and what levels should these factors be for y to reach an optimal value. The relevance of the setting to the ILP problem we are considering here is evident from Fig. 1.

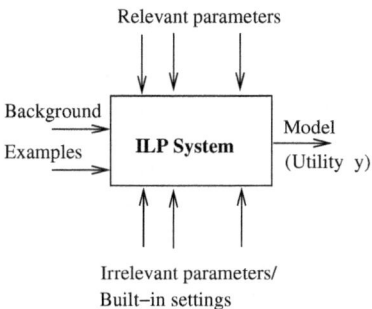

Fig. 1. An system engineer's view of an ILP system. We are assuming here that "Background" includes syntactic and semantic constraints on acceptable models. "Built-in settings" are the result of decisions made in the design of the ILP system. An example is the optimisation function used by the system.

There are a wide variety of techniques developed within the area of experimental design: we will be concentrating here on some of the simplest, based around the use of regression models. Specifically, using designed variations of input variables, we will use a stepwise linear regression strategy to identify variables most relevant to the ILP system's output response. This resulting linear model, or response surface, is then used to change progressively the values of the relevant variables until a locally optimal value of the output is reached. We demonstrate this approach empirically on some ILP benchmarks.

2   Design and Analysis of Experiments

The area broadly concerned with the design of experiments (DOE) deals with devising deliberate variations in the values of input variables, or *factors*, and analysing the resulting variations in a set of one or more output, or *response*, variables. The objectives of conducting such experiments are usually: (a) Understand how variation or uncertainty in input values affects the output value. The goal here is the construction of robust systems in which a system's output is affected minimally by external sources of variation; and (b) Maximise (or minimise) a system's performance. In turn, these raise questions like the following: which factors are important for the response variables; what values should be given to these factors to maximise (or minimise) the values of the response variables; what values should be give to the factors in order that variability in the response is minimal, and so on.

In this paper, we will be restrict ourselves to a single response variable and the analysis of experimental designs by multiple regression. It follows, therefore, that we are restricted in turn to quantitative factors only. Further, by "experimental design" we will mean nothing more than a specification of points from the factor-space, in order that a statistically sound relationship between the factors and the response variable can be obtained.

2.1   Screening Using Factorial Designs

We first consider designs appropriate for screening. By this, we mean deciding which of a set of potentially relevant factors are really important, statistically speaking. The usual approach adopted is what is termed a 2-level factorial design. In this, each factor is taken to have just two levels (encoded as "-1" and "+1", say)[2], and the effect observed on the response variable of changing the levels of each factor. It is evident that with k factors, this will result in 2^k experiments (these may need to be repeated in case there is some source of random variation in the response variable). For example, with $k = 2$, we will perform 4 experiments, each of which will yield a value for y. We are then able to construct a regression model relating the response variable to the factors:

$$y = b_0 + b_1 x_1 + b_2 x_2 + b_3 x_1 x_2$$

The model describes the effect of each factor $x_{1,2}$ and interactive effect $x_1 x_2$ of the two factors on y.[3] Further, it is also the case that with a 2-level full factorial design only linear effects can be estimated (that is, the effect of terms like x_i^2 cannot be obtained: in general, a nth order polynomial will require $n + 1$ levels for each factor). In this paper, we will use the coefficients of the regression model to guide the screening of parameters: that is, parameters with coefficients significantly different from 0 will be taken to be relevant.

Clearly, the number of experiments required in a full factorial design constitute a substantial computational burden, especially as the number of factors increase. Consider, however, the role these experiments play in the regression model. Some are necessary for estimating the effects of each factor (that is, the coefficients of $x_1, x_2, x_3, \ldots$: usually called the "main effects"), others for estimating the coefficients for two-way interactions (the coefficients of $x_1 x_2$, $x_1 x_3$, $\ldots$) , others for three-way interactions ($x_1 x_2 x_3$, $\ldots$) and so on. However, in a screening stage, all that we wish to do is to identify the main effects. This can usually be done with fewer than the 2^k experiments needed for a full factorial design with k factors. The result is a 2-level "fractional" factorial design. Figure 2 below illustrates a 2-level fractional factorial design for 3 factors that uses half the number of experiments to estimate the main effects (from [11]).

---

[2] One way to achieve the coded value x of a factor X is as follows. Let X^- and X^+ be the minimum and maximum values of X (these are pre-specified). Then $x = \frac{X - (X^+ + X^-)/2}{(X^+ - X^-)/2}$.

[3] Interaction effects happen if the effect of a factor, say X_1 on the response depends the level of another factor X_2.

Expt.	x_1	x_2	x_3	y
E1	-1	-1	-1	...
E2	-1	-1	+1	...
E3	-1	+1	-1	...
E4	-1	+1	+1	...
E5	+1	-1	-1	...
E6	+1	-1	+1	...
E7	+1	+1	-1	...
E8	+1	+1	+1	...

Expt.	x_1	x_2	x_3	y
E2	-1	-1	+1	...
E3	-1	+1	-1	...
E5	+1	-1	-1	...
E8	+1	+1	+1	...

Fig. 2. A full 2-level factorial design for 3 factors (left) and a "half fraction" design (right)

The experiments in the fractional design have been selected so that $x_1 x_2 x_3 = +1$. Closer examination of the table on the right will make it clear that the following equalities also hold for this table: $x_1 = x_2 x_3$; $x_2 = x_1 x_3$; and $x_3 = x_1 x_2$. That is, main effects and interaction terms are said to be *confounded* with each other. Thus, the price for fractional experiments is therefore, that we will in general, be unable to distinguish the effects of all the terms in the full regression model. However, if it is our intention—as it is in the screening stage—only to estimate the main effects (such models are also called "first-order" models), then provided we can ignore interactions, these can be estimated with a table that is a fraction required by the full factorial design. For example, the half fraction in Fig. 2 is sufficient to obtain a regression equation with just the main effects x_1, x_2 and x_3.

We focus here on a linear model that contains the main effects only. Depending on the number of factors, this can be done with a fractional designs of "Resolution III" or above (see [9]). Standard tests of significance can be performed on each of the coefficients $b_1, b_2, \ldots, b_k$ to screen factors for relevance (the null and alternative hypotheses in each case are $H_0 : b_i = 0$ and $H_1 : b_i \neq 0$). In fact, this test is the basis for inclusion or exclusion of factors by stepwise regression procedures. Using such a procedure would naturally return a model with only the relevant factors.

2.2 Optimisation Using the Response Surface

Suppose screening in the manner just described yields a set of k relevant factors from a original set of n factors (which we will denote here as $x_1, x_2, \ldots, x_k$ for convenience). We are now in the position of describing the functional relationship between the expected value of the response variable and the relevant factors, by the "response surface" $E(y) = f(x_1, x_2, \ldots, x_k)$. Usually, f is taken to be some low-order polynomial, like a first-order model involving only the main effects (that is, $b_0 + \sum_{i=1}^{k} b_i x_i$).

The principal approach adopted in optimising using the response surface is a sequential one. First, a local approximation to the true response surface is constructed, using a first-order model. Next, factors are varied along a path that improves the response the most (this is along the gradient to the response surface). Experiments are conducted along this direction and the corresponding responses obtained until no further increase in the response is observed. At this point, a new first-order response surface is constructed, and the process repeated until it is evident that a first-order model is inadequate (or no more increases are possible). If the fit of the first-order model is poor, a more detailed model is then obtained—usually a second-order model, using additional levels for factors—and it stationary point obtained.

2.3   Screening and Optimisation for ILP

We are now in a position to put together the material in the previous sections to state more fully a procedure for screening and optimisation of parameters for an ILP system:

SO: Screen quantitative parameters using a two-level fractional factorial design, and optimise values using the response surface.

ScreenFrac. Screen for relevant parameters using the following steps:

$S1$. Decide on a set of n quantitative parameters of the ILP system that are of potential relevance. These are the factors x_i in the sense just described. Take some quantitative summary of the model constructed by the system—for example, its estimated predictive accuracy—as the response variable y (we will assume here that we wish to maximise the response).

$S2$. Decide on on two levels ("low" and "high" values) for each of the factors. These are then coded as ± 1.

$S3$. Devise a two-level fractional factorial design of Resolution III or higher, and obtain values of y for each experiment (or replicates of values of y, if so required).

$S4$. Construct a first-order regression model to estimate the role of the main effects x_i on y. Retain only those factors that are important, by examining the magnitude and significance of the coefficients of the x_i in the regression model (alternatively, only those factors found by a stepwise regression procedure are retained.

OptimiseRSM. Optimise values of relevant parameters using the following steps:

$O1$. Construct a first-order response surface using the relevant factors only (this is not needed if stepwise regression was used at the screening stage).

$O2$. Progressively obtain new values for y by changing the relevant parameters along the normal to the response surface. Stop when no increases in y are observed.[4]

$O3$. If needed, construct a new first-order response surface. If this surface is adequate, then return to Step $O2$.

$O4$. If needed, construct a second-order response surface. Return the optimum values of the relevant factors using the second-order surface, or from the last set of values from Step $O2$.

---

[4] In practice, this would typically mean that no increases have been observed for some number of consecutive experimental runs.

3   Empirical Evaluation

Our aim here is to demonstrate the utility of the screening and optimisation procedure **SO** that we have described in Section 2.3. We assess this utility by comparing the ILP system when it employs **SO** against the performance of the system with denoted **Default**, in which no screening or optimisation is performed and default values provided for all parameters are used.

For reasons of space, we will only report results on the well-studied ILP biochemical problems concerned with identifying mutagenic and carcinogenic chemicals. A longer version of this paper (available from the first author) has results on other datasets used in the literature. The two datasets we report on here have been described extensively elsewhere (for example, see [7] for mutagenesis; and [8] for carcinogenesis) and we refer the reader to these reports for details. For each application, the input to an ILP can vary depending on the background information used. In our experiments, we report results using the ILP system Aleph with two different subsets of background knowledge for each of the problems (the subsets are termed here as B_{min} and B_{max}). Finally, in this shorter version of our paper, we only present a qualitative comparison of the results. The extended version of the paper contains a more quantitative assessment using appropriate statistical tests.

3.1   Results and Discussion

We present first the results concerned with screening for relevant factors. Figure 3 show responses from the ILP system for experiments conducted for screening four parameters of the system using an appropriate fractional design. For lack of space, we are unable to show the full tabulation here of the sequence of experiments for optimising relevant parameter values using the response surface. The best model obtained from the gradient ascent is compared against the model obtained with default values in Fig. 4. We note the following: (1) Although no experimentation is needed for the use of default values, the model obtained with **ILP+Default** usually has the lowest predictive accuracies (the exception is Carcinogenesis, with B_{min})[5]; and (2) The classification accuracy of **ILP+SO** is never lower than the alternative.

Taken together, these observations provide reasonable empirical evidence for the conjecture that using **SO** is better than using **Default**. We now turn to the most serious implications of these results to current ILP practice. First, the results suggest that default levels for factors need not yield optimal models for all problems, or even when the same problem is given different inputs (here, different background knowledge). This means that using ILP systems just based on default values for parameters—the accepted practice at present—can give misleading estimates of the best response possible from the system. Secondly, the screening results suggest that as inputs change, so can the factors that are relevant (for example, when the background changes from B_{min} to B_{max} in

---

[5] We recall that the regression model obtained for Carcinogenesis (B_{min}) was not very good. This affects the performances of both **ILP+SO**.

Expt.	C	$Nodes$	$Minacc$	$Minpos$	Acc
E1	-1	-1	-1	-1	0.798
E2	-1	-1	$+1$	$+1$	0.612
E3	-1	$+1$	-1	$+1$	0.771
E4	-1	$+1$	$+1$	-1	0.723
E5	$+1$	-1	-1	$+1$	0.771
E6	$+1$	-1	$+1$	-1	0.761
E7	$+1$	$+1$	-1	-1	0.803
E8	$+1$	$+1$	$+1$	$+1$	0.612

Expt.	C	$Nodes$	$Minacc$	$Minpos$	Acc
E1	-1	-1	-1	-1	0.883
E2	-1	-1	$+1$	$+1$	0.845
E3	-1	$+1$	-1	$+1$	0.883
E4	-1	$+1$	$+1$	-1	0.867
E5	$+1$	-1	-1	$+1$	0.883
E6	$+1$	-1	$+1$	-1	0.872
E7	$+1$	$+1$	-1	-1	0.883
E8	$+1$	$+1$	$+1$	$+1$	0.862

(a) Mutagenesis (B_{min})
$Acc = 0.731 - 0.054 \; Minacc - 0.040 \; Minpos$

(b) Mutagenesis (B_{max})
$Acc = 0.872 - 0.011 \; Minacc$

Expt.	C	$Nodes$	$Minacc$	$Minpos$	Acc
E1	-1	-1	-1	-1	0.481
E2	-1	-1	$+1$	$+1$	0.454
E3	-1	$+1$	-1	$+1$	0.439
E4	-1	$+1$	$+1$	-1	0.460
E5	$+1$	-1	-1	$+1$	0.424
E6	$+1$	-1	$+1$	-1	0.490
E7	$+1$	$+1$	-1	-1	0.445
E8	$+1$	$+1$	$+1$	$+1$	0.460

Expt.	C	$Nodes$	$Minacc$	$Minpos$	Acc
E1	-1	-1	-1	-1	0.591
E2	-1	-1	$+1$	$+1$	0.475
E3	-1	$+1$	-1	$+1$	0.564
E4	-1	$+1$	$+1$	-1	0.525
E5	$+1$	-1	-1	$+1$	0.561
E6	$+1$	-1	$+1$	-1	0.490
E7	$+1$	$+1$	-1	-1	0.582
E8	$+1$	$+1$	$+1$	$+1$	0.513

(a) Carcinogenesis (B_{min})
$Acc = 0.456 - 0.013 \; Minpos \; (*)$

(b) Carcinogenesis (B_{max})
$Acc = 0.537 - 0.037 \; Minacc$

Fig. 3. Screening results (procedure $ScreenFrac$ in Section 2.3). Acc refers to the estimated accuracy of the model. The regression model is built using the "Autofit" option provided with [11]. This essentially implements a stepwise regression procedure. The regression equation in (c) is marked with a "*" to denote that significance levels had to be relaxed from the usual levels to obtain a model. This suggests that this model should be treated with caution.

Procedure	(Accuracy,Expts.)			
	Mutagenesis		Carcinogenesis	
	B_{min}	B_{max}	B_{min}	B_{max}
ILP+Default	$(0.755 \pm 0.031, 1)$	$(0.846 \pm 0.026, 1)$	$(0.510 \pm 0.028, 1)$	$(0.504 \pm 0.028, 1)$
ILP+SO	$(0.814 \pm 0.028, 15)$	$(0.883 \pm 0.023, 12)$	$(0.510 \pm 0.028, 15)$	$(0.605 \pm 0.027, 16)$

Fig. 4. Comparison of procedures, based on their estimated accuracies and the number of experiments needed to obtain this estimate. The accuracies are 10-fold cross-validation estimates, for which there is no unbiased estimator of variance [2]. The standard error reported is computed using the approximation in [5].

Mutagenesis, $Minpos$ ceases to be a relevant factor). This means that a once-off choice of relevant factors across all possible inputs can lead to sub-optimal performances from the system for some inputs.

Finally, we note that a controlled comparison of **Default** and **SO** has required us to enforce that the ILP system is the same in all experiments. In practice, we are often interested in controlled comparisons of a different kind, namely, the comparison of two ILP systems with the same input data. Our results suggest that equipping each ILP system with the procedure **SO** could enable a controlled comparison of best-case performances: a practice which has hitherto not been adopted by empirical ILP studies, but whose value is self-evident.

References

1. Bengio, Y.: Gradient based optimisation of hyperparameters. Neural Computation 12(8), 1889–1900 (2000)
2. Bengio, Y., Grandvalet, Y.: No unbiased estimator of the variance of k-fold cross-validation. Journal of Machine Learning Research 5, 1089–1105 (2004)
3. Box, G.E.P., Wilson, K.B.: On the Experimental Attainment of Optimum Conditions. Journal of the Royal Statistical Society, Series B (Methodological) 13(1), 1–45 (1951)
4. Bratko, I., Muggleton, S.H.: Applications of Inductive Logic Programming. Communications of the ACM 38(11), 65–70 (1995)
5. Breiman, L., Friedman, J.H., Olshen, R.A., Stone, C.J.: Classification and Regression Trees. Wadsworth, Belmont (1984)
6. Dzeroski, S.: Relational Data Mining Applications: An Overview. In: Dzeroski, S., Lavrac, N. (eds.) Relational Data Mining, pp. 339–360. Springer, Berlin (2001)
7. King, R.D., Muggleton, S.H., Srinivasan, A., Sternberg, M.J.E.: Structure-activity relationships derived by machine learning: The use of atoms and their bond connectivities to predict mutagenicity by inductive logic programming. Proc. of the National Academy of Sciences 93, 438–442 (1996)
8. King, R.D., Srinivasan, A.: Prediction of rodent carcinogenicity bioassays from molecular structure using inductive logic programming. Environmental Health Perspectives 104(5), 1031–1040 (1996)
9. Montgomery, D.C.: Design and Analysis of Experiments, 5th edn. John Wiley, New York (2005)
10. Srinivasan, A.: Four Suggestions and a Rule Concerning the Application of ILP. In: Lavrac, N., Dzeroski, S. (eds.) Relational Data Mining, pp. 365–374. Springer, Berlin (2001)
11. Steppan, D.D., Werner, J., Yeater, R.P.: Essential Regression and Experimental Design for Chemists and Engineers (1998),
   http://www.jowerner.homepage.t-online.de/download.htm

Don't Fear Optimality: Sampling for Probabilistic-Logic Sequence Models

Ingo Thon

Katholieke Universiteit Leuven
{firstname.lastname}@cs.kuleuven.be

Abstract. One of the current challenges in artificial intelligence is modeling *dynamic environments* that change due to the actions or activities undertaken by people or agents. The task of inferring hidden states, e.g. the activities or intentions of people, based on observations is called filtering. Standard probabilistic models such as Dynamic Bayesian Networks are able to solve this task efficiently using approximative methods such as particle filters. However, these models do not support logical or relational representations. The key contribution of this paper is the upgrade of a *particle filter* algorithm for use with a *probabilistic logical* representation through the definition of a proposal distribution. The performance of the algorithm depends largely on how well this distribution fits the target distribution. We adopt the idea of logical compilation into Binary Decision Diagrams for sampling. This allows us to use the optimal proposal distribution which is normally prohibitively slow.

1 Introduction

One of the current challenges in artificial intelligence is modeling dynamic environments that are influenced by actions and activities undertaken by people or agents. Consider modeling the activities of a cognitively impaired person [1]. Such a model can be employed to assist people, using common patterns to generate reminders or detect potentially dangerous situations, and thus can help to improve living conditions. To realize this, the system has to infer the intention or the activities of a person from features derived from sensory information. The typical model used in such processes are Hidden Markov Models (HMM) and their generalizations like factorial HMMs, coupled HMMs or Dynamic Bayesian Networks (DBN). These models can represent the intentions and/or activities with a hidden state. Estimating the hidden state distribution based on a sequence of observation in such model, called *filtering*, is the task we are focusing on paper.

Algorithms that perform efficient and exact inference in single state models like HMMs are well known. Also, for factorial HMMs and coupled HMMs efficient approximative algorithms exist that exploit structural properties [2] and for DBNs particle filters [3] present a good alternative.

However, recent research has shown that in many activity modeling domains, relational modeling is not only useful [4] [5] but also required [6]. Here, transitions between states are factored into sets of probabilistic logical conjectures that allow a dynamic number of random variables, which makes the translation into a standard Dynamic Bayesian Network impossible.

L. De Raedt (Ed.): ILP 2009, LNAI 5989, pp. 226–233, 2010.

Our **contributions** are: First we show how hidden state inference problems can be formulated through Causal Probabilistic Time Logic (CPT-L). CPT-L was introduced previously, together with the inference and learning algorithms for the fully observable case [7]. We use a logical compilation approach to implement efficient sampling from the *optimal proposal distribution*. The proposal distribution is a key component of the particle filter algorithm *Sequential Importance Resampling* (SIR), we want to use in this paper for solving the filtering problem.

Logical compilation has gained lot of interest in the past few years for speeding up inference and learning in probabilistic logics, especially compilation into Binary Decision Diagrams (BDD) [8] [9] [7] annotated with probabilities and its variants Zero Suppressed Decision Diagrams [7] [10] and Algebraic Decision Diagrams [11]. In this work we show as second contribution how a BDD is generated to represent the proposal distribution. Finally we will show as third contribution how the generated BDDs can be used to sample models of the represented formula (the states) according to the underlying distribution.

Related Work: Most sequential SRL models restrict themselves to a single atom per time point [12] or one probabilistic choice, e.g. outcome of an action. We are only aware of the following three exceptions: The Simple Transition Cost Models [13] proposed by Alan Fern. These models allow the specification of costs for transitions, which can be used to represent probabilities, but they have to be equal over all transitions. In Biswas et al.[4] the authors learn a Dynamic Markov Logic Network, but translate to DBNs for inference. Even though this is not a problem in general, it requires a state space with a fixed size that is known in advance. In Zettlemoyer et. al [14], the hidden states are defined by means of weighted FO-Formula the hypothesis. This approach requires mutually exclusive hypotheses, which are hard to construct and it is unclear whether they can deal with functors.

After introducing some terminology we proceed by introducing the CPT-L model. Afterwards, we specify the components of the algorithm that samples from the filtering distribution. Finally we discuss experimental result and conclude.

1.1 Preliminaries

Logical representation: A *logical atom* is an expression $p(t_1, \ldots, t_n)$ where p is a predicate symbol of arity n. The t_i are terms, built from constants, variables and complex terms. Constants are denoted by lower case symbols and variables by upper case symbols. Complex terms are of the form $f(t_1, \ldots, t_k)$ where f is a functor symbol and the t_i are again terms. An expression is called ground if it does not contain any variables. A substitution θ maps variables to terms, and $a\theta$ is the atom obtained from a by replacing all variables according to θ. As an example consider the atom $a = p(X_1, c_1, X_2)$ and the substitution $\theta = \{X_1 \mapsto c_2\}$ which replaces the variable X_1 by c_2 as in $a\theta = p(c_2, c_1, X_2)$. A set of ground atoms $\{a_1, \ldots, a_n\}$ is called Herbrand interpretation used to describe complex states and observations.

Reduced ordered binary decision diagrams: A Boolean formula is a formula build from a set of literals $l_1, \ldots, l_n$ and the connectors $\wedge$ (and), $\vee$ (or) and $\neg$ (not). Such a function can be represented using a rooted, directed acyclic graph (cf. Figure 1). The

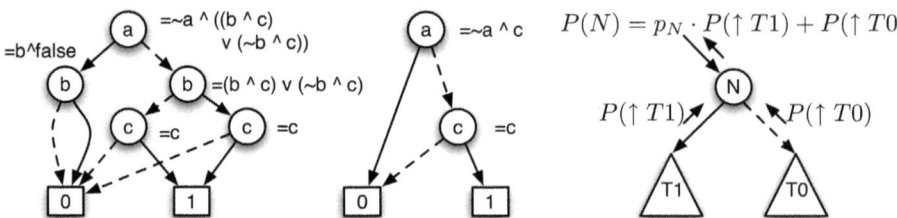

Fig. 1. An ordered Binary Decision Diagram before (left) and after (middle) reducing it. Each node is annotated with the logical formula it represents. For the reduction the c nodes are merged and afterward both b nodes have to be replaced by their child. Calculation of upward probability (right, cf section 3).

terminal nodes correspond to the logical true and false. Each node within the graph corresponds to a literal and has two children one corresponding to assigning the node the value one which is called high-child. The other child corresponds to assigning the value zero called low-child. A reduced ordered binary decision diagram is a decision diagram, where all literals are ordered, isomorphic subtrees are merged and all nodes which have for both branches the same childes are removed. In the following we use the terms BDD and Reduced order BDD synonymously.

2   Model

The model considered is basically a HMM where states and observations are Herbrand interpretations and transition- and observation-probabilities are defined in terms of a probabilistic logic (cf. Fig 2). More formally:

Definition 1. *A* **CPT-L model** *consists of a set of rules of the form*

$$r = (h_1 : p_1) \vee \ldots \vee (h_n : p_n) \leftarrow b_1, \ldots, b_m$$

where the $p_i \in [0, 1]$ form a probability distribution such that $\sum_i p_i = 1$, h_i are logical atoms, b_i are literals (i.e. atoms or their negation).

For convenience we will refer to $b_1, \ldots, b_m$ as $body(r)$ and to $(h_1 : p_1) \vee \ldots \vee (h_n : p_n)$ as $head(r)$. We assume that the rules are range-restricted, that is, all variables appearing in the body also appear in the head.

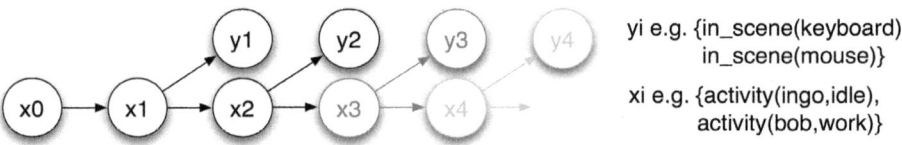

Fig. 2. Graphical representation of an HMM. States and observations are in our case herbrand interpretations.

The interpretation of such a rule is: whenever the body is true at time-point k the rule will cause one of the head elements to be true at time-point $k + 1$.

Consider the following example rules that models the current activity:

$$r(P, X) = a(P, X) : 0.8 \vee a(P, drink) : 0.1 \vee a(P, work) : 0.1 \leftarrow a(P, X).$$
$$od(P) = \qquad ois(can) : 0.7 \vee nil : 0.3 \qquad\qquad \leftarrow a(P, drink).$$
$$ow(P) = \qquad ois(pen) : 0.7 \vee nil : 0.3 \qquad\qquad \leftarrow a(P, work).$$

The first rule states that person P will continue its current activity X with probability 0.8 or switch to one of the activities $work$, $drink$ with probability 0.1. The second and third rule specifies: if someone works/drinks one can observe a pen/can.

The semantics of such a model is given by the following probabilistic constructive process. This stochastic process is assumed to be initialized with the empty interpretation or an interpretation containing a special symbol. Starting from the current state, all groundings $r\theta$ of all rules r, where $body(r)\theta$ holds in the current state are generated. For each of these grounding, one of the grounded head elements $h_1\theta, \dots, h_n$ of r is chosen randomly according to the distribution given by $p_1, \dots, p_n$. We will refer to this choice as selection. The chosen head element is then added to the successor state. A more detailed description can be found in [7].

In this work, we additionally assume that a subset of the predicates are declared as observable whereas the others are unobservable. In the previous example the predicate $a/2$ would typically be unobservable, whereas predicates like $pose/2$, $movement/2$, $object\_in\_scene/1$ $(ois/1)$ would be observable. We assume that all predicates in the head of one rule are either observable or unobservable. A rule is called un-/observable according to the observability of the predicates in the head.

In the rest of the text, x_k denotes the set of all unobservable facts and y_k the observable facts true at time point k. The probability of a hidden state sequence together with a sequence of observations follows directly from the semantics of CPT-L. Note: From the viewpoint of CPT-L the observation y_k of time-point k belongs to the successor state as both are caused by the current state. For readability purposes, we keep indexes of hidden states and observations together. With $x_{l:k}$ we denote the sequence of all states between l and k.

To continue our example assume that the current state is $x_i = \{a(ann, work), a(bob, work)\}$ then the applicable rules are $r(ann, work)$, $r(bob, work)$, $ow(ann)$, $od(bob)$. For each rule a head element is selected and added to the successor state respectively the observation. The state is $x_{i+1} = \{a(ann, work), a(bob, drink)\}$ and the observation is $y_i = \{ois(pen)\}$ for example with probability $(0.8 + 0.1) \cdot 0.1 \cdot (0.7 \cdot (0.3 \cdot 0.7))$.

3   Algorithm

Our goal is to estimate the filtering distribution $p(x_k|y_{1:k})$. This distribution can be used to answer queries like $a(P1, Act), a(P2, Act), P1 \neq P2$ whether *two persons performing the same activity*. Furthermore the distribution can be used for parameter learning with (Stochastic) Expectation Maximization. Exact calculation of this distribution is typically prohibitively slow due to the large structured state space [15]. Therefore

---

Algorithm 1. Number particles N, time step k, proposal distribution π

---

  function SAMPLE$((\pi, p_k, p_o, P))$
    for $i = 1, \ldots, N$ **do**
      $x_k^i \sim \pi(x_k | x_{0:k-1}^i, y_{1,k})$
      $\hat{w}_k^i := w_{k-1}^i \dfrac{p(y_k | x_k^i) p(x_k^i | x_{k-1}^i)}{\pi(x_k^i | x_{0:k-1}^i, y_{0:k})}$
    Normalize weights $w_k^i = \hat{w}_k^i / \sum_j \hat{w}_k^j$ $\overbrace{}^{\text{effective \# particles}}$

    Sample N particles of x_k^i according to w_k^i with weight $1/P$ if $\hat{N}_{thresh} > \underbrace{\left(\sum_i (w_k^i)^2 \right)^{-1}}$

---

we use Sampling Importance Resampling (SIR) [3]: we sample from a proposal distribution and compute importance weights that make up for the difference. In this section we first briefly discuss the mechanics of SIR. Afterwards we alter the original CPT-L algorithm [7] using a BDD to represent the distributions required by SIR. Finally we give the algorithm to sample states from this distribution using the constructed BDD.

3.1 Sampling Importance Resampling (SIR)

The filtering distribution can be approximated by a set of N particles (w_k^i, x_k^i) consisting of a weight w_k^i and a state x_k^i. The weights are an approximation of the relative posterior distribution. This empirical distribution is defined as

$$\hat{p}(x_k | y_{1:k}) = \frac{1}{N} \sum_{i=1}^{N} w_i \delta_{x_k^i}(x_k),$$

where $\delta_{x_k^i}(\cdot)$ is the point mass distribution located at x_k^i.

The SIR algorithm calculates the samples recursively. A single step is described in Algorithm 1. In each step, the particles are drawn from a sampling distribution $\pi(x_k | x_{0:k-1}^i, y_{0:k})$ and each particle's weight is calculated. In principle an arbitrary admissible distribution can be chosen as proposal distribution. A distribution is called admissible if it has probability greater zero for each state, which has probability greater zero for the target distribution. So the correctness does not depend on the choice, but the sample variance does, and thus the required number of particles, largely depends on this choice.

Typical sampling distributions are the transition prior $p(x_k | x_{k-1}^i)$, a fixed importance function $\pi(x_k | x_{0:k-1}^i, y_{0:k}) = \pi(x_k)$, or the transition posterior $p(x_k | x_{0:k-1}^i, y_{0:k})$.

3.2 Optimal Proposal Distribution

The transition prior $p(x_k | x_{k-1})$ is often used as proposal distribution for SIR as it allows for efficient sampling. Using the transition prior means, on the other hand, that the state space is explored without any knowledge of the observations which makes the algorithm sensitive to outliers. While this nonetheless works well in many cases, it is problematic in discrete, high dimensional state spaces when combined with spiked observation distributions. But high dimensional state spaces are common especially

---

Algorithm 2. Generate formula/BDD representing $\sum_{x_k^i} P(y_k, x_k^i | x_{k-1}^i)$

1: Initialize $f := \top$, $I_{max} = \emptyset$ the "maximal" successor state
2: Compute applicable ground rules $\mathbf{R}_k = \{r\theta | body(r\theta)$ is true in x_{k-1}^j, r unobservable$\}$
3: **for** all rules $(r = (p_1 : h_1, ..., p_n : h_n) \leftarrow b_1, ..., b_m)$ in $\mathbf{R}_k$ **do**
4:     $f := f \wedge (r.h_1 \vee ... \vee r.h_n)$, where $r.h$ denotes the proposition obtained by concatenating
        the name of the rule r with the ground literal h resulting in a new propositional
        variable $r.h$ (if not $h_i = nil$).
5:     $f := f \wedge (\neg r.h_i \vee \neg r.h_j)$ for all $i \neq j$
6:     $h_i \leftarrow r.h_i$; $I_{max} = I_{max} \cup h_i$
7: Compute applicable ground observation $\mathbf{S}_k = \{r\theta | body(r\theta)$ is true in I_{max}, r observable$\}$
8: **for** all observations $(r = (p_1 : h_1, ..., p_n : h_n) \leftarrow b_1, ..., b_m)$ in $\mathbf{S}_k$ **do**
9:     $f := f \wedge ((r.h_1 \vee ... \vee r.h_n) \leftrightarrow (b_1, ..., b_n))$, for $h_i \in I_{y_{t+1}}$
10:     $f := f \wedge (\neg r.h_i \vee \neg r.h_j)$ for all $i \neq j$
11: **for** all facts $l \in I_{y_{t+1}}$ **do**
12:     Initialize $g := false$
13:     for all $r \in \mathbf{S}_k$ with $p : l \in head(r)$ do $g := g \vee r.l$
14:     $f := f \wedge g$

---

in relational domains. It can be shown that the proposal distribution $p(x_k^i | x_{k-1}^i, y_k)$[1] together with weight update $w_k^i := w_{k-1}^i P(y_k | x_{k-1}^i)$ is optimal [3] and does not suffer from this problem.

BDD construction: To sample from the proposal distribution and update the weight efficiently we build a BDD that represents $P(y_k | x_{k-1}^i)$. The algorithm (shown as Algorithm 2) is a modification of the algorithm presented in previous work [7]. The algorithm builds a BDD representation of a formula which computes the joint probability of all possible selections that result in a transition for which the following four conditions hold. The transition (a) starts at x_{t-1}^i (line 2) and (b) goes over to a valid successor state x_t. In x_t it (c) generates the observation y_t (line 8-14) using (d) the observable rule applicable in x_t. Each node of the generated BDD $r : h_i$ corresponds to selecting (for one rule) the head h_i or not, as dictated by the probability p_i.

BDD sampling: Sampling a path according to the p_i from the root of this BDD to the terminal node with label 1 corresponds to sampling a value from $p(x_k | x_{k-1}^i, y_k)$. However, in most cases, sampling paths naively according to the p_i's will yield a path ending in the 0-terminal, that will then have to be rejected. Notably this would correspond to sampling from the transition prior. Therefore at every node, when choosing the corresponding subtree, we base our choice not only on its probability, but also on the probability of reaching the 1-terminal through this subtree. This corresponds to conditioning the paths such that we get a valid successor state together with the observation y_k. We call this probability upward probability because of the way it is computed. The upward probability of the root node corresponds to the probability of reaching the 1-terminal, i.e., $P(y_k | x_{k-1}^i)$. The computation starts by initializing the values in the leafs with the

---

[1] Here it is crucial, to realize that the observation y_k is the observation generated by the state x_k.

label of the leaf. Afterwards, the probabilities are propagated upward as illustrated in Fig. 1 [2]. The subtree is then chosen in every node according to

$$\frac{p_N \cdot P(\uparrow T1_N)}{P(\uparrow T0_N) + p_N \cdot P(\uparrow T1_N)} = \frac{p_N \cdot P(\uparrow T1_N)}{P(\uparrow N)} \qquad \text{respectively:} \quad \frac{P(\uparrow T0_N)}{P(\uparrow N)}.$$

4 Experiments

To evaluate our algorithm, we recreated the model of Biswas et al [4], according to the parameters specified in their work. There a person is observed during writing, typing, using a mousing, eating, and so on. The computer has multiple cues to classify the activities. Two examples are the pose of the person or whether an apple is observed in the scene. As the observation distribution is fairly smooth and had nowhere zero mass the transition-prior is expected to perform well as proposal distribution.

For the experiments we sampled 5 sequences of length 10. Afterwards we run the particle filter algorithm with the exact proposal distribution and the transition prior using 100 particles. For the optimal prior each run took less then a minute on a MacBook Pro 2.16 Ghz. The transition prior was approximately 5 times faster. In Fig 3 the effective number of particles (cf. Alg 1) divided by the runtime in ms is plotted. The horizontal axis is the sequence length. Even though not significant the optimal performed on average better. In toy example with spiked observation distribution the transition prior typically lost all particles in a few steps.

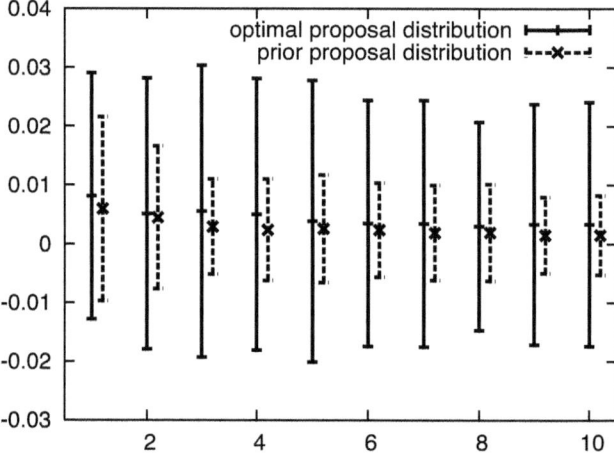

Fig. 3. Effective number of particles divided by runtime in dependence on sequence length

---

[2] For the reader familiar with [8] [9] the use here and in [7] is a bit different. The former is the choice of a literal being true or false, whereas the latter represents whether one of the head gets chosen. Using the backward probability of [9] instead of the upward, the sampling generalizes.

5 Conclusions and Future Work

We propose a novel way of sampling from a joint distribution in the presence of evidence by the means of BDDs. We show that the final system allows more efficient filtering than using the transition prior in relational domains. An advantage of our complete system is that the final algorithms are very intuitive as it builds on well established algorithm like SIR. We plan to extend our filtering algorithm towards more elaborate techniques like for example Rao-Blackwellized Particle Filters, and Online Stochastic EM. Finally, we will investigate the use of our technique of sampling from a BDD also for non-sequential probabilistic logics, as well as for standard DBNs.

Acknowledgments. For discussion we are grateful to (in temporal order) K. Kersting, N. Landwehr, L. De Raedt, and B. Gutmann. Special thanks to K. Driessens for approximating a native speaker.

This work was supported by the FWO project: Relational action and activity learning and the GOA/08/008 project "Probabilistic Logic Learning".

References

1. Pollack, M.E.: Intelligent technology for an aging population: The use of AI to assist elders with cognitive impairment. AI Magazine 26(2), 9–24 (2005)
2. Landwehr, N.: Modeling interleaved hidden processes. In: ICML (2008)
3. Doucet, A., Defreitas, N., Gordon, N.: Sequential Monte Carlo Methods in Practice (Statistics for Engineering and Information Science), 1st edn. Springer, Heidelberg (June 2001)
4. Biswas, R., Thrun, S., Fujimura, K.: Recognizing activities with multiple cues. In: Elgammal, A., Rosenhahn, B., Klette, R. (eds.) Human Motion 2007. LNCS, vol. 4814, p. 255. Springer, Heidelberg (2007)
5. Natarajan, S., Bui, H., Tadepalli, P., Kersting, K., Wong, W.K.: Logical hierarchical hidden markov models for modeling user activities. In: Železný, F., Lavrač, N. (eds.) ILP 2008. LNCS (LNAI), vol. 5194, pp. 192–209. Springer, Heidelberg (2008)
6. Sridhar, M., Cohn, A.G., Hogg, D.C.: Learning functional object-categories from a relational spatio-temporal representation. In: ECAI (2008)
7. Thon, I., Landwehr, N., De Raedt, L.: A simple model for sequences of relational state descriptions. In: Daelemans, W., Goethals, B., Morik, K. (eds.) ECML PKDD 2008, Part II. LNCS (LNAI), vol. 5212, pp. 506–521. Springer, Heidelberg (2008)
8. De Raedt, L., Kimmig, A., Toivonen, H.: Problog: A probabilistic Prolog and its application in link discovery. In: ICJAI (2007)
9. Ishihata, M., Kameya, Y., Sato, T., Minato, S.: Propositionalizing the EM algorithm by BDDs. In: ILP (2008)
10. Minato, S.: Compiling bayesian networks by symbolic probability calculation based on zero-suppressed BDDs. In: AAAI (2007)
11. Chavira, M., Darwiche, A., Jaeger, M.: Compiling relational bayesian networks for exact inference. Int. J. Approx. Reasoning 42(1-2), 4–20 (2006)
12. Kersting, K., De Raedt, L., Gutmann, B., Karwath, A., Landwehr, N.: Relational sequence learning. In: De Raedt, L., Frasconi, P., Kersting, K., Muggleton, S.H. (eds.) Probabilistic Inductive Logic Programming. LNCS (LNAI), vol. 4911, pp. 28–55. Springer, Heidelberg (2008)
13. Fern, A.: A simple-transition model for relational sequences. In: IJCAI (2005)
14. Zettlemoyer, L.S., Pasula, H.M., Kaelbling, L.P.: Logical particle filtering. In: Probabilistic, Logical and Relational Learning - A Further Synthesis (2008)
15. Boyen, X., Koller, D.: Tractable inference for complex stochastic processes. In: UAI (1998)

Policy Transfer via Markov Logic Networks

Lisa Torrey and Jude Shavlik

University of Wisconsin, Madison WI, USA
ltorrey@cs.wisc.edu, shavlik@cs.wisc.edu

Abstract. We propose using a statistical-relational model, the Markov Logic Network, for knowledge transfer in reinforcement learning. Our goal is to extract relational knowledge from a source task and use it to speed up learning in a related target task. We show that Markov Logic Networks are effective models for capturing both source-task Q-functions and source-task policies. We apply them via demonstration, which involves using them for decision making in an initial stage of the target task before continuing to learn. Through experiments in the RoboCup simulated-soccer domain, we show that transfer via Markov Logic Networks can significantly improve early performance in complex tasks, and that transferring policies is more effective than transferring Q-functions.

1 Introduction

The transfer of knowledge from one task to another is a desirable property in machine learning. Our ability as humans to transfer knowledge allows us to learn new tasks quickly by taking advantage of relationships between tasks. While many machine-learning algorithms learn each new task from scratch, there are also *transfer-learning* algorithms [13] that can improve learning in a *target task* using knowledge from a previously learned *source task*.

In reinforcement learning (RL), an agent navigates through an environment, sensing its state, taking actions, and trying to earn rewards [12]. The *policy* of the agent determines which action it chooses in each step. An agent performing RL typically learns a *value function* to estimate the values of actions as a function of the current state, and its policy typically is to take the highest-valued action in all except occasional *exploration* steps.

In complex domains, RL can require many early episodes of nearly random exploration before acquiring a reasonable value function or policy. A common goal of transfer in RL is to shorten or remove this period of low performance. Recent research has yielded a wide variety of RL transfer algorithms to accomplish this goal [13]. In one category of methods, RL agents apply a source-task policy or value function at some point(s) while learning the target task. Approaches of this type vary in the representation of the source-task knowledge and in the timing and frequency of its application.

Madden and Howley [6] learn a set of rules to represent a source-task policy, and they use those rules only during exploration steps in the target task. Fernandez and Veloso [2] use the original representation of the source-task policy, and

L. De Raedt (Ed.): ILP 2009, LNAI 5989, pp. 234–248, 2010.
© Springer-Verlag Berlin Heidelberg 2010

give the target-task agent a three-way choice between using the current target-task policy, using a source-task policy, and exploring randomly. Croonenborghs et al. [1] learn a relational decision tree to represent the source-task policy, and use the tree as a multi-step action (an *option*) that may be chosen in place of a target-task action in any step.

Our own work in this area has contributed several relational methods, in which the knowledge transferred is at the level of first-order logic, and is extracted from the source task with inductive logic programming (ILP). Using ILP [8], we transfer several types of relational models. In one recent approach [15], the transferred model is a first-order finite-state machine that we call a *relational macro*, and it represents a successful generalized source-task plan.

In this paper, we propose transfer via a statistical-relational model called a Markov Logic Network (MLN). An MLN combines first-order logic and probability [9], and is capable of capturing more source-task knowledge than a macro can. With experiments in the simulated soccer domain RoboCup [7], we show that MLN transfer methods can significantly improve initial performance in complex RL tasks.

2   Reinforcement Learning in RoboCup

In one common form of RL called Q-learning [12], the value function learned by the agent is called a Q-function, and it estimates the value of taking an action from a state. The policy is to take the action with the highest Q-value in the current state, except for occasional exploratory actions taken in a small percent ϵ of steps. After taking an action and receiving some reward (possibly zero), the agent updates its Q-value estimates for the current state.

Stone and Sutton [11] introduced RoboCup [7] as an RL domain that is challenging because of its large, continuous state space and non-deterministic action effects. Since the full game of soccer is quite complex, researchers have developed several simpler games within the RoboCup simulator.

In M-on-N BreakAway (see Figure 1), the objective of the M reinforcement learners called *attackers* is to score a goal against $N-1$ hand-coded *defenders* and a *goalie*. The game ends when they succeed, when an opponent takes the ball, when the ball goes out of bounds, or after a time limit of 10 seconds. The learners receive a +1 reward if they score a goal and 0 reward otherwise. The attacker who has the ball may choose to move (ahead, away, left, or right with respect to the goal center), pass to a teammate, or shoot (at the left, right, or center part of the goal).

Figure 2 shows the state features for BreakAway, which mainly consist of distances and angles between players and the goal. They are shown in logical notation since we perform transfer learning in first-order logic; our basic RL algorithm uses grounded literals in a fixed-length feature vector. Capitalized atoms indicate typed variables, while constants and predicates are uncapitalized. The attackers (labeled *a0*, *a1*, etc.) are ordered by their distance to the agent in possession of the ball (*a0*), as are the non-goalie defenders (*d0*, *d1*, etc.).

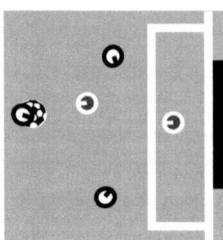

distBetween(a0, Player)
distBetween(a0, GoalPart)
distBetween(Attacker, goalCenter)
distBetween(Attacker, ClosestDefender)
distBetween(Attacker, goalie)
angleDefinedBy(topRight, goalCenter, a0)
angleDefinedBy(GoalPart, a0, goalie)
angleDefinedBy(Attacker, a0, ClosestDefender)
angleDefinedBy(Attacker, a0, goalie)
timeLeft

Fig. 1. Snapshot of a 3-on-2 BreakAway game. The attacking players have possession of the ball and are maneuvering against the defending team towards the goal.

Fig. 2. The features that describe a Break-Away state in their first-order logical form, where variables are capitalized

Our basic RL algorithm uses a $SARSA(\lambda)$ variant of Q-learning [12] and employs a support vector machine (SVM) for Q-function approximation [5]. It relearns the SVM Q-function after every batch of 25 games. The exploration rate ϵ begins at 2.5% and decays exponentially over time. Stone and Sutton [11] found that discretizing the continuous features into Boolean interval features called *tiles* is important for learning in RoboCup; following this approach, we add 32 tiles per feature.

Agents in the games of 2-on-1, 3-on-2, and 4-on-3 BreakAway take between 1000 and 3000 training episodes to reach a performance asymptote in our system. Differences in the numbers of attackers and defenders cause substantial differences in optimal policies, particularly since there is a type of player entirely missing in 2-on-1 (the non-goalie defender). However, there remain strong relationships between BreakAway games of different sizes, and transfer between them should improve learning.

3   Markov Logic Networks

The Markov Logic Network (MLN) is a model developed by Richardson and Domingos [9] that combines first-order logic and probability. It expresses concepts with first-order rules, as ILP does, but unlike ILP it puts weights on the rules to indicate how important they are. While ILP rulesets can only predict a concept to be true or false, an MLN can estimate the probability that a concept is true, by comparing the total weight of satisfied rules to the total weight of violated rules. This type of probabilistic logic therefore conveys more information than pure logic. It is also less brittle, since world states that violate some rules are not impossible, just less probable.

Formally, a Markov Logic Network is a set of first-order logic formulas F, with associated real-valued weights W, that provides a template for a Markov network. The network contains a binary node for each possible grounding of each predicate of each formula in F, with groundings determined by a set of constants C.

Edges exist between nodes if they appear together in a possible grounding of a formula. Thus the graph contains a clique for each possible grounding of each formula in F.

The classic example from Richardson and Domingos [9] follows. Suppose the formulas are:

$$\text{Weight } 1.5: \forall y \ \text{Smokes}(y) \Rightarrow \text{Cancer}(y)$$
$$\text{Weight } 0.8: \forall y, z \ \text{Friends}(y, z) \Rightarrow (\text{Smokes}(y) \Leftrightarrow \text{Smokes}(z))$$

These rules assert that smoking leads to cancer and that friends have similar smoking habits. These are both good examples of MLN formulas because they are often true, but not always; thus they will have finite weights. Given constants *Anna* and *Bob* that may be substituted for the variables y and z, this MLN produces the ground Markov network in Figure 3. (Note that the convention for capitalization is opposite here from in ILP; variables here are lower-case and constants are upper-case.)

Let X represent all the nodes in this example, and let $X = x$ indicate that among the possible worlds (the true/false settings of those nodes), x is the actual one. The probability distribution represented by the Markov network is:

$$P(X = x) = \frac{1}{Z} \ exp \sum_{i \in F} w_i n_i(x) \qquad (1)$$

Here Z is a normalizing term, w_i is the weight of formula $i \in F$, and $n_i(x)$ is the number of true groundings of formula i in x. Based on this equation, one can calculate the probability of any node in the network given *evidence* about the truth values of some other nodes.

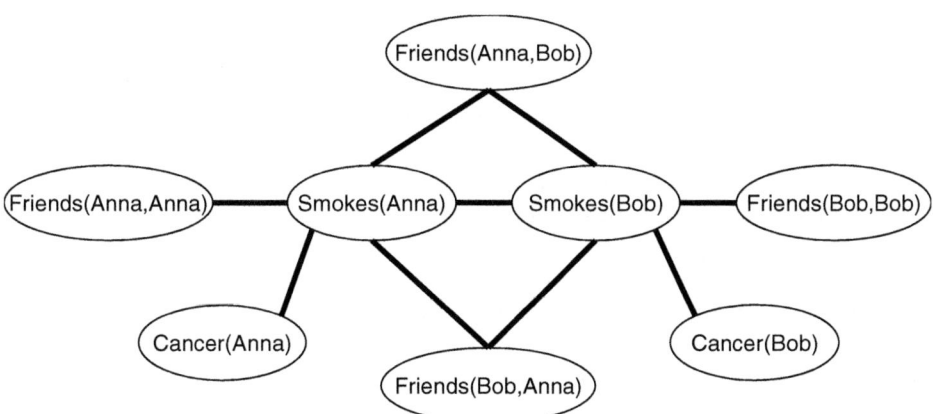

Fig. 3. The ground Markov network produced by the MLN described in this section. This example and this image come from Richardson and Domingos [9]. Each clique in this network has a weight (not shown) derived from the formula weights.

Given a set of positive and negative examples of worlds, appropriate formula weights can be learned rather than specified manually. There are several algorithms for weight learning; the current state-of-the-art is a method called *preconditioned scaled conjugate gradient* [4]. This is the default algorithm in the Alchemy software package [3], which we use for our experiments.

We learn formulas with ILP, and then assign weights to them with Alchemy. The type of formulas we learn determines the type of source-task knowledge captured by the MLN. The following sections describe two possible types.

4   MLN Q-Function Transfer

MLN Q-function transfer [14] is a transfer method that learns an MLN to express the source-task Q-function relationally, and allows target-task agents to use it for an initial demonstration period. This allows the target-task agents to avoid the slow process of random exploration that traditionally occurs at the beginning of RL.

This method uses an MLN to define a probability distribution for the Q-value of an action, conditioned on the state features. It chooses a source-task batch and uses its training data to learn an MLN Q-function for transfer. The choice of which source-task batch has an impact, as we will discuss.

In this scenario, an MLN formula describes some characteristic of the RL agent's environment that helps determine the Q-value of an action in that state. For example, assume that there is a discrete set of Q-values that a RoboCup action can have (*high*, *medium*, and *low*). In this simplified case, one formula in an MLN representing the Q-function for BreakAway could look like the following:

> IF   distBetween(a0, GoalPart) > 10
> AND   angleDefinedBy(GoalPart, a0, goalie) < 30
> THEN levelOfQvalue(move(ahead), high)

The MLN could contain multiple formulas like this for each action. After learning weights for the formulas from source-task data, one could use this MLN to infer, given a target-task state, whether action Q-values are most likely to be high, medium, or low. Note that Q-values in RoboCup are continuous rather than discrete, so I do not actually learn rules classifying them as high, medium, or low. Instead, the algorithm discretizes the continuous Q-values into bins that serve a similar purpose.

Table 1 gives the algorithm for MLN Q-function transfer. The sections below describe the steps of this algorithm in more detail.

4.1   Learning an MLN Q-Function from a Source Task

The first step of the MLN Q-function transfer algorithm in Table 1 is to divide the Q-values for an action into bins, according to the procedure in Table 2. The training example Q-values could have any arbitrary distribution, so it uses

Table 1. Algorithm for MLN Relational Q-Function Transfer

INPUT REQUIRED
A set of batches $B = (b_1, b_2, ...)$ to consider for transfer
The Q-function Q^b for each batch $b \in B$
The set of games $G(b)$ that trained the Q-function for each batch $b \in B$
A parameter ϵ determining distance between bins
A demonstration-period length D
A validation-run length V

CREATE Q-VALUE BINS   // This is a hierarchical clustering procedure.
For each batch $b \in B$
    For each source-task action a
        Determine $bins(b, a)$ for action a in batch b using Table 2
        (Provide inputs $G(b)$ and ϵ)

LEARN FORMULAS   // This accomplishes MLN structure learning.
For each batch $b \in B$
    For each source-task action a
        For each $bin \in bins(b, a)$
            Let $P = \emptyset$ // These will be the positive examples.
            Let $N = \emptyset$ // These will be the negative examples.
            For each state s in a game $g \in G(b)$
                If s used action a and $Q_a^b(s)$ falls into bin
                    Set $P \leftarrow P \cup g$ // Examples that fall into the bin are positive.
                Else if s used action a and $Q_a^b(s)$ does not fall into bin
                    Set $N \leftarrow N \cup g$ // Examples that fall outside the bin are negative.
            Learn rules with Aleph to distinguish P from N
            Let $M(b, a, bin)$ be the ruleset chosen by the algorithm in Table 3
        Let $M(b, a)$ be the union of $M(b, a, bin)$ for all bins

LEARN FORMULA WEIGHTS
For each batch $b \in B$
    For each source-task action a
        Learn MLN weights $W(b, a)$ for the formulas $M(b, a)$ using Alchemy
        Define $MLN(b, a)$ as $(M(b, a), W(b, a))$
    Define $MLN(b)$ as the set of MLNs $MLN(b, a)$

CHOOSE A BATCH   // Do a validation run in the source task to pick the best batch.
For each batch $b \in B$
    For V episodes: Use $MLN(b)$ as shown in Table 4 to choose actions in a new source-task run
    Let $score(b)$ be the average score in this validation run
Choose the highest-scoring $b^* \in B = argmax_b \; score(b)$

LEARN TARGET TASK
For D episodes: Perform RL but use $MLN(b^*)$ to choose actions as shown in Table 4
For remaining episodes: Perform RL normally

---

a hierarchical clustering algorithm to find good bins. Initially every training example is its own cluster, and it repeatedly joins clusters whose midpoints are closest until there are no midpoints closer than ϵ apart. The final cluster midpoints serve as the midpoints of the bins.

Table 2. Algorithm for dividing the Q-values of an action a into bins, given training data from games G and a parameter ϵ defining an acceptable distance between bins

---

For each state i in a game $g \in G$ that takes action a
    Create cluster c_i containing only the Q-value of example i
Let C = sorted list of c_i for all i
Let m = min distance between two adjacent $c_x, c_y \in C$
While $m < \epsilon$ // Join clusters until too far apart.
    Join clusters c_x and c_y into c_{xy}
    $C \leftarrow C \cup c_{xy} - \{c_x, c_y\}$
    $m \leftarrow$ min distance between two new adjacent $c'_x, c'_y \in C$
Let $B = \emptyset$ // These will be the bins for action a.
For each final cluster $c \in C$ // Center one bin on each cluster.
    Let bin b have midpoint $\bar{c}$, the average of values in c
    Set the boundaries of b at adjacent midpoints or Q-value limits
    Set $B \leftarrow B \cup b$
Return B

The value of ϵ should be domain-dependent. For BreakAway, which has Q-values ranging from approximately 0 to 1, we use $\epsilon = 0.1$. This leads to a maximum of about 11 bins, but there are often less because training examples tend to be distributed unevenly across the range. We experimented with ϵ values ranging from 0.05 to 0.2 and found very minimal differences in the results; the approach appears to be robust to the choice of ϵ within a reasonably wide range.

The second step of the MLN Q-function transfer algorithm in Table 1 performs structure-learning for the MLN. The MLN formulas are rules that assign training examples into bins. We learn these rules using the ILP system Aleph [10]. Some examples of bins learned for *pass* in 2-on-1 BreakAway, and of rules learned for those bins, are:

> IF   distBetween(a0, GoalPart) ≥ 42
> AND distBetween(a0, Teammate) ≥ 39
> THEN pass(Teammate) has a Q-value in the interval [0, 0.11]

> IF   angleDefinedBy(topRightCorner, goalCenter, a0) ≤ 60
> AND angleDefinedBy(topRightCorner, goalCenter, a0) ≥ 55
> AND angleDefinedBy(goalLeft, a0, goalie) ≥ 20
> AND angleDefinedBy(goalCenter, a0, goalie) ≤ 30
> THEN pass(Teammate) has a Q-value in the interval [0.11, 0.27]

> IF   distBetween(Teammate, goalCenter) ≤ 9
> AND angleDefinedBy(topRightCorner, goalCenter, a0) ≤ 85
> THEN pass(Teammate) has a Q-value in the interval [0.27, 0.43]

From the rules generated by Aleph, our algorithm selects a final ruleset for each action. It does so using an efficient method shown in Table 3 that approximately

Table 3. Algorithm for selecting a final ruleset from a large set of rules. Rules are added to the final set if they increase the overall F measure.

Let S = rules sorted by decreasing precision on the training set
Let $T = \emptyset$ // This will be the final ruleset.
For each rule $r \in S$ // Select rules.
    Let $U = T \cup \{r\}$
    If $F(U) > F(T)$
    Then set $T \leftarrow U$
Return T

---

optimizes for both precision and recall. It sorts rules from highest precision to lowest and greedily adds them to the final ruleset if they improve its F score. The combined rulesets for all the actions form the set of formulas M in the MLN.

The third step of the algorithm learns weights for the formulas using Alchemy's conjugate gradient-descent algorithm, as described in Section 3. The fourth step of the algorithm selects the best batch from among the set of candidates. We found that the results can vary widely depending on the source-task batch from which the algorithm transfers, so we use a validation set of source-task data to select a good batch.

4.2   Applying an MLN Q-Function in a Target Task

The final step of the MLN Q-function transfer algorithm in Table 1 is to learn the target task with a *demonstration* approach, in which the target-task agent executes the transferred strategy for an initial period before continuing with standard RL. During the demonstration period, the target-task learner queries the MLN Q-function to determine the estimated Q-value of each action, and it takes the highest-valued action. Meanwhile, it learns normal Q-functions after each batch, and after the demonstration ends, it begins using those normal Q-functions.

The algorithm in Table 4 shows how to estimate a Q-value for an action in a new state using an MLN Q-function. For each action a, the algorithm infers the probability p_b that the Q-value falls into each bin b. It then uses these probabilities as weights in a weighted sum to calculate the Q-value of a:

$$Q_a(s) = \sum_b p_b E[Q_a|b]$$

where $E[Q_a|b]$ is the expected Q-value given that b is the correct bin, estimated as the average Q-value of the training data in that bin. The probability distribution that an MLN provides over the Q-value of an action could look like one of the examples in Figure 4.

Table 4. Algorithm for estimating the Q-value of action a in target-task state s using the MLN Q-function. This is a weighted sum of bin expected values, where the expected value of a bin is estimated from the training data for that bin.

Provide state s to the MLN as evidence
For each bin $b \in [1, 2, ..., n]$
    Infer the probability p_b that $Q_a(s)$ falls into bin b
    Collect training examples T for which Q_a falls into bin b
    Let $E[Q_a|b]$ be the average of $Q_a(t)$ for all $t \in T$
Return $Q_a(s) = \sum_b (p_b * E[Q_a|b])$

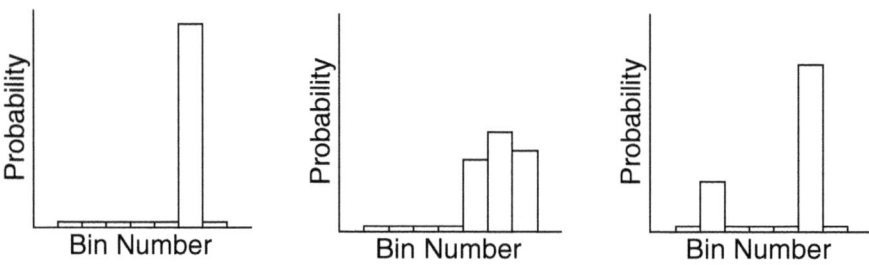

Fig. 4. Examples of probability distributions over Q-value of an action that an MLN Q-function might produce. On the left, the MLN has high confidence that the Q-value falls into a certain bin, and the action will get a high Q-value. In the center, the MLN is undecided between several neighboring bins, and the action will still get a high Q-value. On the right, there is a high likelihood of a high bin but also a non-negligible likelihood of a low bin, and the action will get a lower Q-value (this case suggests methods for intelligent exploration, which could be a direction for future work).

4.3   Experimental Results for MLN Q-Function Transfer

To test MLN Q-function transfer, we learn MLNs from 2-on-1 BreakAway source tasks and transfer them to 3-on-2 and 4-on-3 BreakAway. Figure 5 shows the performance of MLN Q-function transfer in 3-on-2 and 4-on-3 BreakAway compared to standard RL and to our previous approach, macro transfer [15]. Each curve in the figure is an average of 25 runs and has points averaged over the previous 250 games to smooth over the high variance in the RoboCup domain. The transfer curves consist of five target-task runs generated from each of five source-task runs, to account for variance in both stages of learning.

These results show that MLN Q-function transfer is comparable to macro transfer in some cases and less effective in others. In 3-on-2 BreakAway, the area under the curve for MLN Q-function transfer is not significantly different than for macro transfer ($p > 0.05$). In 4-on-3 BreakAway, the area under the curve for macro transfer is significantly higher ($p < 0.05$).

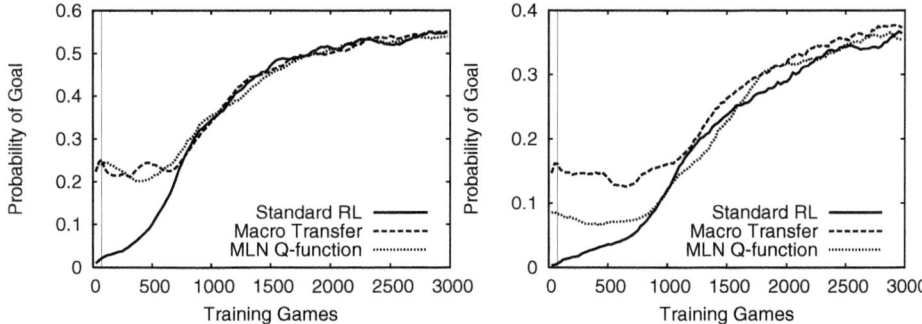

Fig. 5. Probability of scoring a goal in 3-on-2 BreakAway (left) and 4-on-3 Break-Away (right) with standard RL, macro transfer from 2-on-1 BreakAway, and MLN Q-function transfer from 2-on-1 BreakAway. The thin vertical line marks the end of the demonstration period.

5   MLN Relational Policy Transfer

MLN relational policy transfer is a method that learns an MLN to express the source-task policy, and allows target-task agents to use it for an initial demonstration period. This approach is closely related to MLN Q-function transfer, but it has the potential to transfer more effectively by focusing on policy rather than Q-values.

A policy simply determines which action to take given a state, and does not require numeric values to be assigned to actions. Thus instead of needing to create bins for continuous Q-values, MLN policy transfer learns an MLN that simply predicts the best action to take. It is also simpler than MLN Q-function transfer in that it does not need to choose a batch from which to transfer, which was a significant tuning step in the previous method.

Table 5 gives the algorithm for MLN Q-function transfer. The section below describes the steps of this algorithm in more detail.

5.1   Learning and Using an MLN Policy

The first two steps of the MLN policy-transfer algorithm in Table 5 perform structure-learning and weight-learning for the MLN. These steps are similar to those in MLN Q-function transfer. However, the formulas simply predict when an action is the best action to take, rather than predicting a Q-value bin for an action as they do in MLN Q-function transfer.

Each action may have many formulas with different weights. They are learned from examples of actions chosen in the source-task. We use only high-reward source-task episodes since those guarantee good action choices. In 2-on-1 Break-Away, these are games in which the learner scored a goal.

Table 5. Algorithm for MLN Relational Policy Transfer

INPUT REQUIRED
Games G from the source-task learning process
A definition of high-reward and low-reward games in the source task
The demonstration-period length D

LEARN FORMULAS
Let G be the set of high-reward source-task games
For each source-task action a
    Let $P = \emptyset$ // These will be the positive examples.
    Let $N = \emptyset$ // These will be the negative examples.
    For each state s in a game $g \in G$
        If s used action a
            Set $P \leftarrow P \cup s$ // States that use the action are positive.
        Else if s used action $b \neq a$
            Set $N \leftarrow N \cup s$ // States that use a different action are negative.
    Learn rules with Aleph to distinguish P from N
    Let M be the ruleset chosen by the algorithm in Table 3

LEARN FORMULA WEIGHTS
Learn MLN weights W for the formulas M using Alchemy
Define MLN by (M, W)

LEARN TARGET TASK
For D episodes: Perform RL but choose the highest-probability action according to MLN
For remaining episodes: Perform RL normally

Some examples of rules learned for *pass* in 2-on-1 BreakAway are:

    IF   angleDefinedBy(topRightCorner, goalCenter, a0) ≤ 70
    AND timeLeft ≥ 98
    AND distBetween(a0, Teammate) ≥ 3
    THEN pass(Teammate)

    IF   distBetween(a0, GoalPart) ≥ 36
    AND distBetween(a0, Teammate) ≥ 12
    AND timeLeft ≥ 91
    AND angleDefinedBy(topRightCorner, goalCenter, a0) ≤ 80
    THEN pass(Teammate)

    IF   distBetween(a0, GoalPart) ≥ 27
    AND angleDefinedBy(topRightCorner, goalCenter, a0) ≤ 75
    AND distBetween(a0, Teammate) ≥ 9
    AND angleDefinedBy(Teammate, a0, goalie) ≥ 25
    THEN pass(Teammate)

The final step of the MLN policy-transfer algorithm in Table 5 learns the target task via demonstration. During the demonstration period, the target-task learner queries the MLN to determine the probability that each action is best, and it takes the highest-probability action. Meanwhile, it learns normal Q-functions after each batch, and after the demonstration ends, it begins using those normal Q-functions.

5.2   Experimental Results for MLN Policy Transfer

To test MLN policy transfer, we learn MLNs from the same 2-on-1 BreakAway source tasks as before and transfer them to 3-on-2 and 4-on-3 BreakAway. Figure 6 shows the performance of MLN policy transfer in 3-on-2 and 4-on-3 BreakAway compared to standard RL, macro transfer, and MLN Q-function transfer.

These results show that transferring an MLN policy is more effective than transferring an MLN Q-function, and that it can also outperform macro transfer in some cases. In 3-on-2 BreakAway, the area under the curve for MLN policy transfer is significantly higher than for both other transfer approaches ($p < 0.05$). In 4-on-3 BreakAway, it is higher than MLN Q-function transfer but still lower than macro transfer ($p < 0.05$).

6   Comparing an MLN Policy to a Ruleset Policy

The ILP rulesets that we learn for MLN policy transfer could themselves represent a policy, without the addition of an MLN. Here we perform an experiment to determine whether the MLN provides any additional benefit.

In order to use rulesets as a policy, we need a way to decide which rule to follow if multiple rules recommending different actions are satisfied in a state. To do this, we assign each rule a score. The score of a rule approximates the probablity that following the rule will lead to a successful game, as estimated from the source-task data. At each step in the target task, we have our agents check all the rules and take the action recommended by the highest-scoring satisfied rule.

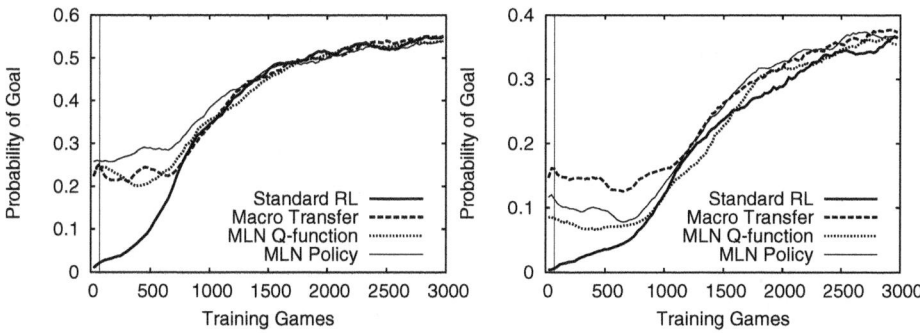

Fig. 6. Probability of scoring a goal in 3-on-2 BreakAway (left) and 4-on-3 BreakAway (right) with standard RL, macro transfer from 2-on-1 BreakAway, MLN Q-function transfer from 2-on-1 BreakAway, and MLN policy transfer from 2-on-1 BreakAway. The thin vertical line marks the end of the demonstration period.

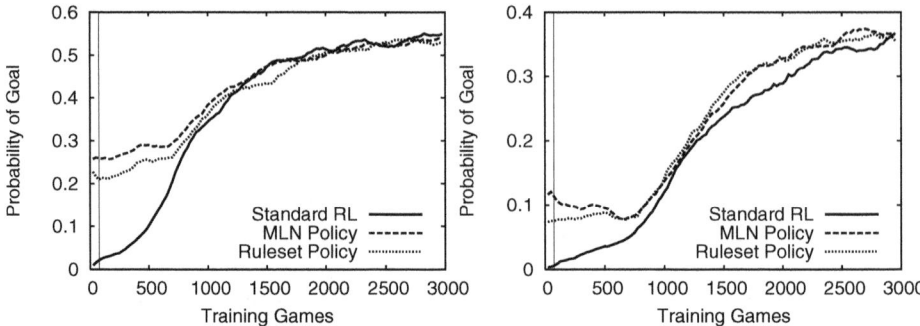

Fig. 7. Probability of scoring a goal in 3-on-2 BreakAway (left) and 4-on-3 BreakAway (right) with standard RL, regular MLN policy transfer from 2-on-1 BreakAway, and ruleset policy transfer from 2-on-1 BreakAway. The thin vertical line marks the end of the demonstration period.

Figure 7 shows the performance of this approach in 3-on-2 and 4-on-3 Break-Away, compared with standard RL and regular MLN policy transfer. The area under the curve for rulesets is significantly less than for MLNs in 3-on-2 Break-Away ($p < 0.05$). Thus MLNs can provide an additional benefit over ILP alone. In some cases, they may be comparable; the areas are not significantly different in 4-on-3 BreakAway ($p > 0.05$).

7   MLN Policies with Action Sequence Information

MLN policy transfer assumes the Markov property, in which the action choice depends only on the current environment and is independent of previous environments and actions. However, it need not do so; the MLN formulas for action choices could use such information. Here we examine the benefit of doing so by adding predicates to the ILP hypothesis space that specify previous actions. We add predicates for one, two, and three steps back in a game. Like macros, this approach allows transfer of both relational information and multi-state reasoning.

In these experiments, Aleph only chose to use the predicate for one previous step, and never used the ones for two and three previous steps. This indicates that it is sometimes informative to know what the immediately previous action was, but beyond that point, action information is not useful.

Figure 8 shows the performance of multi-step MLN policy transfer in 3-on-2 and 4-on-3 BreakAway, compared with standard RL and regular MLN policy transfer. The area under the curve for the multi-step version is significantly less than for the regular version in 3-on-2 BreakAway ($p < 0.05$) and is not significantly different in 4-on-3 BreakAway ($p > 0.05$). These results suggest that adding action-sequence information does not improve MLN policy transfer.

The Markov property appears to be a valid assumption in the BreakAway domain. While action patterns do exist in 2-on-1 BreakAway policies, and macro

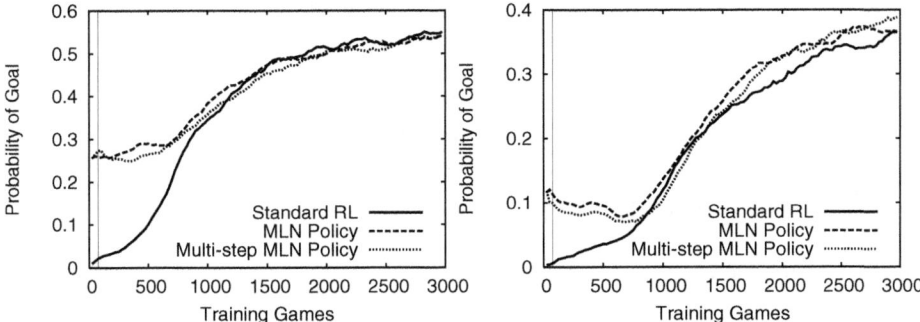

Fig. 8. Probability of scoring a goal in 3-on-2 BreakAway (left) and 4-on-3 BreakAway (right) with standard RL, regular MLN policy transfer from 2-on-1 BreakAway, and multi-step MLN policy transfer from 2-on-1 BreakAway. The thin vertical line marks the end of the demonstration period.

transfer takes advantage of them, there is apparently enough information in the current state to make action choices independently in MLN transfer. A multi-step MLN policy is therefore unnecessary in this domain, though it could be helpful in different domains where the Markov property does not hold.

8   Conclusions and Future Work

We propose algorithms for transfer in reinforcement learning via Markov Logic Networks and evaluate them in a complex domain. In MLN Q-function transfer, we represent the source-task Q-function relationally with an MLN. In MLN policy transfer, we represent the source-task policy with an MLN.

Transferring a policy with an MLN is a more natural and effective method than transferring a Q-function. Rulesets expressing a policy can be demonstrated effectively as well, but using an MLN to combine the rulesets provides additional benefits. An MLN captures complete enough information about the source task that adding knowledge about actions previously taken provides no additional benefit.

MLN policies outperform relational macros in some transfer scenarios, because they capture more detailed knowledge from the source task. However, they can perform worse in more distant transfer scenarios; in these cases they are likely capturing *too much* detail from the source task. This is a phenomenon that we call *overspecialization*, which is related to overfitting, but is specific to the context of transfer learning.

Future work in this area could focus on revision of a transferred model after the initial demonstration episodes. Our methods currently revert to standard RL, but they could instead learn by incrementally revising the source-task knowledge. Applying MLN knowledge in ways other than demonstration may also be effective.

A related area of potential work is MLN-based relational reinforcement learning. Domains like RoboCup could benefit from relational RL, which would provide substantial generalization over objects and actions. The main challenge to overcome in performing relational RL in such a complex domain is the computational cost of learning MLN structures and weights.

Acknowledgements

This research is supported by DARPA grants HR0011-07-C-0060 and FA8650-06-C-7606.

References

1. Croonenborghs, T., Driessens, K., Bruynooghe, M.: Learning relational skills for inductive transfer in relational reinforcement learning. In: Blockeel, H., Ramon, J., Shavlik, J., Tadepalli, P. (eds.) ILP 2007. LNCS (LNAI), vol. 4894, pp. 88–97. Springer, Heidelberg (2007)
2. Fernandez, F., Veloso, M.: Probabilistic policy reuse in a reinforcement learning agent. In: AAMAS (2006)
3. Kok, S., Singla, P., Richardson, M., Domingos, P.: The Alchemy system for statistical relational AI. Technical report, University of Washington (2005)
4. Lowd, D., Domingos, P.: Efficient Weight Learning for Markov Logic Networks. In: Kok, J.N., Koronacki, J., Lopez de Mantaras, R., Matwin, S., Mladenič, D., Skowron, A. (eds.) KDD 2007. LNCS (LNAI), vol. 4702, pp. 200–211. Springer, Heidelberg (2007)
5. Maclin, R., Shavlik, J., Torrey, L., Walker, T.: Knowledge-based support vector regression for reinforcement learning. In: IJCAI Workshop on Reasoning, Representation, and Learning in Computer Games (2005)
6. Madden, M., Howley, T.: Transfer of experience between reinforcement learning environments with progressive difficulty. AI Review 21, 375–398 (2004)
7. Noda, I., Matsubara, H., Hiraki, K., Frank, I.: Soccer server: A tool for research on multiagent systems. Applied Artificial Intelligence 12, 233–250 (1998)
8. De Raedt, L.: Logical and Relational Learning. Springer, Heidelberg (2008)
9. Richardson, M., Domingos, P.: Markov logic networks. Machine Learning 62, 107–136 (2006)
10. Srinivasan, A.: The Aleph manual (2001)
11. Stone, P., Sutton, R.: Scaling reinforcement learning toward RoboCup soccer. In: ICML (2001)
12. Sutton, R., Barto, A.: Reinforcement Learning: An Introduction. MIT Press, Cambridge (1998)
13. Torrey, L., Shavlik, J.: Transfer learning. In: Soria, E., Martin, J., Magdalena, R., Martinez, M., Serrano, A. (eds.) Handbook of Research on Machine Learning Applications. IGI Global (2009)
14. Torrey, L., Shavlik, J., Natarajan, S., Kuppili, P., Walker, T.: Transfer in reinforcement learning via Markov Logic Networks. In: AAAI Workshop on Transfer Learning for Complex Tasks (2008)
15. Torrey, L., Shavlik, J., Walker, T., Maclin, R.: Relational macros for transfer in reinforcement learning. In: ICML (2007)

Can ILP Be Applied to Large Datasets?

Hiroaki Watanabe and Stephen Muggleton

Imperial College London, 180 Queen's Gate, London SW7 2AZ, UK
{hw3,shm}@doc.ic.ac.uk

Abstract. There exist large data in science and business. Existing ILP systems cannot be applied effectively for data sets with 10000 data points. In this paper, we consider a technique which can be used to apply for more than 10000 data by simplifying it. Our approach is called Approximative Generalisation and can compress several data points into one example. In case that the original examples are mixture of positive and negative examples, the resulting example is ascribed in probability values representing proportion of positiveness. Our longer term aim is to apply on large Chess endgame database to allow well controlled evaluations of the technique. In this paper we start by choosing a simple game of Noughts and Crosses and we apply mini-max backup algorithm to obtain database of examples. These outcomes are compacted using our approach and empirical results show this has advantage both in accuracy and speed. In further work we hope to apply the approach to large database of both natural and artificial domains.

1   Introduction

There exist large data in science and business. Although Inductive Logic Programming (ILP) [2] has been tackling challenging problems [1], existing ILP systems cannot be applied effectively for data sets with 10000 data points unfortunately. A natural approach for handling such large data is to simplify it by reducing the amount of information. We could compress several data points into one example although the resulting example would need to capture proportion of positiveness especially when the original examples are mixture of positive and negative examples. Such a data compression technique is expected to be smoothly integrated into ILP frameworks since it is a part of domain knowledge. The purpose of this study is to investigate if we could achieve high predictive accuracies with simplified data in Probabilistic ILP (PILP) [5].

In this paper, our new technique called Approximative Generalisation is formally introduced in Section 2 first. It characterises PILP not only from an uncertainty but also from a *non-deterministic* point of view. Then we empirically study Approximative Generalisation in Noughts and Crosses domain in Section 3. Brief discussions conclude this paper in Section 4.

L. De Raedt (Ed.): ILP 2009, LNAI 5989, pp. 249–256, 2010.
© Springer-Verlag Berlin Heidelberg 2010

2   Approximative Generalisation

2.1   Learning from Specialised Examples

In a standard ILP setting [3], we search the set of hypotheses, H, which satisfies the entailment relation:

$$BK \cup H \models E \tag{1}$$

where E is a set of given examples, and BK is background knowledge. We propose to consider a set of specialised examples, E', which is a specialisation of E associated with BK under entailment as follows.

$$BK \cup E \models E' \tag{2}$$

Then E' satisfies the following relation.

Theorem 1. *Let BK, H, E' be background knowledge, a set of hypotheses, and a set of examples. If (1) and (2) are held, the following entailment relation is also held.*

$$BK \cup H \models E'$$

Proof. From (1), $BK \cup H \models BK \cup E$. From (2), $BK \cup E \models E'$ is held. Therefore $BK \cup H \models E'$.

Theorem 1 shows that the original concept can still be learned with the specialised examples. Now we define Approximative Generalisation as follows.

Definition 1 (Approximative Generalisation). *A set of hypotheses H' is called Approximative Generalisation of E if $BK \cup H \models BK \cup H' \models E'$ such that $BK \cup E \models E'$.*

Example 1. Let us assume the following BK and E.

$$BK = \{human(s), \forall X \; mortal(X) \rightarrow need\_grave(X)\} \quad E = \{mortal(s)\}$$

From $BK \cup E$, we obtain $E' = \{need\_grave(s)\}$. Now $BK \cup \neg E'$ is:

$$BK \cup \neg E' = \{human(s), \neg need\_grave(s), \forall X \; mortal(X) \rightarrow need\_grave(X)\}$$

$$\models \{\exists X (human(X) \wedge \neg need\_grave(X))\} \underset{def}{=} \neg H'.$$

This is equivalent to $H' = \forall X(human(X) \rightarrow need\_grave(X))$. Then H can be $H = \{human(X) \rightarrow mortal(X)\}$ since $\{mortal(X) \rightarrow need\_grave(X)\} \cup \{human(X) \rightarrow mortal(X)\} \models \{human(X) \rightarrow need\_grave(X)\}$.

This example shows that there exists a case such that $BK \cup H' \not\models E$ although $BK \cup H \models BK \cup H' \models E'$.

2.2 Numerically Approximating Examples by Surjection

We further characterise the specialised examples by introducing surjective functions from E to E' in order to transfer the associated Boolean labels of positive or negative examples. Let us consider a class of projections, *surjective* functions, from E to E' as follows.

Definition 2. *A function* $f : E \rightarrow E'$ *is surjective if and only if for every* $e' \in E'$ *there is at least one* $e \in E$ *such that* $f(e) = e'$.

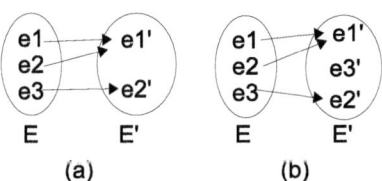

Fig. 1. (a) Surjective and (b) non-surjective functions

Example 2. In Fig. 1, Figure (a) is a surjective function whereas Figure (b) is a non-surjective function.

Now non-deterministic example is defined as follows.

Definition 3 (Non-deterministic Examples). *Let* E^+ *is a set of given positive examples and* E^- *a set of given negative examples. For* $E = E^+ \cup E^-$ *and a surjective function* f, *a non-deterministic example* , $l : f(e)$, *is a labelled example in which* l *is defined as:* $l = \frac{|f_{E^+}(e)|}{|f_E(e)|}$ *where* $|f_{E^+}(e)|$ *is a number of positive examples surjected onto* $f(e)$ *and* $|f_E(e)|$ *is a number of examples surjected onto* $f(e)$.

For example in Fig. 1, if we assume $e1 \in E^+$ and $e2 \in E^-$, we obtain $0.5 : e1'$.

In PILP, probabilistic examples can capture such a degree of truth in probability. *Now our task is to compute the approximative generalisation of E, H', with non-deterministic examples.* Further discussions on Approximative Generalisation can be found in [7].

3 Example: Noughts and Crosses Domain

In the previous section, we extend the standard ILP setting by adding the two computations: logical specialisation and numerical approximation. In this section, we empirically show such an extension can realise learning from large datasets in PILP.

3.1 Generating Never-Lose Sequences of Plays

We show an empirical result of the new framework in Noughts and Crosses domain where our study shows a more relational representation requires less number of examples. Noughts and Crosses, also called Tic Tac Toe, is a two-person, perfect information, and zero-sum game in which two players, *Nought* (O) and *Cross* (X) take turns marking the space of a 3×3 grid. Nought goes first and the player who succeeds in placing three respective marks in a horizontal, vertical or diagonal row wins the game. Although the setting and the rule are simple, there exist 9! (= 362880) possible ways of placing noughts and crosses on the board without regarding to winning combinations.

Noughts and Crosses game is known that there exist a never-lose strategy for each player [6]. That is, the game is always draw if two players know the optimal strategy. However, it becomes a probabilistic game once one side plays randomly. Let us assume that Nought plays in never-lose strategy whereas Cross plays randomly. Under such a probabilistic setting, we study (1) if a Machine Learning algorithm can obtain the never-lose strategy by cloning the behaviour of Nought and (2) how we can reduce the number of examples by changing knowledge representations from propositional one to relational one via surjective functions.

Before starting Machine Learning, we perform a retrograde analyse of Noughts and Crosses game in order to generate *never-lose* sequences of plays. In two player zero-sum game theory, mini-max is a rule which always selects a decision to *minimise the maximum possible loss.* By "maximum possible loss" we mean one players assumes the opponent always takes his best action which results the maximum loss to the other side. Mini-max algorithm evaluates the game in *forward* direction from the initial plays to the ends. In the Noughts and Crosses case, let us assume Nought plays first. Mini-max rule suggests Cross to always select the play which maximises Nought's loss whereas Nought to select the action to minimise such a loss.

Unfortunately mini-max criteria cannot force a player to win, however, an alternation of mini-max called *mini-max backup* algorithm [6] can do. It is originally developed for retrograde analysis of chess endgames in which the algorithm evaluate the player's actions and board positions in *backward* direction from the ends of the game to the initial plays. The key idea is that we only starts from Nought-won end-positions of the game and generate only sequences of predecessors which *never reach to losing end positions*. We apply this idea to generate the database of all the *never-lose* sequences of plays.

3.2   Two Logical Representations

We study two logical representations of Noughts and Crosses game. A natural way to express the 3×3 board is in the following atom:

$$board(p0, p1, p2, p3, p4, p5, p6, p7, p8)$$

where the term p_i is either 1, 2, or 0 to express *nought*, *cross*, and *empty* respectively as shown in Figure 2. The atom, $board(2, 1, 0, 0, 0, 0, 0, 0, 0)$, expresses the state of the board (b) of Figure 2. Language $\mathcal{L}_1$ is defined as: (a)Predicate: board/9 and (b)Terms: 0,1,2. We also introduce a different relational language, $\mathcal{L}_2$. Figure 3 shows 6 relations in the board. The atoms, $corner(mark, p_i)$,

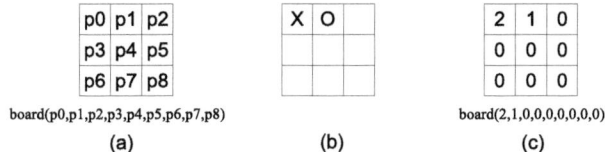

board(p0,p1,p2,p3,p4,p5,p6,p7,p8)   board(2,1,0,0,0,0,0,0,0)

(a)   (b)   (c)

Fig. 2. Logical representation of a state of the 3 × 3 board. (a) shows the mappings between the locations of the grids and arities of the atom. (b) is a state of the board whose logical expression is shown in (c).

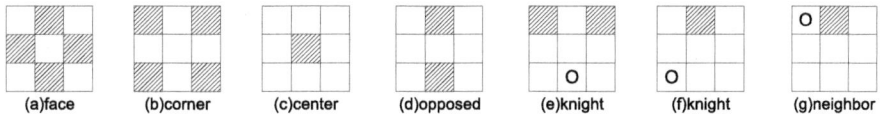

(a)face   (b)corner   (c)center   (d)opposed   (e)knight   (f)knight   (g)neighbor

Fig. 3. Relations between grids of the game board

$face(mark, p_i)$, and $center(mark, p_i)$ take the term "$mark$" (either 1 for *nought* or 2 for *cross*) and p_i (i = 0,...,8) to express the *mark* being placed at p_i. The atoms, $opposed(p_i, p_j)$, $knight(p_i, p_j)$, and $neighbor(p_i, p_j)$ represent the relations between two grids, p_i and p_j. These relations are static and do not essentially depend on the plays, however, we only describe them when any *placed* mark has such relations. More precisely, the figures, (e), (f) and (g) in Figure 3, show the relative grids from the placed noughts. If any mark is placed in the shadowed grids, the associated relations are expressed. Language $\mathcal{L}_2$ is defined as follows.

- Predicate: corner/2, face/2, center/2, opposed/2, knight/2, neighbor/2
- Terms: 1, 2, p_0, p_1, p_2, p_3, p_4, p_5, p_6, p_7, p_8,

For example, the board (b) in Figure 2 can be expressed in the conjunctions of the form as $corner(2, p_0) \wedge face(1, p_1) \wedge neighbor(p_0, p_1)$. Note that $\mathcal{L}_2$ expresses all the marks of *nought* and *cross* on the 3×3 boards even after the specialisation since the second arity of $corner/2$, $face/2$, and $center/2$ tells the grid locations although $\mathcal{L}_2$ cannot express the locations of the empty grid at all.

Logical specialisation from the representation in language $\mathcal{L}_1$ to in language $\mathcal{L}_2$ can be expressed in a logic program. A part of such background knowledge is as follows.

```
corner(X,p0)    :- board(X,_,_,_,_,_,_,_,_), X != 0.
face(X,p1)      :- board(_,X,_,_,_,_,_,_,_), X != 0.
neighbor(X,Y)   :- board(X,Y,_,_,_,_,_,_,_), X != 0.
```

3.3 Probabilistic Logic Automaton

We introduce Probabilistic Logic Automaton (PLA) [7] as a probabilistic logic. Intuitively, PLA is a logical extension of Probabilistic Automaton each of whose node can be an interpretation of an existentially quantified conjunction of literals (ECOL).

Definition 4 (Logical State). *Let L be a first-order language for describing ECOLs, F. A logical state q is a pair (n, F) where $n \in \mathcal{N}$ is the name of the logical state.*

Two logical states, (n_1, F_1) and (n_2, F_2), are treated as different if $n_1 \neq n_2$ even if F_1 and F_2 are logically equivalent. Each edge can be associated with disjunctions of ground actions.

Definition 5 (Logical Edge). *A logical edge is (a) a directed edge between two logical states and (b) associated with a set of ground atoms called logical actions.*

We introduce three probability distributions. Probabilistic transition function, $T : S \times \Sigma \to 2^S$ defines probabilistic state transitions from S to 2^S via Σ which is a set of ground atoms for describing logical actions. S_0 is a probability distribution over a set of initial logical states. We assign probability distribution over a set of logical actions, $B = \{b_{ij}(a_k)\}$, which is a set of probability distributions of taking logical action a_k during the state transition from state n_i to n_j. Now PLA is defined in the following six tuple.

Definition 6 (Probabilistic Logical Automaton). A probabilistic automaton is a 6-tuple

$$PLA = (S, \Sigma, T, S_0, B, G)$$

where S is a finite set of the logical states, Σ is a set of ground atoms for describing logical actions, $T : S \times \Sigma \to 2^S$ is a probabilistic transition function, S_0 is a probability distribution over a set of initial logical states, B is a set of probability distributions over logical actions at each edge, and G is a set of accept states.

In PLA, an input is a chain of observations of situations and actions as follows.

Definition 7 (Logical Sequence). *A logical sequence: $o_1 a_1 o_2 a_2 ... a_{n-1} o_n$ is an input of PLA in which o_i is ECOLs and a_i is a ground atom.*

Intuitively o_i may be viewed as an observation of a situation at time i and a_i is the action taken at time i. For example, a Nought-won game in a logical sequence is shown in Figure 4 where each state of the board is expressed as a logical node and actions of the players are attached to the directed edges.

3.4 Machine Learning of Noughts and Crosses Game

We study Machine Learning of Noughts and Crosses game next. We randomly sampled *never-lose* sequences of plays from the database and expressed in PLA based on the language $\mathcal{L}_1$. Let us call this example in $\mathcal{L}_1$ as E_1. Then the logical contents in each node in E_1 is specialised by a logic program a part of which

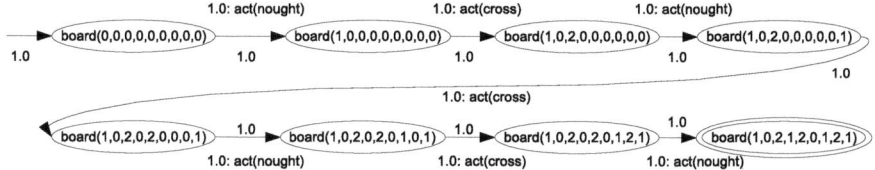

Fig. 4. A positive example of a never-lose sequence of plays in Probabilistic Logic Automaton

is shown in the previous section. This specialised examples are called E_2. Note that E_1 is a set of positive example whereas E_2 is a set of non-deterministic examples with $l = 1.0$.

We learn the Nought strategy both in $\mathcal{L}_1$ and $\mathcal{L}_2$ by *Cellist* system [7] which provides (a) topology learning via Stirling Numbers of the second kind-based topology sampling algorithm [7], (b) specialisation of ECOLs by adopting Plotkin's lgg algorithm [4,7], and (c) EM algorithm for estimating probabilistic parameters.

We tested 12 sample sizes, (5 , 10, 15, 20, 25, 30, 35, 40, 45, 50, 55, 60) for the training examples. Regarding the empirical results, we evaluated predictive accuracies of the generated PLA models using 100 Nought-won test examples. For each number of sample sizes, we calculated the average error of the learned models and plotted in those figures. The best model results 92.3% predictive accuracy (0.077 error) when $m = 55$ in language $\mathcal{L}_2$.

The results[1] are shown in Figure 5 and Figure 6 in which how the error, ε, is decreased by increasing the number of non-deterministic examples. Clearly, the knowledge representation in $\mathcal{L}_2$ shows better predictive accuracies for all the sizes of training data.

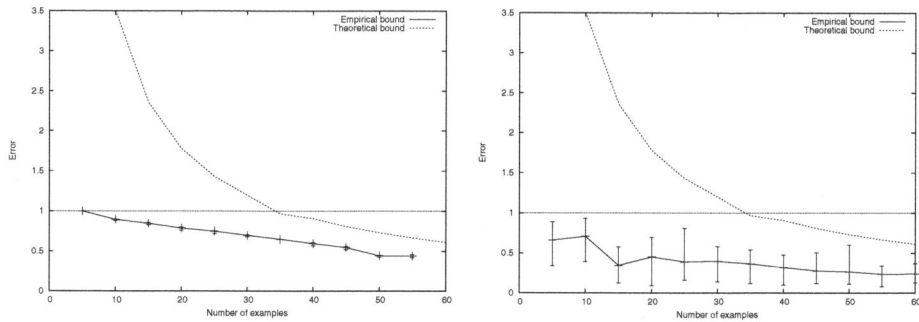

Fig. 5. Predictive Error Rates in $\mathcal{L}_1$ **Fig. 6.** Predictive Error Rates in $\mathcal{L}_2$

---

[1] The theoretical bounds shown in Figure 5 and Figure 6 are based on our average-case sample complexity analysis [7]. The theoretical bound expresses the delay bound; more than 34 examples are required in theory.

4 Conclusion

In this paper, we present Approximative Generalisation for tackling large datasets in PILP. Our approach can compress several data points into one example. If the given examples are mixture of positive and negative examples, the proportion of positiveness is captured in probabilistic examples. We empirically confirmed that our technique has advantage both in accuracy and speed in Noughts and Crosses domain even though the compressed examples are more relational than the original examples. This aspect should encourage PILP to tackle more relational applications.

In Approximative Generalisation, the data compression is smoothly encoded in background knowledge. Approximative Generalisation should also be discussed from an Active Learning point of view since we might need to test several data compressions repeatedly for finding better predictive accuracies.

In future work, we hope to apply our technique to large database of both natural and artificial domains including Chess End Game database to allow well controlled evaluation of our technique.

References

1. King, R.D., Whelan, K.E., Jones, F.M., Reiser, P.K.G., Bryant, C.H., Muggleton, S.H., Kell, D.B., Oliver, S.G.: Functional genomic hypothesis generation and experimentation by a robot scientist. Nature 427, 247–252 (2004)
2. Muggleton, S.H., De Raedt, L.: Inductive logic programming: Theory and methods. Journal of Logic Programming 19(20), 629–679 (1994)
3. Nienhuys-Cheng, S.H., de Wolf, R.: Foundations of Inductive Logic Programming. LNCS (LNAI), vol. 1228. Springer, Heidelberg (1997)
4. Plotkin, G.: Automatic Methods of Inductive Inference. PhD thesis, Edinburgh University, UK (1971)
5. De Raedt, L., Kersting, K.: Probabilistic inductive logic programming. In: Ben-David, S., Case, J., Maruoka, A. (eds.) ALT 2004. LNCS (LNAI), vol. 3244, pp. 19–36. Springer, Heidelberg (2004)
6. Thompson, K.: Retrograde analysis of certain endgames. ICCA Journal 9(3), 131–139 (1986)
7. Watanabe, H.: A Learning Theory Approach for Probabilistic Relational Learning (DRAFT). PhD thesis, Imperial College London, UK (2009), http://www.doc.ic.ac.uk/~hw3/thesis.pdf

Author Index

GPSR Compliance

The European Union's (EU) General Product Safety Regulation (GPSR) is a set of rules that requires consumer products to be safe and our obligations to ensure this.

If you have any concerns about our products, you can contact us on ProductSafety@springernature.com

In case Publisher is established outside the EU, the EU authorized representative is:

Springer Nature Customer Service Center GmbH
Europaplatz 3
69115 Heidelberg, Germany

Batch number: 09474024

Printed by Printforce, the Netherlands